Building Relationships

Building Relationships

Online Dating and the New Logics of Internet Culture

Dawn Shepherd

LEXINGTON BOOKS
Lanham • Boulder • New York • London

Published by Lexington Books
An imprint of The Rowman & Littlefield Publishing Group, Inc.
4501 Forbes Boulevard, Suite 200, Lanham, Maryland 20706
www.rowman.com

Unit A, Whitacre Mews, 26-34 Stannary Street, London SE11 4AB

British Library Cataloguing in Publication Information Available

Library of Congress Cataloging-in-Publication Data
The hardback edition of this book was previously cataloged as follows:

Names: Shepherd, Dawn, 1971- author.
Title: Building relationships : online dating and the new logics of internet culture / Dawn Shepherd.
Description: Lanham : Lexington Books, 2016. | Includes bibliographical references and index.
Identifiers: LCCN 2016002992 (print) | LCCN 2016005883 (ebook)
Subjects: LCSH: Online dating.
Classification: LCC HQ801.82 .S534 2016 (print) | LCC HQ801.82 (ebook) | DDC 306.730285--dc23
LC record available at http://lccn.loc.gov/2016002992

ISBN 9781498508575 (cloth : alk. paper)
ISBN 9781498535830 (pbk : alk. paper)
ISBN 9781498508582 (Electronic)

∞™ The paper used in this publication meets the minimum requirements of American National Standard for Information Sciences Permanence of Paper for Printed Library Materials, ANSI/NISO Z39.48-1992.

Printed in the United States of America

Contents

Introduction

A Minneapolis father learns his teenage daughter is pregnant after she receives a promotional flier for baby-related items from Target. The man reports his concern to a store manager: receiving such promotions might accelerate his daughter's sexual activity. The manager apologizes. When a representative from the retailer later follows up to express regrets once more, he learns that the young woman was indeed pregnant. She had received a high "pregnancy prediction score" based on the retailer's analysis of her buying habits and was sent customized discounts as a result.[1] That same year, Facebook directed the news feeds of nearly seven hundred thousand site members to include items with a majority of either positively or negatively worded items in order to test the effects of "emotional contagion."[2] These much-discussed cases are examples of the use of big data[3] and computational processing to influence behavior. In the case of the former, Target used what it knew about customers' buying habits to pinpoint an important juncture for altering purchasing tendencies: the birth of a child.[4] In the case of the latter, Facebook manipulated the content shown to its users in order to determine how that content affected viewers' moods. This phenomenon exists within a larger condition of informationalization, "a specific form of social organization in which the generation, processing, and transmission of information become the fundamental sources of productivity and power."[5] When considered in terms of big data and computational processing, informationalization as an organizing logic for social relations has significant implications for understandings of new media.

These technologies are especially prevalent on the Web,[6] where they have infiltrated Americans' everyday lives through search engines, online shopping recommendations, and personalized Web radio stations. Facebook, Google, Netflix all use information they gather about our tastes, interests, and habits—and the tastes, interests, and habits of others—to inform their interactions with us. For example, Facebook makes friend suggestions based on a combination of factors, including friends two users have in common. Likewise, Netflix's Cinematch uses millions of movie ratings to predict the viewing choices of its members. Even Google uses a complex and highly guarded set of algorithms to match users with information based on choices both they and other users have made in the past. We make meaning of the world through these processes, though we rarely do so consciously, and it is easy to understand how we might begin to outsource meaning making to technologies without realizing it. However, it is important that we understand the implications of this outsourcing. *Building Relationships: Online Dating and the New Logics of Internet Culture* contributes to the growing body of literature that seeks to understand this phenomenon of new media in terms of systems of procedural and cultural logics.

With the growing popularity of mass customization and user-generated content on the Web, user-application exchanges have changed in recent years.
Although the activity of looking for an address or buying a book online may not feel different for the user, the processes that enable those activities—and the information made available by them—have transformed significantly. The Internet of the late twentieth century generally relied on simple one-to-one retrieval in which a user requested information and a browser displayed it. This relationship was transactional, meaning each exchange between the user and the application was isolated from any past and future exchanges and any exchanges other users might have with the application. In contrast, the Internet of the early twenty-first century offers a variety of user-site relationships in which sites record vast amounts of data on users and their behavior, going beyond a cookie deposited on a hard drive. Although they are called by many names—code, procedure, protocol, algorithm—the logics that manage these processes online have become increasingly prevalent and increasingly complex, and they have gained additional attention from scholars outside computer science and engineering, particularly in regard to their relationship to human behavior.

THE REGIME OF COMPUTATION

In order to elucidate the recent proliferation of informationalization, it is necessary to examine a complex web of relationships—between users and technologies, between technologies and the larger cultural contexts in which they function, between individuals and those same contexts. In *My Mother Was a Computer*, Hayles discusses what she calls the Regime of Computation—a tradition that views computational processes as elemental to all function, be it human, natural, or technical. She notes that "the pervasiveness and importance of computing in contemporary culture continue to reinforce the idea that computation is a fundamental process,"[7] and recently the humanities and social sciences have taken up the examination of computation.

Early in Internet studies, Lessig pointed out the double meaning of code in this context—as legal code regulating the behavior of individuals and organizations and as source code regulating the behavior of site or program elements.[8] Likewise, Galloway focuses attention protocol on the relationships among power, distributed networks, and computers and the computational processes that enable them. This understanding of protocol is native to the current historical moment; however, it is rooted in previous uses and understandings of the term. *Protocol* is the organizing code, the "management style," that, along with digital computers and distributed networks, defines "a new apparatus of control that has achieved importance at the start of the new millennium."[9] This new attention represents both means and opportunity for examining contemporary computer-mediated computational processes and their relationship to previous human practices.

Scholars in rhetorical studies have taken an approach that seeks to account for "the real" in The Regime of Computation. As Packer and Wiley note in their introduction to *Communication Matters*, a collection of works from the 2009 North Carolina State University symposium on materialism in communication and rhetoric, "Following a long period of focusing almost exclusively upon the problematics of language, meaning, and representation, there has been a renewed interest in establishing means for making claims in the real. . . . [T]here are now attempts to set aside epistemological debates in order to compose new ontologies that enable theoretical innovations and political intervention."[10] In rhetorical studies in particular, that attention has turned to studying procedures and procedurality, such as "computational literacy" of software engineers[11] and "object-oriented rhetoric."[12] One especially productive new area of research centers on Bogost's *procedural rhetoric*, or "a new and promising way to make claims about *how things work*."[13] Like Bogost's original work, most studies of procedural rhetoric have examined videogames;[14] others have expanded the concept to include additional computational systems, such as cybertexts[15] and electronic narratives.[16]

BIG DATA AND COMPUTATION IN SURVEILLANCE AND RISK ASSESSMENT

Not surprisingly, researchers have focused on how informationalization plays an increasing role in surveillance and risk assessment, especially algorithmic surveillance, or "surveillance technologies that make use of computer systems to provide more than the raw data observed."[17] Research on algorithmic surveillance tends to center on how facial analysis, movement recognition, and biometric technologies are used to observe, record, and analyze the passage of bodies through space and time. Although the research acknowledges potential positive implementations of these technologies, such as eliminating corruption and bias,[18] most focuses on the negative repercussions. Computational processing and aggregated data are used to determine levels of access in a number of arenas, including who receives priority when routing calls in customer-service call centers,[19] who is allocated increased bandwidth when accessing the Internet,[20] and who has the right to travel on certain roadways.[21] By controlling access to both public and private resources, these technologies are implicated in both the privileging and marginalizing of populations and "must be at the center of any attempt to conceptualize the formation, maintenance and experience of social and geographical inequalities within contemporary capitalist societies."[22]

Some have noted that algorithmic surveillance is closely related to informationalization in management of mobility, and the same processes can simultaneously enable homo- and heterogeneity. Using the example of Vindigo, a location-aware city guide with maps, movie listings, and other local information, Andrejevic argues that by presenting users with distinctive views of their own paths through space and time, their "[s]ubjection to interactive surveillance becomes the precondition for orientation and, at the same time, a form of individual expression."[23] Likewise, Packer notes that in military arenas data and computation are used to construct combat zones that see (CTS), and risk is identified not with the individual but rather with the very activity of mobility.[24] This reification is akin to "categorical suspicion"[25] or the use of crime-reporting information and postal codes to label neighborhoods as dangerous.[26]

Scholars frequently observe that informationalization is implicated in perpetuating dominant ideologies and limiting deliberation. Indeed, one significant obstacle to online deliberation is the standardized character of interactions. With content filtering and personalization, Internet users are isolated. In many discussion venues online, members are able to review both contributors and their contributions, and those ratings are often used to rank submissions automatically. As a result, the best-liked users/content will likely be the most viewed. Under the best of circumstances, the better written, researched, or argued pieces will make their way to the top. Under the worst, the voice of the majority will drown out dissent, limiting the opportunity for deliberation.[27] Similarly, technologies such as collaborative filtering conceal themselves while making suggestions and are "undoubtedly an expression of power, not of someone having power over someone else, but of software making choices and connections in complex and unpredictable ways in order to shape the everyday experiences of the user."[28] This development is expressed through algorithms that construct Facebook news feeds, Google search results, and Amazon recommendations, to name a few.

THE NETWORKED SUBJECT

The proliferation of new media technologies affects humans' understanding of self, identity, and subjectivity. Chirography, print, photography, and film, for example, have all been associated with transformations in identity formation in the past. More recently, personal computers and the Web have been connected with multiple subjectivities and networked selves that enable *cycling through*,[29] or the "ability to alternate quickly among several identities."[30] Poster notes that identity in modernity is theorized in two contradictory ways, as consisting of consciousness but also strongly associated with the body. In the late eighteenth century, state use of media for documenting citizens expanded to include the body through use of dossiers and passports, for example. In the twentieth century, people began associating consumption with identity; clothing with logos or popular culture references (e.g., the catchphrase from a favorite television program) became a way of conveying identity. Therefore, contemporary identity "both captures individuality and yet exists in the forms of external traces."[31] Likewise, Barad argues that subjects do not exist outside "intra-action" with the material world but rather "*in relation to their mutual entanglement; they don't exist as individual elements.*"[32] Hayles, following Chun, notes that metaphors used to orient individuals (e.g., desktop, trashcan) come with expectations that convert people into users but that this conversion "into the machinic system does not require [their] conscious recognition of how [they are] being disciplined by the machine to become a certain kind of subject."[33] These complexities become increasingly prevalent through the use of digital computers in distributed networks.

The relationship between sites and their visitors has changed since the early days of the Web, and contemporary computational processing models allow for sites to "know" their users—based on the pages they visit, products they purchase, and media they consume. Further, this knowledge is no longer site-specific; it follows users across sites, as anyone who has seen an advertisement for shoes on a news site after shopping for new running shoes on Amazon can attest. Although it may not feel that way, identity in such systems is "*anonymous but descriptive.*"[34] In other words, although these web technologies gather, compile, and store a host of information on users, individual identity does not matter. Indeed, Internet data on search histories, usage patterns, and buying habits—or offline purchasing and mobile device location—are more significant than traditional demographics or real names. This phenomenon is closely tied to the shift in market research from use of demographics to psychographic information, a move from understanding consumers in terms of statistical categories (e.g., age, gender) toward relying on attitudes and values (e.g., political affiliation, religion).[35] This new logic is rooted in consumption and what John Cheney-Lippold calls "new algorithmic identity."[36] These instruments of subjectification are increasingly hidden while simultaneously further integrated with consumption and, in the case of online dating, (re)production.

THE CASE FOR ROMANTIC MATCHMAKING

I have chosen online dating as the cornerstone for this case study because it offers both a low point of entry and the most fertile ground on which to operate. Since matchmaking is a long and generally familiar tradition, online dating supplies an opportunity to make explicit connections that are immediately recognizable to most any reader. At the same time, it is not so familiar so as to become transparent. In addition, offering a critique of the billion-dollar online dating industry[37] will supplement the body of current, primarily social scientific, research. Next, examining the relationship between online dating and contemporary marriage provides an opportunity to position marriage not as an unchanging institution but as a historical construction. Most significantly, considering informationalization through romantic matchmaking and the establishment of procreative relationships allows for a deeper understanding of the interplay between these technologies and biopolitical regimes of power.

Recently scholars, in the social sciences particularly, have begun to focus attention on the Internet and romantic relationships. Most of this work has fallen into one of two categories: 1) studies examining the role of Internet communication in the maintenance of existing—online-only and offline-native—relationships and 2) those that investigate the Internet as a medium for establishing new romantic relationships.

Most of the work to date on the maintenance of relationships via the Internet has dealt with issues of interpersonal communication, some of which has been focused on relationships that were established and maintained online. For example, Wildermuth and Vogl-Bauer surveyed 202 members of an Internet discussion group about their online romances and found that participants seemed to use increasingly rich communication media, perhaps moving from a discussion group to text-based chat to voice over IP (VOIP), as they knew their relationship partners longer and better.[38] Additionally, Pauley and Emmers-Sommer studied thirty-six individuals who were involved in online-only relationships in order to understand better uncertainty reduction and self-disclosure strategies. While they found a relationship between the perceived possibility of meeting the online-relationship partner in the future and uncertainty reduction, they did not find a conclusive relationship between media richness and intimacy.[39] Use of Internet technologies in the maintenance of relationships established offline has received scholarly attention as well. Most notably, Sidelinger et al. surveyed 123 undergraduates who were in romantic relationships about how they use e-mail and instant message to maintain those relationships. They found a number of significant correlations, including one between participants' satisfaction with their relationship and their communication satisfaction.[40]

Online dating has become a significant object of study of late, isolated not so much as a set of practices or technologies but rather almost as a condition. Research into online dating, with its reliance on models related to the "marriage market" and its focus on issues of presentation, mimics research in marketing effectiveness and branding. Some studies use economic models to examine the connection between the Internet and romantic relationships. Heino, Ellison, and Gibbs used telephone interviews to examine Match.com in terms of social comparison theory and the marriage market, or the tendency of singles to search for the maximum cost-benefit exchange within the relationship.[41] They noted the complexity of the market metaphor in online dating: the ability to search profiles is convenient and empowering, but "shopping" for a partner denies the importance of emotional connection. Hitsch, Hortaçsu, and Ariely employed economic modeling to examine mate preferences, specifically using the Gale-Shapley algorithm to account for preferences in categories such as age and education.[42]

At present, much like studies of the online marriage market, presentation and self-disclosure has been another prominent theme in online dating research. Gibbs, Ellison, and Heino studied the relationship between anticipation of in-real-life (IRL) interactions and self-disclosure. They found that the expectation of future face-to-face meetings affected users' self-disclosure and level of honesty.[43] Ellison, Heino, and Gibbs used phone interviews with thirty-four online dating participants in order to understand the relationship between projection of an "ideal self" and realistic self-presentation. Participants seem to balance these conceptions in their profiles and used the knowledge they gained from self-reflection in order to evaluate the profiles of others.[44] Toma, Hancock, and Ellison used cross-validation to verify profile accuracy. They found that although men had a tendency to be dishonest about their height, and women about weight, people were most honest about their current relationship status and relationship history.[45] Yurchisin, Watchravesringkan, and McCabe's ethnographic study explored how eleven individuals used online dating sites to experiment with identity and how they used both online and offline feedback from other users to actualize possible selves.[46]

One study of online dating that stands out is Arvidsson's work on Match.com. Arvidsson's examination includes both a textual analysis of user profiles and a cursory examination of the site's functionality.

> Match.com caters to a symbol analytical labour which lacks both the time and freedom to pursue its basic reproductive and intimate needs. But it does this in ways that makes its fantasizing and communicative investments of affect evolve within a branded space, which in turn makes it directly economically productive. The basic biopolitical condition of this class of symbol analysts—the mediatization of their lifeworld and the mobility and flexibility of their productive condition—is positioned as a source of surplus value.[47]

In his work, Arvidsson raises a number of issues that require further inspection, and this book examines the political and economic implications in detail.

One observation from my own work—that romantic matchmaking mirrors online shopping—bolsters my belief that romantic matchmaking offers an appropriate case study. I analyzed the *About Me and Who I'm Looking For* section of fifty Match.com user profiles for users' descriptions of themselves and of their perfect match, and I found that the apparent social action in practice in the dating profiles seemed to run counter to the expected goal of online dating. The majority of profiles examined dedicated less than a third of the *About Me and Who I'm Looking For* to describing the user's perfect match. In addition, only a quarter of those who described their perfect match used the expected character allotment to depict the personality of that match.[48] Based on the profiles studied, the exigency for participating in online dating may be as simple as maintaining cultural standards. Although it was surprising that the users did not devote equal proportion of characters to describing themselves and whom they would like to meet, the results were in line with use of social-networking sites, as some critics have labeled that phenomenon "ego-casting."[49] Perhaps most striking, the Match.com profile-page layout examined closely resembles product-page layouts for online retailers, such as Amazon, and I discuss online dating site layouts, functionality, and user experience later in this book.

In order to account for romantic matchmaking's intricacies, an "apparatus form of media analysis"[50] that considers user experience, site functionality, and cultural context as a historically situated, contingent set of relations informs this project. It is hard to imagine that a person would engage exclusively in online dating as a means for meeting a mate. In fact, it is nearly impossible, as the stated goal of online dating is to facilitate face-to-face meetings that culminate in romantic relationships. It makes sense, then, that online dating exists in a complex of active and passive activities for meeting others, like flirting with coworkers or other patrons in a bookstore or restaurant, attending a singles mixer at a local coffee shop, or going on a blind date arranged by friends. In addition, people might meet potential romantic partners by participating in other Web-based activities such as playing a massively multiplayer online role-playing game (MMORPG), conversing about music on a discussion forum, or arguing politics on social media.

Likewise, online dating is a collection of technologies with varying functionalities. On one level, users experience Match.com, for example, primarily as a search tool; they can design searches and look for other users based on any number of physical characteristics; economic, employment, and education factors; or hobbies and tastes. Beyond that, users encounter the site as a communication tool through which they can exchange information with others who are registered with the service. They can express interest in a range of ways, some more explicit than others. For instance, Match.com members can set their preferences so that they see when other users have viewed their profile, making the mere act of viewing a covert way to express interest. They also can "wink" at one another. When users wink at someone to whom they are attracted, the other user receives a notification of that interest. Finally, they can send free-form messages through the site. In many ways, these features are designed to mimic offline communication techniques. If a person is interested in someone she sees on a bus or in a restaurant, she can let him catch her looking at him, give him a smile, or walk over to him and start a conversation. As discussed earlier, online dating sites can also function as a personal presentation tool. A user can set up a profile that includes a personal narrative, self-categorization in searchable classifications, and descriptions of his favorite places, his background, his ethnicity and education. In fact, since users are not required to contact or respond to others, they can use the site for the sole purpose of constructing an online self or engaging in identity play. And this complex of functionalities is associated with one site, Match.com. As we will see later in this book, the processes used by Match.com to connect users are markedly different from those employed by eHarmony.com or OkCupid.com.

While online dating does make an excellent case study, it also includes one significant shortcoming related to sexuality and gender. Most online dating sites privilege "traditional" romantic unions, especially marriage, and in doing so have generally excluded or marginalized LGBTQ users. In addition, since one goal of this project is to position online dating historically, the second chapter focuses on the position of romantic matchmaking in the context of marriage. While I do discuss the issue of marriage equality and touch on matchmaking for establishing otherwise restricted relationships, such as the use of classified advertising by gay men in the late nineteenth century, much of the analysis takes a heteronormative perspective. Similarly, this discussion primarily relies on a cisgender outlook. First, much like the challenge for sexuality, the focus on marriage lends itself to a female-male gender binary. In addition, when discussing use of online dating sites, it is less confusing to use feminine and masculine pronouns to discuss profile viewing, for example.

APPARATUS, ARCHIVE, AND ANALYSIS

Galloway discusses protocol in "its political sense, [as] a pseudo-ideological force that has influence over real human lives."[51] He argues that protocol is strongly related to Deleuze's control, a regime of power that relies on information technologies to observe and manage people's movements, and Foucault's idea of biopolitics, or the use of quantified data for the management of population. One example Galloway provides that is especially meaningful here is collaborative filtering's use of "powerful algorithms to determine and at the same time reflect the identity of the user."[52]

> Collaborative filtering is therefore an extreme example of the protocological organization of real human people. Personal identity is formed only on certain hegemonic patterns. In this massive algorithmic collaboration, the user is always suggested to be like someone else, who, in order for the system to work, is already like the user to begin with! Collaborative filtering is a synchronic logic injected into a social relation.[53]

Elsewhere Galloway calls for "a viable critique of collaborative filtering," noting that "examination of the de-politicization of algorithms will help."[54] Such an endeavor requires a layered approach. As a logic of social organization, informationalization must be situated within larger cultural and historical contexts. As a process closely linked to information technology, such as the microprocessor,[55] informationalization should be examined in terms of its procedurality. Therefore, *Building Relationships* relies on Packer's apparatus-centered approach to position online dating within a constellation of historical, cultural, social, juridical, and technological relations.

AN APPARATUS FORM OF MEDIA ANALYSIS

Packer's approach to examining communication technologies is rooted in Agamben's apparatus and Williams's mobile privatization. Agamben, following Foucault, defines apparatus as "a set of practices and mechanisms (both linguistic and nonlinguistic, juridical, technical, and military) that aim to face an urgent need and to obtain an effect that is more or less immediate."[56] He proposes two expansive categories, one of living beings and one of "apparatuses in which living beings are incessantly captured."[57] The subject lies in the space between living beings and apparatuses. Indeed, without the process of subjectification, apparatuses do not function: "Apparatus, then, is first of all a machine that produces subjectifications, and only as such is it a machine of governance."[58] Williams identifies mobile privatization with a number of developments that came together in the mid-twentieth century to enable the proliferation of televisions:

> Socially, this complex is characterised [sic] by two apparently paradoxical yet deeply connected tendencies of modern urban industrial living: on the one hand mobility, on the other hand the more apparently self-sufficient home . . . [that] served an at-once mobile and home-centered way of living.[59]

For Williams, television served as a fulcrum for these seemingly contradictory forces. As mobile privatization emerged, the television allowed for the single-family home to move to the center of individuals' everyday lives while simultaneously keeping them connected to the outside world.

Building on Agamben and Williams, Packer defines the apparatus as "a strategically organized network of discursive and nondiscursive elements brought together to address problems resulting from specific forms of knowledge,"[60] and he offers the cigarette as one example. The cigarette apparatus produces the smoker subject, a process that in turn both structures the allocation of personal resources and desires and creates opportunities to "unsmoke or reform" the smoker subject through public-health programs.[61] He goes on to point out that, although Williams includes television programming in his discussion of mobile privatization, content is not Williams's primary concern. Rather, it is one point in a larger assemblage, including the expansion of consumer goods, increased automobility, and radio broadcasting. Packer suggests that we also shift our focus from ideological critique, noting that mobile privatization is a process of "demassing": "It does not operate merely through ideological mystification but by unmaking forms of social movement and organization whose shared spatial and temporal, not to mention class, experience was deemed dangerous."[62] In order to move beyond a focus on content and ideology, Packer's archive offers a way of examining these complex, contingent relations.

According to Packer, "the archive of the apparatus comprises the sign, signifier, and technologies of inscription and maintenance. It also obviously would need to encompass insight into the power relations enacted through communication and the communication structured by power relations"[63] as well as "the plans and theories in use at the time which allow for an understanding of how such an arrangement of forces might come into being."[64] To that end, he offers five areas of investigation.

First, we must understand the notion of authority—who decides what is true, as well as where and how issues of truth are decided. According to Packer, *determinators* are "organizations, institutions or credentialized experts" who "have been given, granted, or taken authority to make truth claims regarding specific phenomena," and he recommends that we consider the knowledge formations that enable the apparatus as well as the "games of truth" that validate them.[65] He discusses his experience with citizen band (CB) radio and the challenge of studying a medium that leaves no traces, that offers no "content" to study. Building the archive of the CB radio within a "disciplined mobility"[66] apparatus required determining who had the authority to define the bounds of appropriateness for the technology, both in terms of its deployment and its users' behavior. In addition, the archive had to include formal deliberations for determining jurisdiction as well as how stakeholders made claims to dominion over the technology—and how those claims were then used to support a particular articulation of the apparatus.[67] In the case of the CB radio and disciplined mobility, that meant examining not only FCC debates and regulations but also user manuals surrounding its use. For online dating, that means building an archive that includes religious, legislative, and juridical arguments, as well as "popular knowledge and expertise"[68] that legitimate relationship formations (i.e., "traditional" marriage), intervene into the conduct of the married couple, or seek to reform the single subject.

Second, we are to identify *free-floating legitimators*,[69] terms that are used within multiple discourses to validate various actions. Elsewhere, Packer uses the example of *safety* as a free-floating legitimator for understanding how certain populations (e.g., motorcyclists, young people) are identified, and in some instances constructed, as dangerous based on assessments of their potential for loss: "The claim that some activity, product, or form of conduct is unsafe automatically legitimates public concern, media worthiness, litigation, and governmental involvement. The space for public debate about nearly any topic is limited not by what is the just, the good, or the democratic, but rather by what is safe."[70] In the case of online dating, compatibility serves as a free-floating legitimator that enables and justifies processes of subjectification.

Third, we must account for "the full compliment [sic] of competing discourses in their combative play of force" and "how such force was exerted, thwarted, and redirected."[71] He suggests that by studying these techniques of power—how approaches for managing conduct succeeded or failed, how they were (re)focused or (re)invented—we can develop a clearer understanding of how power relations are enacted. In his own work on CB radio, Packer points to how the technology was used in sanctioned ways for state (e.g., military and police operations) and business or civil (e.g., delivery and ambulance fleet management) interests as well as for resistant (e.g., trucker protests) purposes.[72] In the chapters that follow, I examine how romantic matchmaking, and online dating in particular, has both reified and unsettled longstanding beliefs about marriage.

Fourth, we are to consider the role of media and communication technologies in the apparatus. Packer labels these components as *articulators* or *con-*

nectors, and they make available the tools, infrastructure, and procedures that empower the apparatus. In Packer's work, the CB radio serves as an articulator in the disciplined mobility apparatus. Within mobile privatization, it transforms the relationship between home, work, and the automobile by adding a two-way means of communication to the in-car radio's broadcast reception. Through this transformation, "the private sphere of the automobile became a public sphere that could be and was used to initiate political protest and to foster noncommercial forms of (mass) communication."[73] In this context, online dating works as an articulator that both stabilizes marriage during this time of transition and reforms risky single subjects.

Fifth, the process of subjectification is foundational to the apparatus; this process "works on life itself, what it means, how it should be lived, and who can manage it (biopolitics)."[74] Packer recommends considering not only areas such as psychiatry but also popular discourse, such as self-help publications. His CB radio research included personal narratives and popular culture stories that present models for novel subjectivity as well as how-to manuals and hitchhiking advice. Packer suggests paying close attention to "cases in which a subject is turned into an object."[75] When considering online dating, processes of subjectification occur through a variety of mechanisms. Online dating success stories give singles a subject position with which to identify; the process of online dating itself reconstitutes single subjects as marriage-minded adults, regardless of their individual aims.

ONLINE DATING AS ARTICULATOR

Of Packer's five components of the archive, understanding the workings of articulators is the most challenging. The archive must include both a mechanism to account for the mostly invisible processes that power articulators and a vocabulary for discussing them. To that end, I adapt Bogost's procedural rhetoric framework for studying videogames to the processes at work in online dating.

Procedural Rhetoric

Bogost joins rhetoric with procedurality, "a way of creating, explaining, or understanding processes" in order to develop a style of analysis.[76] Although he focuses on the study of videogames, procedural rhetoric offers an excellent frame for examining persuasive processes at work in any number of human-computer interactions, including Web applications such as online dating. Procedurality is not exclusive to computers, and Bogost points out that procedures rule interactions we engage in daily. Three conditions of procedurality—possibility, situatedness, and concealment[77]—are particularly relevant here. First, procedurality determines what is and is not possible. For example, a network of affordances and constraints governs our transactions with retailers. We stand in line to make a purchase, pay for items we have chosen, and then take possession of them. Even when it seems procedure is being broken (Bogost offers the example of cashiers allowing the return of items after a store's grace period has expired), that is not the case. "[T]hey are mustering new processes—for example, a process of promoting repeat business, or for preventing a commotion—and seamlessly blending them with the procedure for product returns."[78] These procedures, such as return policies, not only restrict behavior but also permit it. As Bogost points out, we might not even consider returning an item if stores did not have return policies. Second, procedurality is situated, culturally determined and contingent. If we return to the retail example, the manner in which customers wait is governed by decorum. Standing in a line that forms in front of a cashier is typical in the United States. In other countries, waiting is structured differently, with customers forming small crowds in front of or around the attendant, for example. These procedures may feel "flexible and porous, but they are crafted from a multitude of protracted, intersecting cultural processes."[79] Third, procedurality is generally "nonobvious,"[80] either as a result of intentional concealment or of process complexity. Military procedures might be purposefully hidden for national security reasons; the code that enables a word-processing program is obscured by the interface through which the user accesses it, both to simplify use and to protect proprietary information. Regardless of intention, these examples of procedurality are not obvious to the public or the casual user. While some moments of procedurality, such as the system of processes involved in passing through airport security, are visible and actively experienced, most of the time its operation is not visible to us.

When combined with rhetoric, procedurality becomes *procedural rhetoric*, "the art of persuasion through rule-based representations and interactions rather than the spoken word, writing, images, or moving pictures."[81] Through this lens, "rules of behavior, the construction of dynamic models"[82] function rhetorically; they construct reality, informing behavior and constituting identities. By way of illustration, Bogost offers *The McDonald's Videogame*, in which players manage four enterprises: a cattle farm in the developing world, a slaughterhouse, a restaurant franchise, and the company's home office. This game "mounts a procedural rhetoric about the necessity of corruption in the global fast food business, and the overwhelming temptation of greed, which leads to more corruption."[83] He compares the game to publications that make similar arguments (e.g., Eric Schlosser's *Fast Food Nation*). Procedural rhetorics, and the unique variety of engagement they afford, shed light on online dating sites, the subjectivities and relationship formations permitted (and denied) by them, and the cultural assumptions they rely on for theorizing compatibility and compelling use.

Challenges to Studying Procedural Rhetorics

While studying procedural rhetorics opens up new ways to think about everyday interactions with technology, analyzing computational processing brings with it some challenges. First, computational code is hidden within black boxes, or systems that obscure their functionality.[84] As discussed earlier, this concealment works both to simplify interactions with users and to protect patented software. According to Bogost, this challenge can be understood in terms of black- and white-box analysis, two distinct analytical processes from software engineering and testing.[85] In the former, developers and testers observe a program's inputs and outputs to identify possible methods or bugs based on suppositions about the software. In the latter, developers and testers identify actual approaches or bugs based on available information about the software.[86] These terms generally describe activities undertaken by the software's producer (e.g., in-house quality-assurance testing, contracted usability testing). Still, researchers and critics can seek out available information for conducting their own analyses. Bogost offers examples from videogames, such as descriptions of *The Sims* architecture from presentations given at conferences or documentation of common components in the software that powers *Grand Theft Auto*. He also recommends "[p]ublicly documented hardware and software specifications [and] software development kits" as ways into the study of computational processing.[87]

Second, studying code raises issues of "procedural authorship."[88] Bogost likens the call to "open the box" of videogame engines to some literary critics' desire to make suppositions about an author's intent. He reminds us that through dissemination texts become distanced from their authors. Especially important to the study of computational processing, the distancing enables a number of different "readings." In other words, a videogame or online dating is "coauthored" by the game/site and the player/user. This understanding of procedural authorship, combined with an approach rooted in white-box analysis, lays the foundation for my approach to examining procedural rhetorics.

A PROCEDURAL RHETORICAL APPROACH TO STUDYING ARTICULATORS

There are three recursive phases required for building the artifact in my approach to procedural rhetorical analysis:

1. Gathering information on application functionality
2. Engaging with the user experience[89]
3. Developing a framework for analyzing both the application functionality and the user experience

As Bogost discussed, conducting a procedural rhetorical analysis relies on incomplete knowledge of the application's code. Since many of these applications are produced by for-profit enterprises that provide limited or no access to the processes and engines that power their software, this requires a bit of detective work. Bogost mentions reviewing software development kits, presentations at academic and trade conferences, hardware and software specifications, and so on.

Likewise, my first step in preparing to examine online dating sites as procedural rhetorics was to gather information about site functionality from all available sources. In addition to searching for software documentation and public presentations as suggested by Bogost, I looked for patent documents, and Website articles as well as published works of or interviews with company representatives to garner information about how the sites worked. As table I.1 shows, procedural rhetorical analysis requires gathering a wide variety of materials with the aim of constructing an archive. When studying online dating site eHarmony.com, I reviewed the company's 2004 patent document that describes the technical procedures of its romantic matchmaking system; *Falling in Love for All the Right Reasons: How to Find Your Soul Mate*, a self-help book in which founder Neil Clark Warren discusses the characteristics measured by the site both to determine users' likelihood for relationship satisfaction and to make matches between users; and pages on the company's Website that describe the matchmaking process to its users. For my analysis of Match.com, I relied on popular press interviews with site executives and engineers, coverage in technical journals, and com-

Table I.1. Sources of Information on Site Functionality

	eHarmony.com	Match.com	OkCupid.com
Patents and software specifications	X		
Initial account setup	X	X	X
Member profile page(s)	X	X	X
Member dashboard	X	X	X
Communication processes	X	X	X
Published articles and interviews with site representatives		X	X
Books written by site executives	X		X
Technical journal articles		X	
Site pages (e.g., About, FAQ) or blogs	X	X	X

pany Web pages. With OkCupid.com, I used information published on the site and company blog, interviews with site executives, and cofounder Christian Rudder's *Dataclysm: Who We Are (When We Think No One's Looking)*.

Since procedural rhetorics are expressed through the interaction of the user and the application, it is important to engage with the user experience. With a videogame, that means participating in and documenting game play. For my study of online dating, I set up accounts and reviewed matching procedures and communications processes for eHarmony.com, Match.com, and OkCupid.com.[90] That included answering eHarmony.com's nearly three-hundred questions on twenty-four screens in the initial account setup as well as reviewing and documenting the layout and functionality of the member dashboard and Guided Communication process. In the case of Match.com, the initial account setup required answering nearly sixty questions, with approximately two-hundred variables total, on eighteen screens. The OkCupid.com process involved answering and ranking the importance of nearly fifty questions, each on its own page. I also familiarized myself with both the profile page and member dashboard. In all instances, I documented processes by taking screenshots.

After gathering information on site functionality and engaging with the user experience, this approach necessitates developing a framework for analyzing procedural rhetorics and discussing results. Although not all videogames are the same, there are some consistencies across games within the same genre or game system. Likewise, Web applications with similar outcomes likely share commonalities that will inform the development of a framework for analysis. As with my study, this framework may be discovered while conducting research during the first phase but may not demonstrate its potential as a framework until the second phase. The 2004 eHarmony.com patent document offered both a precise vocabulary and specialized framework for the analysis.

The eHarmony patent breaks its matchmaking system into three *stages*: data preparation, matching, and communications.[91] Of course, the patent document frames all three of these stages in terms of the role of the site—that is, preparing empirical data, classifying and then matching complementary users, facilitating user contact—rather than how the user engages with the site. However, as we discussed earlier, procedural rhetorics are expressed through site functionality, user experience, and how the two interact. The patent's framework provided me with an excellent means for analyzing system utilities and user encounters, but stages connotes a linear trajectory. The system initially circulates through data preparation, matching, and communications sequentially, but some processes do recur after an account has been established. In addition, the user experience is not as straightforward, and user and site activities in the same stage do not necessarily take place simultaneously. As a result, I found *flow* a more appropriate term for describing processes, both in terms of site operation and user practice. This term also harkens back to Williams's examination of television, in which he identifies contemporary television arrangement as flow. For Williams, flow means moving seamlessly between programming units as a technique for holding viewers' attention. In addition, flow underscores the importance of continuity to the Web application user experience.[92] Since other online dating sites follow similar procedures, I applied this framework to my analysis of Match.com and OkCupid.com as well.

Online dating's layers of user experience produced additional complications that required further delineation. Users are both working to attract the attention of others and identifying people whom they find attractive, engaging in both self-presentation and other-evaluation, sometimes simultaneously. This division needed discrete naming for these two primary roles, presenter and evaluator, when describing or analyzing Web pages or activities that bring them together. For example, a profile page can be discussed from two user perspectives; the information on the profile page presents User A (the owner of the profile, or presenter) but is evaluated by User B (the viewer of the profile, or evaluator). The processes in which users engage are complex, and this demarcation simplified discussion of them.

IN THIS BOOK

This book employs Packer's apparatus approach to media analysis to examine online dating. Using that framework, chapter 1 surveys the contingent social, cultural, familial, economic, and technological relationships that make up the marriage apparatus. In particular, I focus on the practice of romantic matchmaking as part of that apparatus and a) define it in terms of three conditions—intermediation, mediation, and automation—and b) position it in relation to regimes of power. Intermediation describes the condition in which the marriage match is facilitated by active human decision-making and is closely associated with sovereign power configurations in which kin relations play an important role in how societies are organized, resources are managed, and power is exercised. Like intermediation, mediation is a condition that relies on human decision-making. However, that decision-making no longer belongs to a human intermediary who introduces the potential couple but rather initiates from individuals who engage in their own matching practices using communication technologies as mediators. Mediation is most closely associated with disciplinary power configurations in which institutions must, after changes that include urbanization and mobility, govern at a distance. Automation is a condition in which matching is outsourced to nonhuman operators, specifically computer-mediated matching that usually relies on algorithmic processing. Although this matchmaking may be based in human decision-making, such as social-scientific research, that human decision-making is not visible in the interactions between the matchmaker and the matched. Automation is most closely associated with governmental power configurations and control societies in highly mobile, networked cultures that require management of population.

Chapter 2 uses Bogost's procedural rhetoric to analyze online dating as a formal apparatus through in-depth exploration of computational processes and logics that enable the functionality of three online dating sites—eHarmony.com, Match.com, and OkCupid.com. Grounding this inquiry in procedurality allows me to move beyond issues of representation to examine subjectivities in terms of the persuasive processes that enact them. Specifically, I examine the data preparation, matching, and communications flows from the perspective of both user experience and site functionality. Since the kind of information that is employed to match users reveals much of how sites conceive of subjects, I pay special attention to the data preparation logics of each site, which operate during the initial account setup phase. In addition, I focus on two components of procedurality, possibility (permission-denial) and concealment (transparency-opacity), as they play out in these online dating sites. This chapter both models a process for extending Bogost's framework to applications other than videogames and lays the groundwork for examining online dating's role in contemporary processes of subjectification in the next chapter.

Chapter 3 studies the implications for subjectification identified by the previous chapter's descriptive analysis. Just as procedurality defines permission-denial and transparency-opacity, so is it situated. Procedural rhetoric offers a opportunity to consider online dating in terms of "the coupling of a cultural process and its computational representation," especially "taking a set of cultural systems apart to see what logics motivate their human actors.[93] Using enthymematic procedures and thought processes, I further explicate the cultural assumptions that inform eHarmony.com, Match.com, and OkCupid.com. I argue that compatibility operates as a free-floating legitimator that serves to compel online dating use and to buttress "traditional" relationship formations.

Chapter 4 examines online dating success stories from eHarmony.com, Match.com, and OkCupid.com as confessional discourse. I rely on Foucault's discussion of the confession's transformation during the seventeenth century, when confessions of the flesh shifted from a singular focus on detailed descriptions of physical acts to an inclusive cataloging of all sex-related deed and thought. He links this transformation to the emergence of biopolitics, a new form of power that uses quantified data—such as birth and marriage rates or mortality statistics—and works at the level of population. Online daters are expected to share vast amounts of demographic and lifestyle information. And although romantic matchmaking is an ancient tradition, online dating is a set of technologies and practices unique to the contemporary moment. Indeed, online dating, much like the confession, works as a technique of power. While online dating success could be defined in a variety of ways (e.g., length of relationships established, number of dates facilitated, percentage of dates that lead to sexual intercourse), most sites define success as marriage. Today, when the question could be not only when or whom to marry but also whether to marry, the increasing prominence of online dating requires our attention. I argue that these success stories, in conjunction with sites' functionality, work to reify "traditional marriage" during this period of destabilization and transformation by converting dangerous single subjects to the (re)productive couple.

Chapter 5 uses debates surrounding marriage equality to examine the relationship between marriage and family in contemporary control societies. In "Postscript on the Societies of Control," Gilles Deleuze contrasts Foucault's disciplinary societies and their reliance on enclosure with control societies that use "free-floating control" to govern.[94] He highlights five important institutions of confinement—prison, hospital, factory, school, family—that are threatened as control societies emerge. He goes on to discuss some early control techniques put into place by institutions (e.g., prison systems' use of electronic monitoring for nonviolent offenders). Deleuze's "Postscript" previews many early adoptions of control apparatuses that have become pervasive in our culture. Although he explicitly names the family as an "interior in crisis,"[95] Deleuze does not return to it in "Postscript." As the struggle over same-sex marriage shows, an expansive definition of family may indicate its sustainability as a social institution, but marriage's instability may threaten that sustainability. I argue that although it appears a fitting response to reinforce marriage's boundaries, online dating's (and control's) synchronic logics are incompatible with the historically diachronic logic of marriage.

The final chapter returns to the larger discussion of informationalization. I discuss human decision-making in contrast with computer logics and argue that the labels we use to name many of the technologies so common on the Internet—search engine, recommender system, collaborative filtering—are not sufficiently descriptive. In order to shed light on the tension between user experiences on the Web and the underlying processes that enable them, I propose a new term, *matching technologies*, that describes the combination of big data, computational processing, and the values and cultural assumptions built into informationalization. I also offer a brief look at another matching technology, Netflix's Cinematch, and recommend areas for further study.

NOTES

1. Charles Duhigg, "How Companies Learn Your Secrets," *New York Times*, February 16, 2012, www.nytimes.com/2012/02/19/magazine/shopping-habits.html.

2. Robinson Meyer, "Everything We Know about Facebook's Secret Mood Manipulation Experiment," *The Atlantic*, June 28, 2014, last modified September 8, 2014, www.theatlantic.com/technology/archive/2014/06/everything-we-know-about-facebooks-secret-mood-manipulation-experiment/373648/.

3. I am using the term as defined by danah boyd and Kate Crawford. "There is little doubt that the quantities of data now available are often quite large, but that is not the defining characteristic of this new data ecosystem. . . . Big Data is less about data that is big than it is about the capacity to search, aggregate, and cross-reference large data sets." "Critical Questions for Big Data," *Information, Communication, & Society* 15, no. 5 (2012), 662–79, 665.

4. Duhigg, "How Companies Learn Your Secrets."

5. Manuel Castells, Shujiro Yazawa, and Emma Kiselyova, "Insurgents Against the Global Order: A Comparative Analysis of the Zapatistas in Mexico, the American Militia and Japan's AUM Shinrikyo," *Berkeley Journal of Sociology* 40 (1995–1996), 21–59, 21n.

6. Throughout this book, I use *Internet* and *Web* interchangeably.

7. N. Katherine Hayles, *My Mother Was a Computer: Digital Subjects and Literary Texts* (Chicago: University of Chicago Press, 2005), 30.

8. Lawrence Lessig, *Code: And Other Laws of Cyberspace*, Version 2.0 (New York: Basic Books, 2006).

9. Alexander R. Galloway, *Protocol: How Control Exists after Decentralization*. Cambridge (MA: MIT Press, 2004), 25.

10. Jeremy Packer and Stephen B. Crofts Wiley, *Communication Matters: Materialist Approaches to Media, Mobility, and Networks* (Hoboken: Taylor and Francis, 2011), 4.

11. Annette Vee, "Understanding Computer Programming as Literacy," *Literacy in Composition Studies* 1, no. 2 (2013), 42–64.

12. Scot Barnett, "Toward an Object-Oriented Rhetoric: A Review of *Tool-being: Heidegger and Metaphysics of Objects* and *Guerrilla Metaphysics: Phenomenology and the Carpentry of Things* by Graham Harman," *Enculturation* 7 (2010), enculturation.net/toward-an-object-oriented-rhetoric; James J. Brown Jr., "The Decorum of Objects," *Clinamen*, September 12, 2011, clinamen.jamesjbrownjr.net /2011/09/12/the-decorum-of-objects/.

13. Ian Bogost, *Persuasive Games: The Expressive Power of Videogames* (Cambridge, MA: MIT Press, 2007), 125 (emphasis original).

14. Matt King, "Procedural Rhetorics /Rhetoric's Procedures: Rhetorical Peaks and What It Means to Win the Game," *Currents in Electronic Literacy* (2010), currents.dwrl.utexas.edu/2010/king_procedural_rhetorics_rhetorics_procedures; Molly Storment, "Using Games to Teach Procedural Rhetoric," *HASTAC: Humanities, Arts, Science, and Technology Alliance and Collaboratory*, October 22, 2011, www.hastac.org/blogs/mhstorment/2011/10/22/using-games-teach-procedural-rhetoric.

15. Kevin Brock, "One Hundred Thousand Billion Processes: Oulipian Computation and the Composition of Digital Cybertexts," *Technoculture: An Online Journal of Technology in Society* 2 (2012), tcjournal.org/drupal/vol2/brock.

16. James J. Brown Jr., "Crossing State Lines: Rhetoric and Software Studies," in *Rhetoric and Digital Humanities*, edited by Jim Ridolfo and William Hart-Davidson (Chicago: University of Chicago Press, 2015), 20–32.

17. Lucas D. Introna and David Wood, "Picturing Algorithmic Surveillance: The Politics of Facial Recognition Software," *Surveillance and Society* 2, no. 2–3 (2004), 177–98, 181. See also Stephen Graham and David Wood, "Digitizing Surveillance: Categorization, Space, and Inequality," *Critical Social Policy* 23 (2005), 227–48; David Lyon, "Technology vs. 'Terrorism': Circuits of City Surveillance since September 11th," *International Journal of Urban and Regional Planning Research* 27, no. 3 (2003), 666–78.

18. Graham and Wood, "Digitizing Surveillance"; Introna and David Wood, "Picturing Algorithmic Surveillance."

19. David Beer, "Power through the Algorithm? Participatory Web Cultures and the Technological Unconscious," *New Media & Society* 11, no. 6 (2009): 985–1002; Stephen Graham, "Software-Sorted Geographies," *Progress in Human Geography* 29 (2005): 562–80; Graham and Wood, "Digitizing Surveillance."

20. Graham, "Software-Sorted Geographies."

21. Graham, "Software-Sorted Geographies"; Graham and Wood, "Digitizing Surveillance."

22. Graham, "Software-Sorted Geographies," 575.

23. Mark Andrejevic, "Monitored Mobility in the Era of Mass Customization," *Space and Culture* 6, no. 2 (2003), 132–50, 142.

24. Jeremy Packer, "Becoming Bombs: Mobilizing Mobility in the War on Terror," *Cultural Studies* 20, no. 4–5 (2006), 378–99, 392.

25. Andrejevic, "Monitored Mobility," 136.

26. Roger Burrow and Nick Ellison, "Sorting Out Places? Towards a Social Politics of Neighbourhood Information," *Information, Communication, & Society* 7, no. 3 (2004), 321–36.

27. Azi Lev-On and Bernard Manin, "Happy Accidents: Deliberation and Online Exposure to Opposing Views," in *Online Deliberation: Design, Research, and Practice*, ed. Tom Davies and Seeta Pena Gangadharan (Stanford: CSLI Publications, 2009), 105–22.

28. Beer, "Power through the Algorithm?" 997.

29. Sherry Turkle, *Life on the Screen: Identity in the Age of the Internet* (New York: Touchstone, 1995); Paul C. Adams, *The Boundless Self: Communication in Physical and Virtual Spaces* (Syracuse: Syracuse University Press, 1998).

30. Adams, *The Boundless Self*, 100.

31. Mark Poster, *Information Please: Culture and Politics in the Age of Digital Machines* (Durham, NC: Duke University Press, 2006).

32. Karen Barad, *Meeting the Universe Halfway: Quantum Physics and the Entanglement of Matter and Meaning* (Durham, NC: Duke University Press, 2007), 33.

33. Hayles, *My Mother Was a Computer*, 61.

34. Galloway, *Protocol*, 69 (italics original).

35. John Cheney-Lippold, "A New Algorithmic Identity: Soft Biopolitics and the Modulation of Control, *Theory, Culture, & Society* 28, no. 6 (2011), 164–81.

36. Ibid., 165.

37. Nick Paumgarten, "Looking for Someone: Sex, Love, and Loneliness on the Internet," *New Yorker*, July 4, 2011, www.newyorker.com/magazine/2011/07/04/looking-for-someone.

38. Susan M. Wildermuth and Sally Vogl-Bauer, "We Met on the Net: Exploring the Perceptions of Online Romantic Relationship Participation," *Southern Communication Journal* 72, no. 3 (2007), 211–27.

39. Perry M. Pauley and Tara M. Emmers-Sommer, "The Impact of Internet Technologies on Primary and Secondary Romantic Relationship Development," *Communication Studies* 58, no. 4 (2007), 411–27.

40. Robert J. Sidelinger, "Couples Go Online: Relational Maintenance Behaviors and Relational Characteristics Use in Dating Relationships," *Human Communication* 11, no. 3 (2006), 341–56.

41. Rebecca D. Heino, Nichole B. Ellison, and Jennifer Gibbs, "Are We a Match? Choosing Partners in the Online Dating Market" (paper presentation, International Communication Association Annual Conference, New York City, New York, May 2005).

42. Gunter J. Hitsch, Ali Hortaçsu, and Dan Ariely, "Matching and Sorting in Online Dating," *American Economic Review* 100, no. 1 (2010), 130–63.

43. Jennifer Gibbs, Nichole B. Ellison, and Rebecca D. Heino, "Self-Presentation in Online Personals: The Role of Anticipated Future Interaction, Self-Disclosure, and Perceived Success in Internet Dating," *Communication Research* 33 (2006), 152–77.

44. Nichole B. Ellison, Rebecca D. Heino, and Jennifer Gibbs, "Managing Impressions Online: Self-Presentation Processes in the Online Dating Environment," *Journal of Computer-Mediated Communication* 11 (2006), 415–41.

45. Catalina L. Toma, Jeffrey T. Hancock, and Nichole B. Ellison, "Separating Fact from Fiction: An Examination of Self-Presentation in Online Dating Profiles," *Personality Social Psychology Bulletin* 38, no. 8 (2008), 1023–36.
46. Jennifer Yurchisin, Kittichai Watchravesringkan, and Deborah Brown McCabe, "An Exploration of Identity Re-Creation in the Context of Internet Dating," *Social Behavior and Personality* 33, no. 8 (2005), 735–50.
47. Adam Arvidsson, "'Quality Singles': Internet Dating and the World of Fantasy," *New Media & Society* 8, no. 4 (2006), 686–87.
48. Dawn Shepherd, "Match.com, Genre, and the Conception of Self" (paper presentation, Biennial Conference of the Rhetoric Society of America, Seattle, Washington, May 2008.
49. Michael J. Bugeja, "Facing the Facebook," *Chronicle of Higher Education*, January 23, 2006, chronicle.com/article/Facing-the-Facebook/46904.
50. Jeremy Packer, "What Is an Archive? An Apparatus Model for Communications and Media History," *The Communication Review* 13, no. 1 (2010), 88–104, 94.
51. Galloway, *Protocol*, 81.
52. Ibid., 114.
53. Ibid., 114–15.
54. Alexander R. Galloway, "Protocol," *Theory, Culture & Society* 23, nos. 2–3 (2006), 317–20, 320.
55. Manuel Castells, "Rise of the Network Society" (*The Information Age: Economy, Society, and Culture*, Volume I) (Oxford: Blackwell Publishers, 1996).
56. Giorgio Agamben, *What Is an Apparatus? and Other Essays* (Stanford: Stanford University Press, 2009), 8.
57. Ibid., 10.
58. Ibid., 20.
59. Raymond Williams, *Television: Technology and Cultural Form* (New York: Schocken Books, 1975), 19.
60. Packer, "What Is an Archive?" 89.
61. Ibid., 93.
62. Ibid., 94.
63. Ibid., 94.
64. Ibid., 95.
65. Ibid., 100–1.
66. Ibid., 97. Packer defines *disciplined mobility* as "orchestration and governance of technologies of autonomous motorized transport (primarily automobiles and motorcycles) according to the seemingly competing logics of freedom and safety into a well-disciplined machine for efficient and productive movement of peoples and goods."
67. Ibid., 97.
68. Ibid., 95.
69. Ibid., 101.
70. Ibid.
71. Ibid., 101–2.
72. Ibid., 98.
73. Ibid.
74. Ibid., 102.
75. Ibid., 102–3.
76. Bogost, *Persuasive Games*, 3.
77. These are my terms for Bogost's concepts.
78. Bogost, *Persuasive Games*, 6.
79. Ibid., 7.
80. Ibid., 8.
81. Ibid., ix.
82. Ibid.
83. Ibid., 31.
84. Ibid., 62.
85. Ibid.

86. Ibid., 63.
87. Ibid.
88. Ibid., 29.
89. Don Norman, former vice president of Apple's Advanced Technology Group, coined the term *user experience* in 1998. According to Norman, the user experience is "all aspects of the person's experience with the system" and includes such elements as "industrial design graphics, the interface, the physical interaction and the manual." Although formally addressing components such as graphic design and interface fall outside the scope of this project, user experience is a useful term for distinguishing what online dating users do from how the site works. Interview by Peter Merholz, *Adaptive Path*, December 13, 2007, www.adaptivepath.com /ideas/e000862/.
90. Since this project did not involve work with human subjects, my accounts for the three sites studied included minimal information and no photographs. When possible, they were not visible or searchable.
91. J. Galen Buckwalter et al., Method and System for Identifying People Who Are Likely to Have a Successful Relationship, US Patent 6,735,568, filed on August 10, 2000, and issued May 11, 2004, 5.
92. Williams, *Television*, 91. Galloway defines *continuity* as "the set of techniques practiced by Webmasters that, taken as a totality, create this pleasurable, fluid experience for the user" and argues that, since the internet does not employ the same narrative and temporal elements as other media (e.g., television), continuity is what captivates users. *Protocol*, 55. Christian Rudder remarks that the "tension—between the *continuity* of the human condition and the fracture of the database—can make running a website complicated." *Dataclysm*, 13.
93. Ibid.
94. Gilles Deleuze, "Postscript on the Societies of Control." *October* 59 (Winter, 1992), 3–7, 4.
95. Ibid.

Chapter One

Romantic Matchmaking and the Marriage Apparatus

Online dating is a growing part of American life. According to the Pew Research Center, 42 percent of American adults know someone who has participated in online dating, an increase of 11 percent since 2005. Likewise, 29 percent know someone who found a long-term partner or spouse through a dating site, which is double the number of just eight years ago. And although the percentage of American adults who participate in online dating has remained fairly consistent between 2005 and 2013, they are much more likely to meet a person they have connected with through the service.[1] As online dating develops increasing acceptance as a vehicle for establishing procreative relationships, it is important to understand not only how it circulates within contemporary American culture but also how it relates to antecedent matchmaking processes. Following Packer, this chapter positions online dating within a marriage apparatus. Packer's apparatus-based analysis allows for considering matchmaking within a network of relations. The archive for this chapter includes popular-press coverage as well as histories of the social institution of marriage and communication technologies used for matchmaking. In particular, this chapter focuses on religious, legislative, and juridical efforts to assert authority in defining marriage (i.e., determinators), communication technologies' role in (marriage) matchmaking, and the competing discourses at work in those struggles. Here I consider romantic matchmaking alongside marriage and then complicate that connection in subsequent chapters. I analyze matchmaking in terms of three conditions—intermediation, mediation, and automation—that describe a marriage apparatus that includes the relationship among matchmaking, the potential couple, and communication technologies.

ROMANTIC MATCHMAKING: INTERMEDIATION, MEDIATION, AUTOMATION

We can think of romantic matchmaking in terms of intervention into everyday lives and its place in a set of relations we will call the marriage apparatus, a contingent set of relations that includes the cultural practices, material conditions, and linguistic components that seek to define, enable, and constrain marriage. Though the purpose of romantic matchmaking is cast fairly consistently throughout most of history as facilitating marriage, how matchmaking has intervened, and the relationship between the matchmaker and the matched, vary both across histories and among cultures. Contemporary users of online dating may seek a range of relationship formations in addition to marriage (e.g., dating, casual sexual, cohabitating), but the longest-running matchmaking objective is arranging marital unions.

Intermediation

Intermediation describes the condition in which an authority figure or a matchmaker facilitates the marriage match through active human decision-making. It is closely associated with cultural formations in which kin relations play an important role in how societies are organized, resources are managed, and power is exercised. This process can be more or less formalized, depending on the situations in which the matching occurs. For example, the earliest marriages most likely emerged as a way to arrange sex, childrearing, and division of labor as nomadic groups used these unions to establish relationships with other groups. Men, women, and children within the same camp hunted, gathered, and shared their resources reciprocally, and collaboration was rewarded with increased status in the group.[2] Not surprisingly, having children with familial bonds in multiple camps extended cooperative relationships. A prominent member of the community likely facilitated these exchanges. In fact, within conditions that require marital bonds as a mechanism for basic survival, it is difficult to imagine a marriage match that was not intermediated. In other words, in a situation that warrants the broadest possible cooperation, why would mate selection be anything other than a communal act? Early marriage likely relied on a logic of cooperation, and the quality of the match was likely judged based on its potential for facilitating it.

As humanity became less nomadic and began to produce surpluses, cooperation and reciprocation gave way to wealth consolidation. As marriage became the primary means for transferring status and property, it was therefore more formal and restrictive. To put it another way, the earliest marriages were less stabilized, informal arrangements that arose as a mechanism for creating both loose tethers between groups and partner relations for organizing sex and childrearing. By settling land, producing crops, and establishing more permanent housing structures, the first attempts to stabilize marriage emerged. These attempts focused primarily on maintaining and transferring ownership through advantageous connections and were rooted more often than not in local traditions rather than law.[3]

Similarly, in ancient societies across the globe, rule was based primarily on familial relationships and bloodlines. This manifested itself in two important ways. First, claims to authority were made based on ancestry.[4] Second, strong kinship ties were necessary both for asserting rights to authority and to maintaining rule. Whether as a mechanism for ensuring purity of bloodlines or cementing kin relationships, marriage played a key role in the assertion and maintenance of power. Marriage was also important outside the aristocracy, as advantageous matches provided new means of wealth through establishing new business opportunities or boosting agricultural yields.[5] In these conditions, marital choice was highly controlled, though more by familial and cultural traditions and expectations than by laws or religious edicts.

As marriage stabilized, so did matchmaking grow in prominence, and the Jewish tradition is perhaps the oldest and best known. This marriage match is a *shidduch*, and the oldest known agreement is found in the Torah, when Eliezer pairs Abraham's son Isaac with Rebekah. Historically, a *shadchan* (matchmaker) would approach the family when its daughter reached marrying age. After determining the young woman's marriageability based on her family's standing, her religious dedication, and her physical appearance, the *shadchan* would match the young woman with appropriate young men.[6] The parents were then presented with his selections and determined whether the matches were suitable. Upon parental consent, the couple would meet. Once they were acquainted, a determination would be made about marriage. The matchmaker would be paid upon completion of a successful match. This is, of course, in contrast to the less formal Jewish tradition of matchmaking frequently associated with a *yenta*, the Yiddish term for a nosy woman, who takes it upon herself to make matches among young people with whom she is acquainted. Both the professional and informal practices of matchmaking persist today, the former primarily in Orthodox communities.[7]

Other cultural environments also affect the conditions for marriage and matchmaking. For most of history, marriage was a communal concern, a union that included the husband and wife, their parents, their siblings, and their neighbors. The married couple increased in visibility and importance with the rise of the merchant and artisan classes in the early Modern period in Europe, but those conditions were not like contemporary matrimony. Working-class families often sent their children to work in wealthy households, and middle-class families frequently shared their homes with servants or lodgers. Such mingling of familial residences limited the privacy of married couples.[8] With economic shifts that found most people working away from the home, with changes in transportation that enabled people to live farther away from their birthplaces, with juridical and medical changes that allowed women more freedom, a logic of romantic attachment plays a more central role in marriage matches. As that happens, the relationship of the matchmaker to the match changes as well. Most significantly, matchmaking shifts from contracting with the parents of marriageable children to establishing relationships with the marriage-minded individuals themselves.

Another extensive and well-known matchmaking tradition exists in Ireland. Matchmakers in rural Ireland have long been uniting unmarried landowning men with "attractive" (whether that be beautiful or skilled) women in order to ensure the maintenance of family farms and to preserve the larger culture. These men were generally traders, such as livestock dealers, who traveled frequently and knew many families. As rural life changed, and as the concept of marrying for love emerged in the seventeenth century,[9] the need for the matchmaker appeared to wane. In the 1950s, automobiles and dance halls became increasingly prevalent in the countryside, enabling greater potential for matching without professional intermediation.[10] Nevertheless, the annual Lisdoonvarna Matchmaking Festival continues today in County Clare, and Ireland still has at least one rural professional matchmaker, Willie Daly.[11]

Similarly, by the late 1930s, marriage bureaus, companies providing affluent clientele with introduction services, were gaining popularity in England.[12] These intermediaries tended to highlight both the status and wealth of their potential matches and their discretion. One example is the Marriage Bureau, founded by Mary Oliver and Heather Jenner, two former debutantes, in 1939. Personnel met individually with each client and then selected potential matches for them. By the 1950s, the Marriage Bureau had become a fixture of British high society and professed to have facilitated thousands of marriages with only three ending in divorce.[13]

Professional matchmaking persists, and there are two significant holdovers in contemporary services. First, just as the *shadchan* assessed familial status and the Irish matchmaker looked for heartiness, contemporary professional matchmakers rely on both information they gather and their impressions of their clients and potential matches. They depend on face-to-face communication, generally conducting at least one interview and engaging in follow-up after the initial date. Beyond that, some agencies even require supplementary services (e.g., wardrobe consulting, mock dates, home visits) of their clients.[14] Second, the primary purpose of intermediation is still marriage, and it seems that all professional matchmakers require that their clients be marriage minded and seeking "traditional" relationships. In fact, Patti Stanger, founder of the Millionaire's Club and host of Bravo TV's *Millionaire Matchmaker*, prohibits her clients from having sex on the first date and encourages them to wait until they are "in a committed relationship that could lead to marriage."[15] While Stanger's reality television show and online dating's growing popularity may have raised the profile of professional matchmakers in recent years, it accounts for little of contemporary romantic matchmaking, at least in the United States, where cultural mythos values qualities such as perseverance and work ethic over bloodlines and birthright.

Mediation

Like intermediation, mediation is a condition that relies on human decision-making. However, that decision-making no longer belongs to a human intermediary who introduces the potential couple. Instead, individuals engage in their own matching practices using communication technologies as mediators. In addition, although the preferred, or at least sanctioned, match may be "traditional" marriage, mediators are used to secure a variety of kinds of matches, including companionship, sexual liaisons, and same-sex relationships. The shifts mentioned earlier—in economic systems, in transportation and mobility, in the legal position of women—most certainly play a role in when and how mediation takes hold. The development of new technologies, especially the invention of the printing press and the proliferation of the newspaper, significantly affects processes of mediation. Stabilizations and destabilizations of marriage are closely related to conditions of mediation as well. Mediation, then, is most closely associated with cultural formations in which institutions must, after changes that include urbanization and mobility, govern at a distance.

As we saw in the discussion of intermediation, it is important to understand the position of marriage in order to make sense of the condition of mediation. Leading up to the time in which mediation emerges, there are attempts to assert control over the institution of marriage by both governments and religious institutions, and the rise of Christianity serves as an example of the tug-of-war between familial relationships, religious authority, and governmental institutions. As Christianity flourished, and as religion sought to be the centerpiece of everyday life, the Church began to claim dominion over the institution of marriage. Marriage was considered a threat to self-control, and church oversight of marriage (and divorce) and personal behavior was necessary. Since European marriage practices were diverse and highly unregulated, attempts to stabilize them led to some difficult contradictions for the papacy. Prior to the middle of the twelfth century, the Church agreed with the traditionally held belief that only mutual consent and consummation were necessary for a valid marriage. However, the Bishop of Paris subsequently argued that, under this condition, Mary and Joseph would not have had a legally recognized marriage. He suggested that a pledge to wed without consummation was not a marriage but that an exchanging of consensual vows in the present (e.g., "I take you as my husband") was a marriage, regardless of whether the vow was consummated. This belief became church teaching, but it gave birth to an untenable situation. First, if the Church supported the claim made by an individual that he had privately exchanged vows with a woman, then she was committed to that union for life, regardless of whether they had engaged in sexual congress, or whether the vow exchange had ever even taken place. In addition, the Church was now in the position of defending the unions of defiant young people who had wed privately despite their parents' reprisals.[16]

By the Protestant Reformation, religious leaders' stance on the importance of marriage shifted. It was no longer considered a detriment to self-control, and sermons on love between husband and wife and the married couple's right to privacy became increasingly common. At the same time, Luther and his followers encouraged tightening of marriage laws in order to prevent young people from marrying without parental permission. Continuing concern over marriage by consent in the sixteenth century makes two things clear. First, attempts to codify marriage by the Fourth Lateran Council three centuries prior had little effect on actual marital practices in Europe. In addition, it exposed a new concern during that time, the rise of a "new breed of 'masterless' men and women."[17] As young people increasingly worked as wage laborers outside the home, they were subjected less to the restrictions of family or other authority figures.[18] Unmarried women especially were regarded as both a moral and an economic threat, and clergymen and bureaucrats worried that they would fall into ill repute or become wards of the state.

After centuries of dispute over the control of marriage, interests of the church and state began to coalesce as a means for reigning in these dangerous young people. There was a decline in marriage by consent, fueled in no small part by government intervention. More jurisdictions passed laws that either required the presence of witnesses at the exchange of vows or would undo the unions of young couples who married without parental permission.[19] In the mid-sixteenth century, governments across Europe were penalizing sex outside marriage as well, including new laws in both Germany and Switzerland that prohibited consummation of a marriage prior to the wedding and in both England and France that required punishment for unwed mothers, all attempting to further standardize marriage. By the early seventeenth century, a unique marital formation had taken root in Western Europe that included stringent divorce laws, delayed age of marriage, and couples who both were closer to each other in age and lived in homes away from their families.[20] During the eighteenth century marrying for love, what Coontz calls the *love match*, becomes the ideal: "Until the late eighteenth century, most societies around the world saw marriage as too vital an economic and political institution to be left up to the free choice of the two individuals involved, especially if they were going to base their decision on something as unreasoning and transitory as love."[21] It's not surprising that, as marriage was an increasingly private affair, more formalized definition of marriage and stronger divorce laws appeared. The married couple's retreat from the public sphere necessitates increasing juridical intervention

It is during this period of marriage privacy that mediation emerges, and the exemplary romantic-matchmaking mediator was the personal classified advertisement. The first "short ad," the brief ads of tight prose that are the antecedents of modern classified and personal advertising, was a declaration on Easter festival procedures printed in *The Pyes of Salisbury* in 1477. These ads shaped into a means for locating husbands and wives in the 1690s, a mere five decades after the development of the modern newspaper.[22] These early advertisers searched almost exclusively for marriage, and the advertisements were used especially by people without kinship connections for making marital arrangements. The history of personal classified advertisements is similar in the United States, with the first appearing three centuries ago and focusing primarily on marriage. Many of these personals were by lonely farmers left isolated as increasing numbers of young people moved to urban areas.[23] Midway through the nineteenth century, these advertisements became a regular feature, and matrimonial papers, publications with "pages of hopeful brides and bridegrooms, with handy information about the specific contents of bank accounts as well as detailed physical descriptions," appeared.[24]

By the end of the nineteenth century, commentators in Britain were concerned that established notions of courtship were no longer appropriate. There were early experiments with social networks of unmarried people at this time, and in the early twentieth century, entrepreneurial-minded journalists concluded that these ads could be used to facilitate other kinds of social connections, such as friendship and companionship, or relationships based purely in shared interest in an activity. Alfred Barrett's *The Link* was the first publication devoted entirely to companionship advertisements.[25] It was different from its predecessors in the seventeenth and eighteenth centuries because it sought to bring together unmarried individuals for social interactions that were not intended to lead to marriage, and that included both heterosexual and same-sex pairings. Barrett labeled his venture a "social medium," and the tone of the advertisements was much more jovial than earnest.[26]

The Link came under fire from R. A. Bennett, an anti-prostitution campaigner, and Barrett was eventually tried for indecency. He argued that his periodical was not indecent because it was intended to soothe the isolation he saw as endemic of the early twentieth century. In fact, a number of mainstream publications were already trying to serve the population of single people interested in "palling up," the notion of meeting members of the opposite sex for noncommitted social, and sometimes physically intimate, companionship.[27] The claims of indecency leveled at Barrett highlight an anxiety at both the intervention of this new technology and the new relationship formations that were surfacing during the time.

Since their inception, personal advertisements in Great Britain have struggled against their negative characterization as pointless at best and possibly dangerous.[28] Plagued by the threat of fraud—one case included an advertiser who claimed to be a wealthy young widow defrauding four different men of travel expenses—and a lack of respectability, many major American newspapers (e.g., the *New York Times*, *Atlanta Constitution*) stopped printing personals in the late nineteenth to early twentieth centuries.[29] When they returned to prominence in the 1940s, the disreputable legacy remained, no doubt fueled by cases such as "the lonely hearts killers," a couple who robbed and murdered several people with whom they met through an American correspondence club.[30] Perhaps because of their early accusations of indecency, or their adoption for arranging alternative relationships, first as companionship ads and same-sex personals in the early twentieth century and later for arranging swinging and wife-swapping in the 1960s, personal classified advertisements never seemed to catch on as a mainstream use for marriage facilitation.

Around the same time that Barrett was publishing *The Link*, W. T. Stead started the Wedding Ring Circle, an association for facilitating middle-class marriages.

> The Wedding Ring Circle vetted its members and tightly controlled access to all member information. Clients could access two collections housed in the company's office, an album with member photographs and a journal with their thoughts on topics of interest, to learn more about others. Members could correspond with potential mates anonymously until both agreed to reveal their identities.[31]

Contemporary networking services, organizations that facilitate single individuals' meeting one another without matching specific people, work similarly. In the 1980s, a number of such networking services emerged in large cities. Singles clubs held members-only events for people who paid an annual fee. In Washington, DC, for example, Entrees and Turning Point held large dinner or cocktail parties for their members,[32] and Florida's Southern Fantasies arranged themed mixers, designed to appeal specifically either to men or to women.[33] This trend persisted into the twenty-first century. More recently, DinnerWorks arranges dinner parties in popular restaurants to which four men and four women, all single, are invited. At nuts-and-bolts parties the male invitees receive a nut, and females a bolt. The goal is to find the member of the opposite sex with the complementary piece of hardware. In addition, agencies dedicated to singles' travel (e.g., Meet Market Adventures) and speed dating, an event during which individuals have multiple timed dates with up to twenty different members of the opposite sex, have surfaced.[34]

There is one obsolete technology that also merits attention. In the late 1980s, video dating achieved some popularity. With video dating, clients delivered a sort of dating monologue that was videotaped by the social introduction services provider. Some services, such as Washington, DC's Georgetown Connection, worked as video-dating/matchmaking hybrids, with a representative directing clients as to what information to include in their videos ("first name, job, education, interests, and what they're looking for in a relationship") and then selecting videotapes of potential matches for them to view.[35] On the other hand, the country's largest video-dating service, Great Expectations, initially allowed users to comb their local branch's video library without assistance from one of the service's representatives. As Jeffrey Ullman, founder of Great Expectations, put it in 1989, "We believe the best person to select who you'll go out with is you. You know who you want, you know what you want, and there's no so-called expert who's more of an expert about you than you."[36] In addition, their service allowed a number of different kinds of relationship options for its clients (not just marriage) with various levels of commitment.[37] Within three years, and after sixteen years in business, Great Expectations changed its model. Perhaps in response to market conditions, or because of the rising concern over AIDS, the company decided to focus instead on "communicating itself as a service that makes marriage-minded matches."[38] In addition, branches began providing assistance to their clients, as Ullman notes, "Most adults are just not good selecters [sic]. They might choose a blonde with blue eyes and a big bust. She might be wrong for him; he just lets his lust make the decision."[39] The portrayal of the potential wrong match as a sexualized woman seems to indicate that the company's choices were probably at least in part due to changing sexual practices in the larger culture, and the shift in focus to marriage matching may indicate an ideological shift, rather than a choice made to increase customer satisfaction, on the part of the company. The resilient institution of marriage ebbs and flows with significant changes, but the end of the twentieth century marks a significant shift and the beginning of another matchmaking condition.

Automation

Automation is a condition in which matching is outsourced to nonhuman operators, specifically computer-mediated matching that usually relies on machine processing that stands in for human thought processes. Although this matching may be based in human decision-making, such as social-scientific research, that human decision-making is not visible in the interactions between the matchmaker and the matched. Automation began to emerge in the 1960s, but the roots came at the turn of the twentieth century.

As marriage choice became an expectation, marriage facilitation changed as well.

By the 1890s, the middle-class tradition of calling, a courtship practice in which a young woman's parents invite a young man for a supervised visit in their home, had become common.[40] In the custom of calling, parents were able both to maintain control over who was allowed to call and to chaperone the couple during their interactions. In addition, the girl and her family invited callers, and a boy would never call uninvited. By the early twentieth century, calling had given way to dating, and power dynamics shifted. First, dates happened away from home, distancing the parents' control and supervision. In addition, since a young woman's financial standing was generally subordinate to a young man's, the boy—rather than the girl's family—was now in the driver's seat.

Just as spatial territorializations did with intermediation and the prominence of the married couple with mediation, changes in mobility play a significant role in the emergence of automation. The circumstance in which dating emerged is what Williams calls *mobile privatization*. He identifies a number of developments—including the expansion of consumer goods, increased automobility, radio broadcasting—that came together in the mid-twentieth century to enable to emergence of television and "an at-once mobile and home-centered way of living."[41] By the middle of the twentieth century, marriage had become the center of adult life. In the 1960s, 90 percent of people married, and marriage signified the beginning of adulthood.[42]

One of the greatest destabilizations of the mid-twentieth century is the result of the contraception revolution of the 1960s. Before the availability of dependable birth control, women's sexual freedom was limited by the possibility of pregnancy. By 1970, more than half of American women were using some form of birth control, allowing women to disconnect sex from procreation.[43] With the first successful human artificial insemination in the 1950s and the first in vitro fertilization in the 1970s, as well as the growing popularity of adoption, the act of procreation was further separated from sex. Not only does this make it easier for women to manage sexual activity and procreation, it also allows heterosexual couples to choose to have children later in life or not at all, and for same-sex couples to have children.

Advancements in reproductive technologies and the availability of adoption, as well as changes in women's reproductive rights, work alongside other technological, social, and economic conditions to change the status of marriage. Marriage no longer serves as an obligatory part of a "credentialing process that people have to go through to gain adult responsibility and respectability."[44] Young adults are not required to marry in order to live outside the family home or seek employment or educational opportunities.

Finally, the political and legal battle surrounding marriage equality, especially in the United States, is important to this discussion of contemporary marriage and the condition of automation. In 2000, Vermont conferred the same legal status for same-sex civil unions and heterosexual marriage, and the Massachusetts Supreme Court ruled in 2003 that the state's constitution ensures marriage equality for all couples.[45] Within three months, President George W. Bush called for the protection of marriage in his State of the Union address, prompting the mayor of San Francisco to instruct city officials to begin issuing marriage licenses to same-sex couples.[46] In 2015, the US Supreme Court ruled in favor of marriage equality, effectively legalizing same-sex marriage. This debate is emblematic of much of the concern over the state of contemporary marriage. As the love match becomes the dominant marriage paradigm, and its ties to its original purposes—consolidating power and resources, facilitating cooperation, organizing sexual relationships and procreation—are loosened, some want to shore up its place in society.

Though short-lived, computer dating was the first automator. Computer dating used questionnaire responses to match users. Although it did not reach prominence until the 1980s, the first service was founded at Harvard University in 1964. Operation Match used a computer program to connect singles based on their "vital statistics" and responses to "hypothetical, and awkward, situations."[47] Within a year, Lewis Altfest and Robert Ross had launched Project Technical Automated Compatibility Testing (Project TACT). For a fee of five dollars, customers were matched based on their answers to over one hundred questions.[48] It seems that most computer-dating services (e.g., Compatibility Unlimited) connected singles based on a variety of demographic and personality features.[49] Love Quest, a program developed and run by celebrity sex therapist Joyce Brothers and the Sexual Analysis Group of Miami, matched users based on sexual compatibility as well.[50] As Leigh Rothschild said of Love Quest's multidimensional matching system more than a decade ago, "If you're Mary from Miami and you want to be matched with Harry from Miami you might get John from Maryland. If the computer does not feel two people are compatible it will not match them up even if Mary and John are the only people in the computer data base."[51]

Contemporary online dating sites offer a variety of methods for singles to meet potential partners. Some sites focus primarily on their searchability (offering clients access to customizable tools that they can use to search profiles), while others emphasize their efficiency (matching members based on personality assessments). Either way, online dating's most marketable quality seems to be the sheer volume of available singles registered with sites. Popular sites, such as Match.com and eHarmony.com, widely publicize the large number (in the millions) of registered users, and I will discuss online dating at length in the chapters that follow.

ROMANTIC MATCHMAKING AND REGIMES OF POWER

As we have seen, the most prominent contemporary relationship between the romantic matchmaker and the couple is different than in previous iterations. Intermediation describes human intervention into the pairing, first at the behest of parents or other authority and later of the couple themselves. Mediation describes human intervention into their own matching, as people used traditional one-to-many communication technologies to make pairings. Automation describes computational intervention for matchmaking. In this condition, it is not individuals who are matched. Rather, the computational processes match data—preferences, tastes, behaviors, personality profiles—not people.

Understanding romantic matchmaking in terms of the marriage apparatus and conditions for matching draws attention to the relationship between power formations and the institution of marriage. During intermediation, marriage was central to the consolidation of power and accumulation of wealth, but marriage laws were less stabilized. The matchmaker and the couple's parents maintained tight control over the matchmaking process, determining who would be matched with whom and when the couple would be allowed to meet. As marriage became increasingly codified through religious and juridical involvement in its management, and as the married couple became central to daily life, the focus of romantic matchmaking shifted. It is during the condition of mediation, when familial forces exerted less direct control over the match, that romantic matchmaking is characterized as frivolous and potentially subversive. In the contemporary condition of automation, marriage is again facing destabilization as more and more heterosexual couples choose cohabitation and as same-sex couples are granted access to marriage. In automation, individuals may feel they have greater control over their matchmaking, but they are subject to invisible categorizations and calculations that make determinations for them.

Highlighting matchmaking's position in the marriage apparatus offers an ideal opportunity for understanding matching in terms of Foucault's three regimes of power (sovereignty, disciplinarity, and governmentality) and Deleuze's control. Foucault begins *Discipline and Punish* with the juxtaposition of two forms of retribution. The first is a lengthy and graphic description of a public execution in Paris in 1757, including the boiling and tearing of flesh and the eventual quartering of a prisoner. The second is a highly regimented schedule for Parisian prisoners published in 1837. These two examples—the public execution and the timetable—represent sovereignty and disciplinarity, respectively.[52]

Sovereignty is the primary relation to power circulating during the time of feudal monarchy, a system in which nobility of birth determines the ability to reign. Under this relation, the law is an extension of the monarchy. As such, the expectation in sovereignty is complete obedience to all laws. Since the law is an extension of the king himself, transgression is subjected to visible, bodily, and spectacular punishment. Intermediation is likewise concerned with bloodlines and birthright, and it is in this condition that the greatest explicit control is exerted over both the marriage match and the marriage. Matchmaking is associated almost exclusively with the body—the extension or mingling of bloodlines, procreation, physical labor—and the marrying couple had little to no say in their pairing. Once married, the couple remains physically connected to family through close living quarters that made even the most private parts of what we might call the marital body (e.g., sexual intercourse, childbirth) visually and audibly present for the entire household.

Through the seventeenth and eighteenth centuries, as the dominant organization of nation states transitioned from monarchy to representative forms of government, spectacles such as public execution were eventually outlawed. Foucault notes, "The disappearance of public executions marks therefore the decline of the spectacle, but it also marks the slacking of the hold on the body."[53] If sovereignty centers on rights and corporeal intervention onto bodies, then discipline relies on normalization and the management of bodies at a distance. As the second example above demonstrates, disciplinarity applies constant control to the individual docile bodies by exercising "holds on [them] at the level of the mechanism itself—movements, gestures, attitudes, rapidity."[54] The best-known model of this normalization at the level of individual bodies is Bentham's panopticon, the prison design with a central guard tower encircled by banks of cells in which both inspection and surveillance are pervasive.

In disciplinary societies, power is exerted not at the level of the flesh but rather at the level of the individual. In other words, the prisoner is not shackled, but his movements are observed (from the guard tower), his time structured (by the timetable), and his behavior managed (through guidelines and regulations). As the love match emerges and the married couple is physically distanced from the larger family, individuals are allowed increased flexibility in mate selection. Just as marriage was an important instrument of power in sovereignty—through continuing bloodlines, for example—so was marriage central to containment of individuals in disciplinary societies. In mediation, concern shifts from strengthening familial bonds to shoring up the couple's relationship. It is during this formation that we see increasing attempts by both church and state to codify marriage. In addition, sermons began to address the married couple,[55] and magazines such as *Ladies' Home Journal* advised young people on the intricacies of appropriate behavior.[56]

Governmentality moves from the disciplining of individual bodies to the management of population. In this relation to power, Foucault is concerned with "the 'governmentalization' of the state."[57] Foucault understands governmentality as "power that has the population as its target, political economy as its major form of knowledge, and apparatuses of security as its essential technical instrument."[58] In this regime of power, governing becomes the overarching logic of power, and its primary concern becomes management of population through tracking of birth and death rates, fertility, etc. Similarly, Gilles Deleuze names this relation to power *control societies*, a post-disciplinary type of power that rules not through bodily intervention or confinement but through permitting or denying movement.[59] The signature gives way to the password, and individuals become *dividuals*, compilations of data that can be disassembled and reassembled, as needed. In the condition of automation, a marriageable individual is dissected into innumerable discrete categories, pulled together into a pool of similarly identified matching units, and reconstructed as a match. The computational processes, cultural assumptions, and power relations that govern this practice vary based on its purpose. In the remaining chapters, I will explore the condition of automation in greater depth.

NOTES

1. Aaron Smith and Maeve Duggan, *Online Dating and Relationships* (Washington, DC: Pew Research Center's Internet and American Life Project, 2013).
2. Stephanie Coontz, *Marriage, a History: From Obedience to Intimacy, or How Love Conquered Marriage* (New York: Viking, 2005). At times during this discussion I delve into the history of marriage, and I rely primarily on Coontz's work, a social history of marriage that is ideal in its extensive treatment of periods and cultures as well as its keen criticism of widely held assumptions about "traditional" marriage.
3. Ibid.
4. Ibid., 54.
5. Ibid.
6. Tamar Caspi, "Make Me a Match," *The Jewish Advocate*, December 11, 2009, 31.
7. Ibid.
8. Coontz, *Marriage, a History*, 128.
9. Ibid.
10. Jasper Winn, "Mix 'N' Match," *Geographical*, February 2003, 33.
11. Willie Daly, *The Last Matchmaker: The Heartwarming True Story of the Man Who Brought Love to Ireland* (London: Hachette, 2010); Susan Spano, "Her World: Rural Ireland Is in the Green with Bachelors at Matchmaking Festival," the *Los Angeles Times*, June 17, 2001; Winn, "Mix 'N' Match."
12. H. G. Cocks, *Classified: A Secret History of the Personal Column* (London: Random House Books, 2009).
13. Ibid., 124.
14. Ron Geraci, "A Headhunter for My Heart," *Men's Health*, June 2004, 166–71.
15. Hanh Nguyen, "'Millionaire Matchmaker' Offers Tough Love," Zap2It.com, January 22, 2008, www.zap2it.com/blogs/zap-story-pattistanger-millionairematchmaker.
16. Coontz, *Marriage, a History*, 106.
17. Ibid., 136.
18. Ibid.
19. Ibid.
20. Ibid., 136–45.
21. Ibid., 178.

22. Cocks, *Classified.*
23. Laura Schaefer, "Looking for Love Online, or on Paper," the *New York Times*, February 14, 2003, www.nytimes.com/2003/02/14/opinion/looking-for-love-online-or-on-paper.html.
24. Ibid.
25. Cocks, *Classified.*
26. Ibid., 4.
27. Ibid., 16.
28. Ibid.
29. Schaefer, "Looking for Love."
30. Cocks, *Classified*, 131.
31. Ibid., 16–18.
32. Sandy Kyle Bain, "Singles: Standing by Their Clubs," the *Washington Post*, June 4, 1982, www.washingtonpost.com/archive/lifestyle/1982/06/04/singles-standing-by-their-clubs/6959b80e-05f8-4d36-93c8-d0a3da33a8f0/.
33. Charlie Gofen, "Lonely Hearts: Dating Service Opens Its Doors in Inverness," *St. Petersburg Times*, April 25, 1988.
34. Shanda Deziel, "Love Inc.," *MacLean's*, February 14, 2005, 28–34.
35. Carol Krucoff, "…And a Matchmaker," the *Washington Post*, December 11, 1979.
36. Kevin McManus, "Strike a Match," the *Washington Post*, November 17, 1989, www.washingtonpost.com/archive/lifestyle/1989/11/17/strike-a-match/fe6d37b2-643e-4773-a2f7-f2adbd1850ef/.
37. Ibid.
38. Pat Hinsberg, "Romancing the Singles Scene: Video Dating Services Push Marriage in a Bid to Survive" (includes interview with Jeffrey Ullman), *Adweek*, March 11, 1991.
39. Ibid.
40. Coontz, *Marriage, a History*, 199.
41. Raymond Williams, *Television: Technology and Cultural Form* (New York: Schocken Books, 1975), 19.
42. Coontz, *Marriage, a History*, 227.
43. Ibid., 254.
44. Ibid., 276.
45. Ibid., 273.
46. Robert Barnes, "Supreme Court Agrees to Hear Gay Marriage Issue," the *Washington Post*, January 16, 2015, www.washingtonpost.com/politics/courts_law/supreme-court-agrees-to-hear-gay-marriage-issue/2015/01/16/865149ec-9d96-11e4-a7ee-526210d665b4 _story.html.
47. David Rapp, "The Love Machine," *American Heritage*, November/December 2005, 13–4.
48. Nick Paumgarten, "Looking for Someone: Sex, Love, and Loneliness on the Internet," *The New Yorker*, July 4, 2011, www.newyorker.com/magazine/2011/07/04/looking-for-someone.
49. Aaron C. Ahuvia and Mara B. Adelman, "Formal Intermediaries in the Marriage Market: A Typology and Review," *Journal of Marriage and Family* 54, no. 2 (1992): 452–63; Bob Levey, "$15,000 Buys Computer Cold Shoulder," the *Washington Post*, April 14, 1988.
50. Christine Del Sesto, "Looking for Love at First Byte" (includes interview with Leigh Rothschild), the *Washington Post*, June 4, 1987, www.washingtonpost.com/archive/lifestyle/1987/06/04/looking-for-love-at-first-byte/98a8c6d9-96e8-4376-9847-2b685c6739b7/.
51. Ibid., 5 (italics original).
52. Michel Foucault, *Discipline and Punish: The Birth of the Prison*, translated by Alan Sheridan (New York: Vintage Books, 1977).
53. Ibid., 10.
54. Ibid., 137.
55. Coontz, *Marriage, a History.*
56. Cocks, *Classified.*
57. Michel Foucault, *Security, Territory, Population: Lectures at the College de France, 1977–78*, edited by Michel Senellart, translated by Graham Burchell (New York: Palgrave Macmillan, 2007), 109.

58. Ibid., 108–9.
59. Gilles Deleuze, "Postscript on the Societies of Control," *October* 59 (Winter, 1992), 3–7.

Chapter Two

Procedural Rhetorical Analysis of Three Online Dating Sites

According to Galloway, protocol is the organizing principle that manages interactions in networks, and he discusses it on three levels: physical (or network), power, and form (or apparatus). He emphasizes that protocol is "*against interpretation*."[1] Protocol may capture and store information, but it is generally indifferent to that content.

> To follow protocol means that everything possible within that protocol is already at one's fingertips. Not to follow means no possibility. Thus, protocological analysis must focus on the possible and the impossible (the envelope of possibility), not demystification of some inner meaning or "rational kernel" within technology. *Protocol is a circuit, not a sentence.*[2]

Galloway's position aligns with Packer and Wiley's observation that "Following a long period of focusing almost exclusively upon the problematics of language, meaning, and representation, there has been a renewed interest in establishing means for making claims in the real. . . . [T]here are now attempts to set aside epistemological debates in order to compose new ontologies that enable theoretical innovations and political intervention."[3] They call for researchers to focus on logics rather than interpretations—or, in Galloway's terms, on possibility not meaning.

The *physical* media of protocol are associated with the kind of network. Galloway notes that "protocol is not by nature horizontal or vertical, but that protocol is an algorithm . . . whose form of appearance may be any number of different diagrams or shapes."[4] Galloway discusses three varieties of networks: centralized, decentralized, and distributed. The key features of centralized networks are hierarchy and authority, and all components are organized around a central core. He offers the United States judicial system as an example, with the Supreme Court serving as the center. Currently the most common schema is the decentralized network, which works like a "multiplication of the centralized network."[5] The American airport system illustrates this diagram—a web of numerous hubs with their own networks, none of which serves as the center of the larger system. The final network structure, and the one Galloway is most concerned with, is the distributed network. Two examples of distributed networks are the United States interstate highway system and the Internet, both developed contemporaneously for military purposes but have since proved useful for civilian and economic functions as well.[6]

Galloway's power "consider[s] protocol in a political sense, as a pseudo-ideological force that has influence over real human lives."[7] For this discussion, Galloway relies on the work of both Foucault and Deleuze, specifically the former's concepts of biopower and biopolitics and the latter's notion of control. As Galloway notes, Deleuze's control is protocological. In other words, individuals are absorbed by the (distributed) network of control societies. Foucault's biopower and biopolitics are concerned not with the control over death but with "the calculated management of life."[8] Closely related is the conception of biopolitics. As the individual body was the object of previous forms of power, the population is the object of biopolitics, and this system requires statistical information for the management of population. They "are Foucault's terms for the statistical coding, the making statistical, of large living masses, such that any singular life-form within that mass may be compared in its organic nature to the totality."[9] Biopower and biopolitics, like Deleuzean control, are protocological as well.

Galloway associates the formal apparatus with software, and defines it as "the totality of techniques and conventions that affect protocol at a social level, not simply a technical one."[10] Galloway focuses this discussion on the Internet, and he identifies continuity as a significant component. He defines continuity as "the set of techniques practiced by webmasters that, taken as a totality, create this pleasurable, fluid experience for the user." He reminds us that if network is concerned with the physical field of the Internet that is the focus of network administrators, then form is the responsibility of the Webmaster. The Internet is a paradox, hierarchical in its organization and potentially distributed in its use. So while the user may experience the Web as a meander, clicking through a seemingly endless stream of text, image, and sound, the Internet itself relies on stable hierarchies—of transfer protocols, of site architecture. Galloway argues that the reason the Internet is so compelling despite this contradiction is its continuity. Notwithstanding its lack of reliance on narrative- and time-based techniques present in other media, such as film and television, the Web creates a generally fluid movement within/among sites and actions that makes the user feel comfortable. In addition, its integration with other technologies (e.g., applications available for smart phones) creates the seamlessness that extends this feeling of continuity. For example, Apple devotees using Safari can move from phone to tablet to laptop and have the same information available in their browser display across devices.

This conception of continuity is important to the study of online dating because it allows sites to draw on, and circulate within, larger cultural narratives, both old and new, about romantic love, sexuality, and the importance of marriage. Much of that can be seen in how these companies market themselves, and those claims are fairly obvious. For instance, eHarmony.com's television commercials regularly feature couples who met through the service. With the familiar opening bars of a Natalie Cole classic, presumably disheartened viewers are reassured. With eHarmony.com's assistance, *this will be an everlasting love.* On the site, eHarmony boasts both that it is the "#1 Most Trusted Online Dating Site" and that approximately 4 percent of US marriages resulted from its matchmaking.[11]

What is less obvious is how online dating sites rely on those same cultural assumptions—about the necessity of marriage, the importance of marrying for love—in the relationships they have with their users. How do big data and computational processes work together with cultural assumptions in online dating? How do the techniques and conventions, the formal apparatuses, employed by sites constitute users as subjects and influence our understanding of romantic love? In chapter 1, we began to understand romantic matchmaking as part of a marriage apparatus, a set culturally and historically situated contingent relations that exist in three conditions—intermediation, mediation, and automation. In this chapter, I deepen our understanding of automation by providing a descriptive analysis of three online dating sites (eHarmony.com, Match.com, and OkCupid.com) as procedural rhetorics. Building this archive of sites' procedural rhetorics both models processes for extending Bogost's framework to Internet applications and lays the groundwork for examining the role of articulators in the apparatus. Specifically, I analyze the formal apparatuses at work in the sites' initial account setup phase as well as recurrent data preparation, matching, and communications flows. Situating online dating within Galloway's analytic of the formal apparatus sets up the discussion of online dating and contemporary processes of subjectification in the next chapter as well.

EHARMONY.COM

After three years of research and development led by Neil Clark Warren, a clinical psychologist and marriage counselor, and Galen Buckwalter, a research psychologist, eHarmony.com (EH) launched in 2000. The privately held company claims to have had thirty-three million registered users across its history and to be responsible for the marriages of 438 members daily.[12] The site matches users based on their potential for relationship satisfaction and their level of compatibility. After surveying over a thousand successful married couples, the team determined twenty-nine traits happy couples consistently shared. For matching purposes, these characteristics became the site's 29 Dimensions of Compatibility (29 D), and EH divides them into two categories: core traits and vital attributes. The former includes personality characteristics, while the latter centers on background and socialization.[13]

Warren discusses these dimensions in greater detail in *Falling in Love for All the Right Reasons: How to Find Your Soul Mate*, and a brief review elucidates the logics that inform the procedures EH employs. In the book, Warren positions the 29 D in opposition to the "wishy-washy" public understanding of compatibility, positing that "the culture is satisfied with an extremely thin, narrow base of compatibility, usually settling for five main qualities: [appearance, chemistry, a sense of humor, status, and front-end personality]."[14] By characterizing compatibility as something that is fixed and knowable, and further portraying conventional notions as capricious, Warren creates a favorable comparison for his research-based, social-scientific approach to compatibility and EH's matchmaking system. This air of certainty is present in all three of the site's flows as well. While EH organizes traits in two categories, Warren's book further delineates them into four groups: *Screening Dimensions*, *Core Personal Dimensions*, *Skills That Can Be Developed*, and *Qualities That Can Be Developed* (see table 2.1). Either way, the 29 D are intended as a tool for finding a suitable marriage partner.

The seven *Screening Dimensions* are, as the name implies, baseline characteristics. Just as an employer might screen applications to ensure that pros-

Table 2.1. eHarmony.com 29 Dimensions of Compatibility

Category	Dimensions
Screening Dimensions	Good Character, Good Self-Conception, Red Flags, Anger Management, Obstreperousness, Understandings about Family, Family Background
Core Personal Dimensions	Intellect, Energy Levels, Spirituality, Education, Appearance, Sense of Humor, Mood Management, Traditional (or Nontraditional) Personality, Ambition, Sexual Passion, Artistic Passion, Industry, Curiosity, Vitality and Security, and Autonomy (or Closeness)
Skills That Can Be Developed	Communication, Conflict Resolution, and Sociability
Qualities That Can Be Developed	Adaptability, Kindness, and Dominance/Submissiveness

pective employees have a stable job history and no criminal record, Warren encourages readers to separate out "bad" relationship partners. These characteristics are not intended to help identify a specific would-be spouse but rather represent the must-haves (or must-not-haves) for assessing them. For Warren, the central screening dimension is *Good Character*—"integrity, honesty, moral uprightness."[15] In this process, character is both important and fixed, making it the central screening characteristic. The next four screening dimensions—*Good Self-Conception*, *Red Flags*, *Anger Management*, and *Obstreperousness*—are similarly important and difficult to change, though less fixed than a person's character. The final two—*Understandings about Family* and *Family Background*—are important and unlikely or, in the case of a person's background, impossible to change.

Once potential mates have been screened for their suitability, the *Core Personal Dimensions* come into play. These sixteen characteristics are also difficult to change, but they are not either-or propositions. In other words, in Warren's system people either do or do not possess integrity and good moral character. If they do not, they are immediately skimmed off the top, as they do not belong in the pool of conceivable marriage partners. In contrast, ambition or level of artistic passion or spirituality does not preclude individuals from being a potential soul mate; it simply makes them more or less compatible based on the ambition, level of artistic passion, or spirituality of their matches. To put it another way, this is a question of degree rather than quality, and Warren seems to endorse matching based on similarity rather than opposition or complementarity for qualities in this grouping of the 29 D (table 2.1).

The final two categories—*Skills* and *Qualities That Can Be Developed*—are still important, but there is a greater probability for change. The former includes *Communication*, *Conflict Resolution*, and *Sociability*, while the latter is made up of *Adaptability*, *Kindness*, and *Dominance/Submissiveness*.[16] Everyone is capable of developing skills in communication, conflict resolution, and sociability, and Warren recommends that couples have similar styles rather than advocating one over another. However, he treats the qualities that can be developed much like the screening dimensions; it is imperative that all romantic partners be adaptable and kind. Interestingly, dominance/submissiveness is the only quality for which Warren recommends complementarity rather than similarity. He does not encourage complete opposition but rather that "one person is somewhat dominant and the other is slightly submissive."[17] It is this schema, Warren's experience as a marriage counselor and clinical psychologist, and his religious beliefs that inform all of EH's matching and categorization processes as well as the site's interactions with its users.

DATA PREPARATION

By completing EH's initial account setup phase, the user initiates data preparation. This time-consuming flow includes a minimum of twenty-four screens designed to gather data for matching users, and the first three screens collect information typical of online dating sites, such as gender and birth date. Unlike most online dating sites, however, EH does not have new members select usernames, presumably because profiles are not viewable by others unless and until they have been matched by the site. Users are addressed by the site—and presented to others—using their first names. During the initial account setup phase, users can neither skip nor return to any questions, seemingly to prevent them from manipulating their previous answers based on questions asked in later sections. This practice is frequently used in online personality tests, such as those used by employers to determine the forthrightness of potential employees.

Most of the remaining twenty-one screens include questions related to the user's personality. These questions make up the Relationship Questionnaire (RQ), and responses allow EH to develop the personality profile it uses during the matching process. Seventeen of those screens are broken into ten subsections (*About You*, *Self Descriptions*, *Personality Characteristics*, *About Your Feelings*, *Relationship Orientation and Values*, *Important Qualities*, *About Your Personality*, *Your Personal Interests*, *Living Skills*, and *Communication Style*), and I provide example instructions and responses for each in table 2.2. Of the 272 personality questions that make up the RQ, 249 use a Likert scale.

In the first three subsections (*About You*, *Self Descriptions*, *Personality and Characteristics*), users are asked to indicate how well a statement or an adjective describes them. In *About Your Feelings*, users indicate how often they have had certain feelings during the past month. On the *Relationship Orientation and Values* page, they rate their level of agreement with statements. This screen seems designed to determine users' potential for relationship satisfaction, the cornerstone of EH's functionality that is not covered by the 29 D. *Important Qualities* asks them to indicate how important certain characteristics are. The *About Your Personality* page includes nineteen true-false questions, and this seemingly simple page may play an important role in the site's process by addressing attributes used for screening prospective members, such as good character. *Your Personal Interests* and *Living Skills* require users to indicate the level of their interest and their ability in a number of activities and pursuits. Finally, on the *Communication Style* screen, users select how well statements describe them. As with *About Your Personality*, this page seems related to high-importance traits. However, this time the questions address the more variable attributes, especially communication and conflict resolution. The last four pages in the EH initial account

Table 2.2. eHarmony.com Example Questions/Prompts

Section	Example Prompt	Example Responses	Format
About You	Please indicate how well each of the following statements describe [sic] you.	• I do things according to a plan. • I usually stand up for myself.	7-point Likert scale (*not at all* to *very much*)
Self Descriptions	Use the scale below to indicate how well each of the following describes you.	• Clever • Loyal • Genuine • Healthy	7-point Likert scale (*not at all* to *very much*)
Personal Characteristics	Use the scale below to indicate how well each of the following describes you.	• I am satisfied with my emotional development. • I like to be pampered.	7-point Likert scale (*not at all* to *very much*)
About Your Feelings	Please use the scale below to rate how often during the past month you have felt the following ways.	• Happy • Satisfied • Plotted Against	7-point Likert scale (*rarely* to *almost always*)
Relationship Orientation and Values	Please rate how strongly you agree or disagree with the following statements using the scale below.	• I am looking for a long-term relationship that will lead to marriage. • I don't need to have close friendships to be happy.	7-point Likert scale (*absolutely agree* to *absolutely disagree*)
Important Qualities	Use the scale below to indicate how important each of the follow characteristics would be to you in finding a partner for a relationship.	• My partner's fun-loving nature • Having my partner be open with me about how he/she feels about me	7-point Likert scale (*not at all important* to *very important*)
About Your Personality	Please decide whether each of the following statements is a good description of you. For statements that are NOT at all like you, please click "False." For statements that are at least somewhat like you, click "True."	• I enjoy many different types of recreational activities. • I never worry about work-related issues. • I am sometimes tempted to make fun of people behind their backs.	True/False

Your Personal Interests	Please use the scale below to rate your interest in the following things.	• Live music • Backpacking • Family • Maintaining a healthy diet	7-point Likert scale (*none* to *very strong interest*)
Living Skills	Please use the scale below to indicate how skilled you are at the following things.	• Socializing • Helping those who are less fortunate	7-point Likert scale (*not skilled* to *very skilled*)
Communication Style	Please use the scale below to describe how well you believe each of the following words generally describe you.	• I try to resolve the conflict well. • I am passionate/ intense about my position.	7-point Likert scale (*not at all* to *very well*)

setup phase return to standard online dating questions. The *Matching Information* pages collect data related to lifestyle (i.e., smoking, drinking, having children) and location, and the final two screens allow users to upload a photograph and subscribe to the service. Members may complete the RQ and receive matches for free, but they must subscribe to view photographs and communicate with other users.

Significantly, most of the information displayed on the profile page is not gathered during the initial account setup phase. In fact, responding to the twenty-four pages of questions will complete only about 20 percent of the user profile. [18] In order to have a robust account, users must answer additional questions related to interests, values, and interpersonal relationships, as well as upload photographs through the *Complete My Profile* tab on their profile management dashboard. Although user profiles are only partially complete at the conclusion of the RQ, that is when the site's portion of the data preparation flow begins. Once users have finished the initial account setup phase, the system converts their answers to numerical data and establishes a user database. According to the 2004 patent,

> the approximate individual satisfaction index . . . [is] generated from empirical data. The empirical data is [sic] generated from survey completed by different individuals. Each survey includes a plurality of inquiries into matters that are relevant to each individual in forming relationships with other people. The inquiries can have numerical answers. The answers are used in a factor analysis to identify factors that are each a function of one or more individual inquiries. These factors are used in the individual satisfaction estimator and the couple satisfaction estimator. Because the factors are a function of several inquiries, the use of the factors reduces the number of variables considered in generating the approximate individual satisfaction index and the couple satisfaction index. However, the complexity of the relationship between the variables (question answers) is retained in the results because each of the variables are [sic] taken into consideration when generating the factors. [19]

Each user's data are then compared with the site's individual satisfaction estimator (ISE) for the user's match group to classify members based on their potential for relationship satisfaction. Match groups are broad categories based on gender and sexual orientation, and a match group's ISE is the aggregate of the prospective individual satisfaction of its members. After users are classified based on their likelihood for contentment (unlikely, average, good), the site determines their matching viability and proceeds to the matching flow, in which their couple satisfaction estimator is generated in conjunction with other users.

MATCHING

In the matching flow, users are matched with others based on their relationship satisfaction index (RSI) categorization and their match group. Like the match group's ISE, a user's RSI is broken into three classes (unlikely, average, good) of likelihood for happiness in a romantic relationship. In the case of heterosexual matching, for instance, a man in the average RSI classification would be selected for a woman in the same category.[20] The site then uses other factors provided by users to determine the ideal available range of matches, including the data gathered by most online dating sites (age, ethnicity, religion, lifestyle, etc.) and the user's individual preferences associated with them.

> One of the identified candidates is selected. Data for the user and data for the selected candidate is [sic] compared to the approximate satisfaction the user would have with the candidate. This is repeated for each of the identified candidates. The results are studied to identify the candidate and user combinations that would result in the most satisfaction. The user and the identified candidates are then given the option of communicating with one another.[21]

Once matches are filtered for these other elements, the site presents the results to the user in the communications flow.

Significantly, users cannot access the matching flow until its results are delivered to them at the transition into the communications flow. Users may further filter matches by choosing with whom they would like to communicate, but they cannot engage in self-matching. Unlike most online dating sites, which allow users to browse profiles, EH only permits members to access the matches it provides. In other words, the matching flow is concealed from users. Other than making changes to acceptable match criteria (e.g., decreasing the importance of the potential match's age, adjusting location parameters), users cannot activate processes in the matching flow.

COMMUNICATIONS

The first phase of the EH communications flow is consistent with the role taken on by dating sites that offer greater flexibility for interaction between users. This flow begins when the site shares a potential match with the user. Matches are both sent to users via e-mail and available from the site dashboard in the *My Matches* tab. This flow recurs when new candidates who fit within matching parameters join the site or the initial, best set of candidates has been exhausted and less ideal matches (e.g., outside age or location range) are sent.

As figure 2.1 demonstrates, the EH profile page is broken into two columns that use blocks to organize information. The upper left corner includes the profile picture and responses to two questions, "The one thing I am the most passionate about" and "The most important thing I am looking for in a person is." Just to the right is the *Next Steps* block, which offers the evaluator a number of potential actions. In addition to a large button for beginning EH's Guided Communication process, the evaluator is also presented with the opportunity to send an EH Mail or to request a secure call with the presenter. Users must subscribe to the site in order to use all three of those functionalities. The rest of the right column includes the *Something to Talk About* box, which highlights some commonality the presenter and evaluator share (e.g., an interest in travel), an advertising block, and a *Communication Advice* section with links to site-authored articles on dating. Unlike most sites, which offer an ad-free experience to subscribers, EH presents all users with advertisements, regardless of their subscription status. These ads are generally for self-improvement services (e.g., weight-loss programs).

The bulk of the EH profile page includes information about presenters gathered during the optional, post-RQ portion of the initial account setup phase. This large block takes up most of the left column and is organized around three tabs, *About Me* (default), *IceBreakers*, and *Mailbox*. Just to the right of these tabs is a link to the presenter's personality profile. If presenters choose to share it openly with all matches, evaluators can click this link and read the detailed report. *About Me* begins with the presenters' occupation, age, height, ethnicity, religion, whether and how often they drink, whether there are children living with them, and when they and the evaluator were matched. It also includes additional information about presenters, such as interests, how they spend their leisure time, and who has been their biggest influence.

Unlike most dating sites, which allow members to determine how to communicate with others, EH's default contact sequence for the communications flow is its Guided Communication system. All functionality after initializing the communications flow is available only to subscribers, and non-subscribing members cannot view photographs.[22] Guided Communication

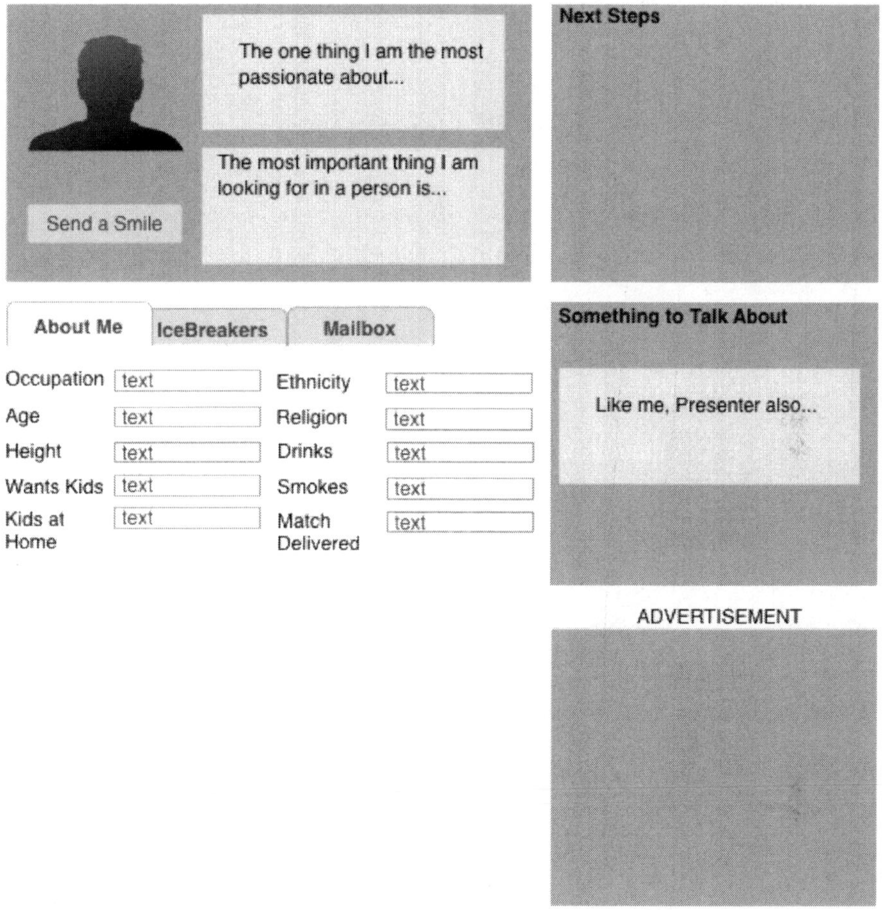

Figure 2.1. eHarmony.com Profile Page

occurs in three sequential steps (table 2.3). First, users each answer five questions of the other's choosing. Next, they trade lists of ten relationship Makes and Breaks. Third, they each respond to three open-ended questions selected by the other. Once a pair has completed these three exchanges, they proceed to EH Mail, which allows users to send messages through the site. Either user may terminate the connection at any point during the Guided Communication process. As the next section shows, EH's communications flow—along with its data preparation and matching flows—builds a user experience that creates a feeling of certainty that strengthens site ethos.

Table 2.3. eHarmony.com Guided Communication Examples

Stage	Example		Format
Quick Questions	If you were taken by your date to a party where you knew no one, how would you respond?		Multiple Choice
Makes or Breaks	Makes	Breaks	Check-boxes
	Chemistry—I must feel deeply in love and attracted to my partner.	Rude—I can't stand someone who is belittling or hateful to people.	
Dig Deeper	What [sic] person in your life has been most inspirational, and why?		Open-Ended Questions
eHarmony Mail	N/A		E-Mail Message

PEOPLE CHANGE, BUT NOT MUCH

> A son talking with his father about a woman the son was considering as a marriage partner. "Isn't she great, Dad? She's so beautiful, and she really motivates me to better myself."
>
> "Yes, I can see that she's very attractive," the father said, stroking his chin, "and I'm sure she can be quite motivating . . . but she seems a bit selfish. Have you noticed that she always wants to do what she wants, and doesn't seem to be interested in what everyone else would like?"
>
> "Oh, yeah, Dad, but she's really flexible. I can talk her into seeing things my way."
>
> "I noticed that she doesn't like to be inconvenienced, either."
>
> "Oh, I know, Dad. But she has a really great attitude. Don't worry, after we're married, she'll change."
>
> The father looked his son straight in the eyes and said, "Yes, son, people change, but not much."[23]

This excerpt from the introduction to the core personal dimensions of *Falling in Love for All the Right Reasons* resembles a Platonic dialogue with the father, presumably a stand-in for Warren, in the role of interlocutor. In it, he warns his son against marrying his selfish girlfriend because "people change, but not much," and the EH experience across all flows is predicated on that notion.

The procedures in the data preparation flow simulate a process of forward momentum while denying the possibility for change. None of the responses made on the twenty-one pages that make up the RQ may ever be changed, or even viewed, after they are submitted to the site. Of course, since EH relies on personality testing to select candidates during the matching flow, this is a necessity in order to gain "authentic" self-impressions of users as they are and always will be. This underlying assumption does not account for multiple subjectivities, even as it depends on them in its users. For example, users of the site must oscillate between a) self-presentation when selecting and uploading photographs of themselves, developing self-descriptions for their profile pages, and composing responses to others during Guided Communication and b) other-evaluation when viewing the profiles of matches and interpreting messages received from them during the communication process. Such oscillations are common, unconscious acts in a culture with pervasive communication technologies. In the case of online dating, they are especially closely tied to mate selection and physical desire, and the fixity created by EH's organizing logics can jettison emotional responses and instead rely on—or rather, give the appearance of relying on—wholly rational social scientific methods and computational processes of the site, all the while ignoring the logics and ideologies that informed the development of those processes. The site's reliance on self-report seems to presume a certain singular truth of identity that will be transparently communicated through the RQ rather than being answered strategically in an effort to increase one's attractiveness to others.

Whereas the matching flow is inaccessible to users, EH makes information about it available to them. It is difficult to say how many users read the *Scientific Matching* and *29 Dimensions of Compatibility* pages on the site—or whether they are familiar with the more than fifty texts available through Amazon.com either authored or coauthored by Warren. Regardless, EH's patented matching system is regularly featured in its advertisements, and the introduction to the RQ highlights it: "Now it is time to complete our comprehensive Relationship Questionnaire and receive your detailed Personality Profile. The results will provide you with insights about yourself and will enable us to find people who are highly compatible with you." Interestingly, the personality profile is linked from the top navigation rather than highlighted in the main content area of a user's profile page. Users may share the results with others, but they are not required to. This report includes five categories (agreeableness, openness, emotional stability, conscientiousness, and extraversion), and each page of the report is broken into a number of subheadings. It is surprising that this information would remain buried, but it makes sense given both the site's paternalistic approach and its stance on identity stability.

Finally, the inflexibility of EH's communications flow deepens the appearance of certainty. By moving couples through a highly regimented Guided Communication process, the site gives the impression that it can protect its users, mitigating ambiguity or deceit, as if they will know their matches better when they begin communicating more openly and/or meet face to face. However, this assumes both self-awareness and honesty that actual matches may not possess. In the most innocent of cases, users may not have reflected sufficiently on their relationship requirements, thus providing misinformation when considering their *Makes* and *Breaks*. In more insidious circumstances, there is nothing to stop manipulative users from answering questions in ways they believe members of their complementary match group would find attractive. EH's communications, matching, and data preparation flows stand in contrast to the Match.com approach discussed in the next section.

MATCH.COM

Originally conceived as an online classified advertising platform called Electric Classifieds, Match.com (Match) was founded by Gary Kremen in 1995. After noticing the popularity of personal ads in the San Francisco area, Kremen made a shift in concept from Internet classifieds to online dating. Match launched in 1996 and was purchased by Ticketmaster Online-City-Search, a subsidiary of InterActiveCorp (IAC) in 1998.[24] Today, it is one of the oldest and best-known online dating sites and serves as the cornerstone of IAC's Match Group, which includes forty brands, such as dating sites Chemistry.com and OkCupid.com as well as popular mobile dating application Tinder. According to Match Group CEO Sam Yagan, the site matches users based on their potential for having a good first date. "The way we can predict whether you're going to have a good first date is whether you're having a lot of communication on the site. That's really what we're trying to maximize, the probability that you're going to get into a conversation with one of the people we recommend to you."[25]

Data Preparation

Match's data preparation flow begins with the initial account setup phase. This flow includes seventeen screens that are designed to enable both site and user processes, and those screens are divided into eight categories: *Basics*, *Appearance*, *Interests*, *Lifestyle*, *Background/Values*, *Get to Know Me*, *About My Date*, and *In My Own Words*. Nearly all of the information gathered during the initial account setup phase requires users either to select from drop-down lists, radio buttons, and check-boxes or to enter information (zip code, for example) into text-boxes with predetermined formats and/or lengths (table 2.4). A handful of questions allow users to write lengthier descriptions, with character counts as the only limitations. The information users share with Match is used, generally speaking, for two purposes. First, the data provided are used to present potential matches to users, both by e-mailing *Top Matches* and by displaying users with similar profiles when a member views another user's page. Second, users may conduct their own searches based on the majority of the data collected.

During the initial account setup phase, users first answer the question, "What brings you here today?" before providing their relationship status, gender, and the desired gender and age range and location (i.e., mile radius from their zip code) of potential matches. On the same page, they then choose whether they are interested in matches who do not include a photo-

Table 2.4. Match.com Example Questions/Prompts

Section	Example Question/Prompt	Format
Basics	• Who are you looking for? (gender and age range) • Where should we search? (e.g., zipcode)	Drop down Fixed-format text; Radio button
Appearance	• Which of these best describes your body type?	Drop down
Interests	• What sort of sports and exercise do you enjoy? • What's the last thing you read? • The kinds of movies I like	check-box(es) Character-limited text Image selection
Lifestyle	• How often do you exercise? • Tell us more [about what you do for a living]. • What is your current annual income? • Which pets do you have (or you like but don't have)?	Drop down Character-limited text Radio button Likert scale (*have, like, no opinion*)
Background/ Values	• Which ethnicities best describe you? • Tell us more [about your faith]. • What's your level of education? • Describe your political views.	check-box(es) Character-limited text Radio button Drop down
Get to Know Me	• Tell us your birth order. • Which comedians make you laugh the most? • If I had 2 weeks off and could take any vacation I desired, I'd:	check-box(es) Radio button Image selection
About My Date	• How tall should [s/he] be? • Hair Color • Ethnicity • Salary Range • Alcohol [Use] • Has kids?	Drop down check-box(es)
In My Own Words	• Describe yourself and your ideal match.	Character-limited text

graph before moving on to the next screen, where they indicate the country in which they were born and their astrological sign.

The next two sections gather information about users' physical characteristics and leisure activities. In *Appearance*, users select their height, body type, hair color, and eye color from drop-down lists, and *Interests* screens gather data related to both users and their ideal match's pastimes and tastes. On the first page in the *Interests* section, users select check-boxes with the kinds of sports and exercise they enjoy from a list of twenty-three options. This page also includes two optional text-boxes in which users describe their favorite pastimes and places in more detail. The second page asks users to check interests they hope they and their partner will share and offers two more optional character-limited text-boxes—this time related to "favorite things" and the last book read. The final page in this section asks users to click a radio button for the kind of movies they enjoy, and it is one of three pages that differ significantly from the others in the initial account setup phase. Instead of asking users to select options from check-boxes or drop-down lists, this screen requires users to select an image from eight options, each representing a film genre. Since responses provided on this page are not searchable by other users, they have less to do with sharing information and more to do with what those choices say about respondents. In other words, the questions are part of the site's—but not users'—matching flow.

The next two portions of the initial account setup phase collect data on users' way of life and personal history. The *Lifestyle* section includes two pages. On the first screen, users select from drop-down lists how often they exercise, whether and how often they smoke and drink, their occupation, and whether they have children. They may also provide additional information regarding their employment in an optional text-box. On the second page in this section, users choose their current annual income range from drop downs and indicate on a scale whether they own, like but do not own, or do not have an opinion on seven kinds of pets. The *Background/Values* section includes two pages, and users select their ethnicities and their faith on the first. The former is selected using check-boxes, which allows users to identify with multiple ethnicities; for the latter, users click a radio button. This is unlike EH, which only allows users to choose one ethnicity. They may provide more information about their "roots, heritage, or culture" and/or their faith in the optional text-boxes on the same page. On the second page in this section, users share languages they speak using check-boxes and their politics and their level of education with drop downs.

The *Get to Know Me* section appears to include more questions solely for the site's portion of the matching flow. The first page includes five questions (see table 2.5). The second page asks users to complete the following statement: *If I had 2 weeks off and could take any vacation I desired, I'd . . .* and includes five options. As with the earlier page on favorite film genres, this screen contains photographs representing each of the choices, and none of the questions in this section is searchable by other users. These questions related to sense of humor, values, and social skills are similar to those used by Chemistry.com, another Match Group online dating site that uses personality profiles to match participants.

Unlike EH, Match's data preparation procedures permit users to share a variety of preferences about their potential partners. The *About My Date* section asks users to identify their ideal matches and includes three subsections: *Appearance*, *Background/Values*, and *Lifestyle*. On the *Appearance* screen, users select a preferred height range and mark check-boxes for hair-color, eye-color, and body-type preferences. On the *Background/Values*

Table 2.5. Match.com Get to Know Me Questions and Responses

Question	Responses
Tell us your birth order.	• the middle child • the middle • from a large family • the oldest • the youngest • No answer
If my friend and I ended up at a party where I didn't know anyone else I would . . .	• Walk straight up to someone and introduce myself • Just wait for the host to introduce me • Hang out with my friend • Say to myself, "I'm leaving in an hour, max" • No answer
I'm most drawn to charities that . . .	• Save the planet • Help people • Save animals • Fight illnesses • No answer
Which of these comedians makes you laugh the most?	• Jim Carrey • Chris Rock • Jay Leno • Jon Stewart • Tina Fey • Ellen DeGeneres • No answer
Let's say you got a big bonus. What would you do with it?	• Treat myself to something special • Celebrate with friends and family • Save half and spend the rest • Donate to a cause I really believe in • Reduce my credit card balance • No answer

screen, users choose their potential dates' ethnicity/ies, religion(s), education level(s), and language(s) spoken. The *Lifestyle* subsection is made up of two screens. On the first, users select appropriate occupations, preferred salary ranges, and smoking preferences. On the second page, they indicate drinking preferences and suitable marital and parenting statuses. If users identify preferences, they must specify how important those preferences are by clicking either a "must have" or "nice to have" radio button. They may also choose to indicate when they have no preferences.

Except for the photo-upload option, *In My Own Words* is the final section in the initial account setup phase. Here users are asked to describe both themselves and their ideal match in 200 to 4,000 characters. They also must write a maximum 140-character dating headline that appears both at the top of their dating profile and in the abbreviated view of their profile displayed in the search results of other users. On this page users are also presented the opportunity to double their chances of finding a match by uploading their profile to Chemistry.com.

After users have completed the initial account setup phase, Match appears to complete three primary tasks during the data preparation flow. First, user data are converted to numerical values and stored in a database that is searchable by other users and used by the site to identify potential matches. Next, users are arranged into match groups based on their gender and sexual orientation. Finally, users' self-description (*In My Own Words*) and dating headline are evaluated by the site. Users' profiles will not become active (i.e., they will not appear in search results, and the user cannot view the profiles of others) until they have completed those two elements, submitted them, and had them approved by Match. Once they are approved, the site initiates the matching flow, and both matching and communications flows become available to the user.

Matching

Information on Match's matching flow is less accessible than EH's. Match does not hold a patent for its matching procedures and provides no explicit explanation of them on its site. However, interviews with site executives and engineers provide valuable information on the matching logics employed. In the initial matching flow, the site moves through three phases: screening, filtering, and sorting. Match begins by screening potential matches based on users' "dealbreakers" (e.g., smoking), as indicated by the user during the data preparation flow. Similar to EH's screening dimensions, dealbreakers eliminate incompatible users from the pool of potential matches. Next, the site filters remaining candidates to eliminate those outside the user's preferred age and location ranges. Finally, matches are sorted with both users' preferences expressed on the *About My Date* screens and the intersection between the user's and candidates' interests, lifestyles, and backgrounds.[26]

Subsequent matching flows also rely on user data gathered implicitly. The site seems to employ data-mining methods similar to those used by Google and others to rank results. According to Gould, Match seems to rely on users' viewing behaviors to make recommendations: "[E]xplanations that Match gave me ('You share a birth month!') were simplifications. It generated them after it found a match by observing whose profiles I spent the most time reading and whose profiles others like me have liked, among any number of other factors."[27]

Indeed, Match uses revealed preferences, in combination with stated preferences, to identify similar users, those whose revealed and stated preferences conflict in similar ways and "whose behavior [the user's seems] to resemble."[28] In this process, the site accounts for fifteen hundred variables. According to Match engineer Amarnath Thombre, "These are compared with the variables of others, creating a series of so-called interactions. Each interaction has a score: a numerical expression of shared trait-tolerance."[29] Thombre likens this process to Netflix's Cinematch system, "except that the movie doesn't have to like you back."[30] I would offer one more distinction. When using Netflix, consumers are aware that the site is using their preferences (both media ratings and viewing history) in combination with the preferences of others to make viewing suggestions; a "best guess" at the user's star rating is displayed above the average user rating on each product page. On Match, these procedures are either concealed entirely or masked—claiming that a match is based on one criterion (e.g., a shared birth month) when it is actually more likely based on an aggregate of behaviors and preferences of similar users.

Match users may engage in self-matching through profile searches at any point after their account is activated, and there are two ways to initiate this search functionality. They can use a simple search—based solely on sexual orientation, age and location range, and whether users have a photograph and are online currently—from their account dashboard, or they can match themselves based on more than twenty additional categories and any keywords. The more intricate search preferences are savable, so that the user can rerun them at a later time. The difference between these two is comparable to the difference between simple and advanced searches of popular Web search engines or library databases.

Communications

Once the site has matched users with potential romantic partners, they receive a *Top Matches* e-mail message that includes username, photograph, total number of candidate's photographs available, age and location, and a link to view the profile page for up to twelve matches. The candidates in the message are divided into two categories: *Mutual Matches* and *Members We Selected for You.* The former is made up of candidates who fit the user's preferences and whose preferences the candidate fits; the latter includes additional, less-ideal matches chosen by the site. For an extra fee, premium subscribers are included in the initial message sent to new users who fit within their matching preferences, regardless of whether the premium subscriber meets the preferences of the new user. As long as users are visiting the site regularly, they will receive these messages daily. Users may also view these candidates at any time on the *Matches* page and take a quiz (*Like at First Sight*) in which they are shown a number of photograph arrays containing celebrities, automobiles, dating scenarios and asked to choose which in the group is most attractive. They are then presented with the option of viewing additional matches based on those choices. Similarly, Match includes a *Daily 5* page linked from the main navigation banner. On this page, users are presented with five potential matches and asked, "Does he [or she] spark your interest?" They then respond to the question by clicking a button (yes, no, or maybe) and are presented the next match. It appears that, in addition to the data-mining techniques discussed earlier, Match uses an aggregated system of user rankings similar to Netflix's Cinematch[31] to determine which candidates to present to users.

Although the kind of information gathered during the initial setup phase has changed only slightly, the Match profile page underwent a significant redesign in 2010.[32] Until that time, the profile page functioned primarily as a display of the data gathered during account initiation and its layout resembled that of a typical ecommerce page. The current site design (figure 2.2) includes a few social features that make visible some of the potential connections and similarities between the presenter and the evaluator. The Match profile page begins with a banner that includes a thumbnail photograph of the presenter, his username, and his dating headline on the left and a button to "favorite" him on the right. Just below the username is a menu with three tabs—*His* [or *Her*] *Story*, *Photos*, and *Our History*. The first tab is the default main screen for the profile page. The majority of the page "above the fold" is divided into two columns. The first column takes up two-thirds of the screen horizontally. On the left, there is an enlarged view of one of the presenter's photographs. Just to the right of that, there are thumbnails of additional photographs and more information about the presenter. The second column includes buttons and links for evaluator actions, including e-mailing or instant messaging the presenter, viewing additional profiles of similar users, or forwarding the profile to a friend's e-mail address; neither this column nor the banner changes when the evaluator chooses the *Photos* or *Our History* tabs.

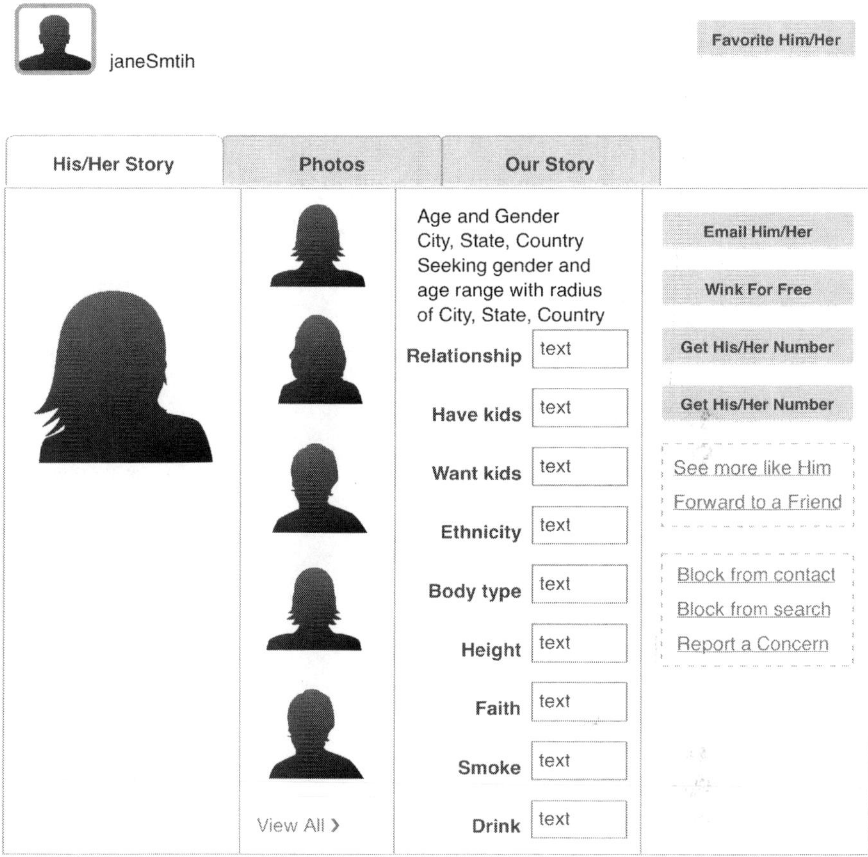

Figure 2.2. Match.com Profile Page

If the evaluator scrolls down on the default profile screen, she will find the presenter's *About Him [Her] & Who He's [She's] Looking For*. This section includes the required self-description gathered during the initial account setup phase as well as information related to his interests, leisure activities, education, etc. The final section of the default page is a table that includes information about appearance, lifestyle, and background/values for both the presenter and his ideal date. Just below the table is the *Ways You Match* check-box. When the evaluator selects it, the table alters to show only the categories on which both agree. Selecting the *Photos* tab on the profile page banner allows the evaluator to see all of the presenter's photographs, and *Our History* contains a record of contact between the evaluator and the presenter. If the two have not been in contact, the page displays a large blue arrow pointing to the evaluator action buttons/links with the message, "Every relationship starts with a first move, [sic] make yours now." After reviewing the profile page, a subscribing evaluator can send the presenter a direct message at any time. If she is not interested in the presenter, she may either return to potential matches identified by the site or engage in another phase of self-matching. Unlike eHarmony, Match users may contact others at any point during the communications flow. In fact, the evaluator calls to action are designed to facilitate communication between as many users as possible. As we saw earlier, facilitating conversation is the cornerstone of Match's approach.

Some [First Dates] Lead Everywhere

Recent Match commercials features current subscribers on first dates. Each of the spots includes voiceover such as the following:

> The world has changed; 1 in 5 relationships now begin on an online dating site. And when it comes to meeting someone great online, there's only one place to go: Match.com. We've led to more dates, more relationships, and more marriages than any other site. Begin something new today.[33]

These advertisements, along with Match's two other recent slogans ("Find Love. Guaranteed." and "It's Okay to Look."), highlight users' ability to take action. Whereas EH's tightly controlled process gives a feeling of certainty, Match's focus on *finding*, *looking*, and *leading to something* engenders an overall atmosphere of user agency that is supported by the site's reliance on user self-matching and open communication.

As is the case with search engines, this feeling of freedom belies the level of control exerted by the site, and the apparent autonomy available to Match users conceals its underlying limitations. For example, both categories (e.g., interests) and their members (e.g., music and concerts) are all predetermined by the site. Users may not add categories or members of categories. If users were only interested in meeting other users born in a particular setting, such as a large metropolitan area or a farming community, they could not search based on that category. Likewise, users could not add interests that people are often passionate about (e.g., comic books, fantasy sports, tabletop gaming, knitting) to that category. Users may be able to conduct profile searches whenever they like—and to include as few or as many variables as they choose—but they are site-specific searches. In other words, users cannot search based on data not gathered by the site, no matter how important they may be to the user. Previously, users could attach MatchWords, user-generated tags, to their profiles and search for them in the profiles of others. That option was removed in the 2010 site redesign. The only flexible search option now available is a keyword search that makes profile essays (e.g., *In My Own Words*) and usernames searchable. However, completing most text-based fields is optional. Only the *About Me & Who I'm Looking For*, which must be completed for account activation, is required.

If EH is a controlling parent, then Match.com is an absentee one. By allowing users to work fairly autonomously on the site, Match has the appearance of being removed from relying on and perpetuating assumptions about gender, race, class, and sexual orientation. Still, by providing this information in the seemingly stable frame of a drop-down list (in the case of categories such as gender, salary range, occupation) or a set of check-boxes (as is the case with ethnicity), Match is complicit in them. As some have pointed out, relying on categorization has significant implications for users as they interact with the site and one another.[34] For example, boyd notes that automatically completing the text-box on a search engine based on frequent searches could have unintended consequences, such as labeling someone with an Arabic name as a terrorist.[35] In the case of Match, requiring users to select race from a menu of predefined options does not account for the complexity of racial or gender identity. What if a user identifies with multiple ethnicities or a gender identity not included in the list? In addition, including such categories further emphasizes them, perhaps adding selection criteria to users' searches they may not have otherwise considered, just as a store's return policy creates the possibility of returning a purchase.

In his discussion of identity theft, Mark Poster notes, "[Contemporary] identity fragments into an aspect of consciousness (an awareness of continuity in time and space) and a complex of media content contained in information machines that combine to define an individual."[36] More and more of how we conceive of ourselves as human is tied up with machines, especially as identity becomes synonymous with information stored in databases. Establishing a Match profile is a time-consuming process. Users go through screen after screen of drop-down lists through which they must record basic demographics and information about their appearance, interests, lifestyle, background, and values. They must then repeat the process regarding the person they would like to meet before they can complete the *In My Own Words* portion of the profile and provide a textual account of who they are. And their profile will not be activated until they provide at least 340 characters. The search and suggestion functionalities available through Match require the conflation of individual identity with media content. Poster notes that identity in post-modernity is "a mixture of materiality and consciousness,"[37] and the reduction of people to the product of a database search is representative of this notion. In fact, as many have noted,[38] Match works very much like online shopping. Users can search for potential partners in the same way that they might search an online retailer for a pair of shoes. They select desired characteristics (size, color, style), receive a list of matches that fit the criteria, and then view images and read descriptions in order to determine which one of the potentials is the perfect fit.

OKCUPID.COM

OkCupid.com (OKC) is a free[39] online dating site cofounded in 2006 by the group of Harvard College graduates who also started TheSpark, a collection of online study guides that eventually became SparkNotes. In 2011 Match's parent company purchased OKC for fifty million dollars, making it part of the company's Match Group collection of dating sites and applications.[40] The site differentiates itself from other online dating sites and includes a number of social features. OKC incorporates forums on which users may discuss issues related to dating and relationships and a broad catalog of both site- and user-authored personality tests. Users are encouraged to share the results of these tests with their social circle, either through e-mail or integration with social-networking sites. The site has approximately 3.5 million users internationally.[41]

OKC also allows users to keep journals and to give one another awards (e.g., *Eye Candy*, *Brilliant Profile*) and offers a number of tools and games. In addition to *MyBestFace*, an application that allows members to upload and compare their profile pictures in order to determine which is the most attractive, OKC includes both single- and multi-player games and two concept maps that visualize results of answers to questions employed during the data preparation flow. The *Match Map* uses question responses to display the states and countries in which a user is most likely to find compatible relationship partners, and *Flowchart to My Heart* presents the path of responses that leads to a date with—or rejection by—the user. Although OKC is billed as an online dating site, individuals are not required to use it for that purpose. Some members seem to join to take the personality tests, and many who join the site for dating appear to maintain active profiles after establishing a relationship so that they may continue to take quizzes, play games, and write and read journals.

Data Preparation

OKC's initial account setup phase includes three required screens, many fewer than Match's seventeen pages and EH's twenty-four. Users begin on a welcome screen that asks them to identify gender (female, male), sexual orientation (straight, gay, bisexual), and relationship status (single, seeing someone, married).[42] On the second page, they enter their birth date, location (city and either country or zip code), and e-mail address. Finally, users select a public username and password and complete a captcha.[43] Clicking the *Done!* button redirects users to a welcome page on which they may upload a photograph, update their profile with additional information, answer questions, or send a message to another user. At this point, members have access to all of the site's functionality, including searching for potential romantic partners, though the site's portion of the matching flow will not be initiated until they have answered fifty *Match Questions*. Users may choose to complete the secondary setup phase immediately or return to it another time. Like Match, OKC's seeming ease of access conceals site complexity.

The *Tell us about yourself* page is the starting point for the secondary setup phase, and it includes areas for user details, desired characteristics for potential partners, and ten "essay" blocks. *My Details* collects information commonly included in online dating user profiles as well as some site-specific data: online status; height and body type; ethnicity; whether and how often the user smokes and drinks; whether the user has children and/or pets; the user's religious affiliation and level of commitment to it; and the user's education level, job, and income. In addition to these categories, OKC's *My Details* contains the user's diet (e.g., vegetarian, kosher) and how strictly it is maintained, frequency of drug use, astrological sign, and languages spoken and level of fluency. Except for ethnicity, all of the *My Details* categories are selected from drop-down lists. Like Match, users identify their ethnicity using check-boxes, allowing for identification with more than one. Although both EH and Match allow users to identify their ethnicity as "other," only OKC allows users to choose not to disclose this information. The *I'm Looking for* section permits the user to report on age range, proximity, relationship status, and sexual orientation and gender of potential partners as well as the kind(s) of relationships (e.g., friendship, short- or long-term dating, casual sexual) the user is interested in. The ten essay sections give users the opportunity to provide additional information about themselves, and there are no word- or character-count requirements for these sections. In addition to *My Self Summary*, they include headings such as *I'm really good at* . . . and *On a typical Friday night I am*. . . . The information provided by users in this section is both viewable and searchable by other users. Unlike EH and Match, all site content, including journal entries and discussion forum posts, can be searched by site members.

As discussed earlier, users need only choose a username and password, enter some basic identifying information, and complete a captcha in order to begin using OKC. However, users do not receive suggestions for potential partners and *Match Search* results until they have answered a minimum of fifty *Match Questions*, an additional process in the data preparation flow (see table 2.6), consisting of the question, a range of answers acceptable to the user, the questions importance to the user, the user's response to the question, and the user's explanation of her answer, if provided. The default state for all questions is public, but users may choose to keep their answers private. Users whose answers are not publicly visible cannot compare their answers to those of their matches.[44]

Matching

OKC's user experience seems to lie somewhere between EH and Match. Like Match, OKC allows its users to search for, view, and contact other users without restriction.[45] Like EH, OKC relies on a kind of compatibility matching, and the site sorts *Match Question* responses based on "how well they divide the population."[46] Questions are also organized by "genre" so that users are presented with different kinds of questions, and by "raciness" so that members who consistently skip sex- or drug-related questions no longer receive them.[47] These questions are the cornerstone of OKC's matching procedures.

After users have met the fifty-question minimum, the site begins the initial matching flow. *Match Percentage* between any two users is based on their responses to—and the importance of—questions they have both answered. For example, if Users A and B both answered the question in table 2.6 ("Which makes a better relationship?"), that question would be included in the set used to determine their compatibility. The site converts users' question ranking to numerical weights: *irrelevant* (0), *a little important* (1), *somewhat important* (10), *very important* (50), or *mandatory* (250). User A is awarded points for answers that fit within User B's parameters, and vice versa; the points awarded are divided by the points available. The two users' satisfaction percentages are multiplied, and the square root of them is then taken. This number is the calculated *Match Percentage* for Users A and B.

Table 2.6. OkCupid.com Match Question Components

Question: Which makes for a better relationship?

Question Component	Potential Responses	Selection Method
User Response	• Passion • Dedication	Radio button
Preferred Response(s) from Other Users (*Answer I'll accept . . .*)	• Passion • Dedication	Check-box (all that apply)
Question Importance (*This question is . . .*)	• Irrelevant • A Little Important • Somewhat Important • Very Important • Mandatory	Radio button
Question Privacy	• Answer this question privately.	Check-box (all that apply)
Explanation	• Explain your answer (optional)	Text-box

This percentage is then adjusted for margin of error,[48] and the result is displayed to both users.

For any pair of users, there are six factors in play for every question: User A's and User B's answers, User A's and User B's preferred answers from other users, and the importance of the question for Users A and B. Table 2.7 provides an example from OKC's *Match Percentage* page. For User A, the first question was very important, and the second was a little important. User B's answers fit within User A's acceptable range for the first question but not for the second. User A's satisfaction with User B's answers would be calculated thus:

$$1(50) + 1(0) = 50/51 = 98\%$$

For User B, the first question was a little important, and the second was somewhat important. User A's answer for the first question does not fit within User B's acceptable range but does for the second. User B's satisfaction with User A's answers would be calculated thus:

$$1(0) + 1(10) = 10/11 = 91\%$$

The match percentage is then calculated by taking the square root:

$$\sqrt{}98\% \text{ x } 91\% = 94\%$$

Users A and B are therefore a 94 percent match.[49]

OKC uses *Match Percentage* to identify candidates for users, but users can also engage in self-matching by searching the site. *Match Search* begins with a simple search that includes options for sexuality, age and location range, when last online, relationship status, and whether a photograph is included in the profile. Participants can also select from an additional twenty advanced search options, including body type, *Dating Persona* test results, and the number of Match Questions answered. Some of these options, such as body type, are only available to subscribers. Unlike Match's search page, which lists all available search criteria, the OKC *Match Search* page uses a small banner for selecting search criteria and the bulk of the page for presenting potential matches. After subsequent matching flows, the site also presents *Quiver* and *Quickmatch*[50] candidates to users. The former is a set of three potential matches delivered on both the *Quiver* page and via e-mail message. When users indicate they are not interested in one (or all) of their matches, the selections are noted and new *Quiver* matches are provided. The latter is available through the *Quickmatch* page and provides a profile with the username stripped out. Users are then asked to rate the profile in order to see the username. Just as Match's *Daily 5* uses profile ranking to refine matches, so

Table 2.7. OkCupid.com Example Match Percentage Calculation

	Question: How messy are you?	Question: Have you ever cheated in a relationship?
Possible Responses	1. Very messy 2. Average 3. Very organized	1. Yes 2. No
User A Response	3	2
User A Preferred Response(s) from Other Users (*Answer I'll accept . . .*)	2 or 3	2
User A Question Importance (*This question is . . .*)	Very Important	A Little Important
User B Response	2	1
User B Preferred Response(s) from Other Users (*Answer I'll accept . . .*)	2	2
User B Question Importance (*This question is . . .*)	A Little Important	Somewhat Important

Source: "Math Percentage," OkCupid.com.

Quickmatch captures additional data on user preferences. OKC explains its matching procedures on the *Support* page and highlights its reliance on mathematical calculations on its blog (*OkTrends*), on its *About* page, in site cofounder Christian Rudder's *Dataclysm: Who We Are (When We Think No One's Looking)*, and in its public relations.

Communications

Once the initial matching flow has been completed, the site sends an e-mail message to users with a photograph, username, location, age, gender, and an excerpt from one of the candidate's essays for multiple matches. The site also presents potential partners in the *You might like . . .* block in left-hand border of the dashboard when the user is logged on to the site. This block persists on subsequent visits to the site and is refreshed with new potential matches with each new page viewed.

As figure 2.3 shows, the top block on the main OKC profile page is divided into three columns. The left column includes information about the presenter: his photograph; his compatibility with the evaluator viewing his profile; his username; his age, gender, sexual orientation, and relationship status; and his location and distance from the evaluator. The next column includes any awards he has received, and the final column includes buttons for evaluator actions (e.g., message, save). The rest of the main presenter profile page is the presenter's *About* section, which is made up of information provided by him on his *Tell us about yourself* page. Each presenter profile contains subpages that are accessible through six additional tabs just above the *About* section. These additional tabs hold photographs, an infographic representing a comparison of the presenter's personality traits (e.g., more trusting, less spiritual) with others, his journal, results of tests he has taken, awards given to him accompanied by awarders' comments, and a comparison of the presenter's and the evaluator's responses to questions.

Like Your Friend in the Real World

According to OKC cofounder, president, and creative director Chris Coyne, the site's matching processes are "software that would be like your friend in the real world":

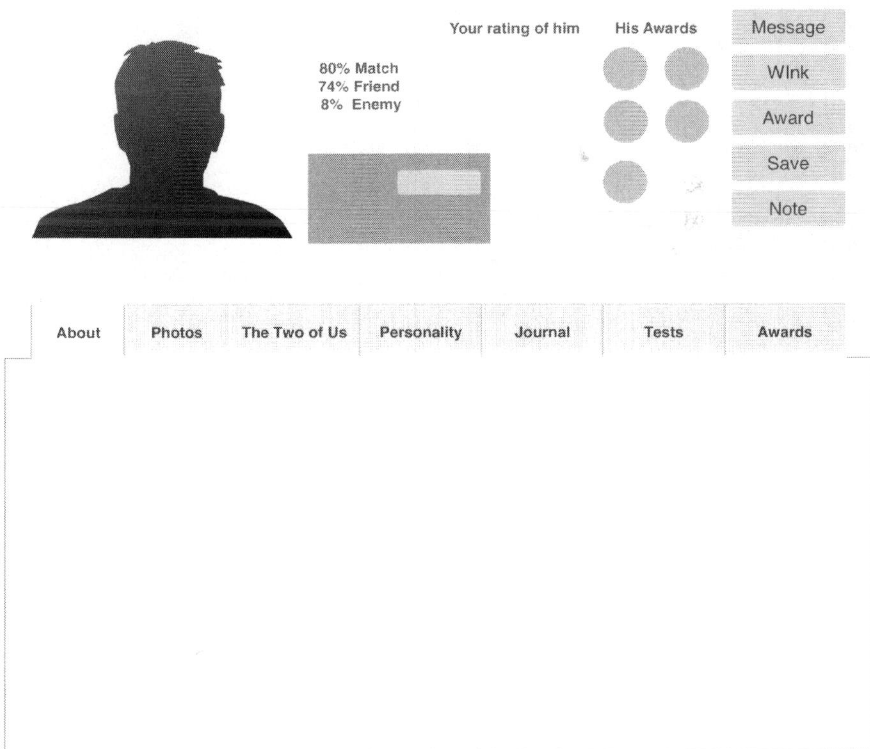

Figure 2.3. OkCupid.com Profile Page

> If I were your friend and I told you that So-and-So would be the perfect date, your response would be to start asking me questions. . . . On the Internet, people will ask—and answer—extremely personal questions.[51]

Indeed, the site is clearly designed to feel fun and friendly, encouraging users to be forthcoming. The color scheme and typography used on the site are playful, and even the way the site refers to itself is tongue in cheek. The site's main landing page has a bright blue background and invites visitors to "Join the best free dating site on Earth"; e-mails sent to users on behalf of the site frequently originate from Staff Robot. Features that encourage users to hang out on the site—quizzes, journals, discussion forums—contribute to the air of sociability.

OKC cofounder and CEO of The Match Group Sam Yagan likens the experience of finding a date through the site to the process offline:

> In the real world dating is social. One girl will call up her girlfriends and say, "Hey, let's all go out tonight." They all get dressed up and they never explicitly say, "Hey, we're looking to get picked up by a guy." They just tell themselves that they're going out to have a good time. No one had ever invited me to join Match.com. No one ever said "go to Match.com, you'll have a great time." Match.com was a very lonely experience. So our vision from the very beginning was to build an online bar, a place where you go with a bunch of friends, you observe other people and decide which ones seem the most appealing to you. Then you gradually and slowly begin to strike up a conversation with them.[52]

This atmosphere instills a level of comfort in certain kinds of users, people who enjoy and make time for socializing. The questions OKC asks of its users do deal with deeply personal and taboo issues, and the site has amassed over eight hundred million answers.[53] The site's original and most prominent quiz, *The Dating Persona*,[54] probes topics such as drug use, promiscuity, and sexual predilections in detail and then suggests that users share their results on their profiles, send them to contacts through e-mail, and post them to social-networking sites. And as with social-networking sites, OKC uses the data it collects for a variety of purposes beyond matching users with one another. In-house researchers use data gathered to conduct investigations into issues related to online dating on *OKTrends*, the site's blog, as well as Rudder's nonfiction bestseller *Dataclysm*. OKC also sells user data for scholarly studies. For example, a team of political scientists from Stanford and Yale used an OKC data set to study the connection between political opinion and social relationship formation.[55]

ONLINE DATING AND THE CONDITIONS OF PROCEDURALITY

Galloway identifies fourteen techniques of continuity, and his final characteristic is that identity on the Internet is "*anonymous but descriptive*,"[56] and it is especially relevant for the discussion of online dating. In other words, although Internet applications store a host of information on users and traffic, the "real identity of those movements is irrelevant."[57]

> Demographics and user statistics are more important than real names and real identities. On the Internet there is no reason to know the name of a particular user, only to know what that user likes, where they shop, where they live, and so on. The clustering of descriptive information around a specific user becomes sufficient to explain the identity of that user.[58]

Significantly, this technique of continuity is in direct contradiction to the user experience in computational systems such as online dating, which make users feel unique and catered to.

As we saw in our earlier discussion, procedurality is a way of understanding processes that includes in it conditions of possibility, concealment, and stituatedness, and procedural rhetoric is a mechanism for making arguments through those processes. By examining online dating sites procedurally, we see both how sites function and how they build relationships with their users—through what those processes permit and deny, by concealing some things and making other things visible, in their connections with culturally determined understandings of compatibility. On one level, EH, Match, and OKC include some significant similarities in their processes. They all use constraints and affordances to help define their relationships with their users. Indeed, all three sites rely heavily on limitations to instill confidence in their users. Early on, EH uses its highly constrained data preparation flow to set an expectation of reliability and trustworthiness. By rigidly structuring interactions between users during the communications flow, EH continues to project an air of confidence. Match's and OKC's data preparation flows may appear to be more flexible, but they too collect specialized information from their users in particular ways. Although the matching and communication flows of Match and OKC are more flexible, it is through that flexibility that those sites make connections to users' larger lives.

All three sites also rely on concealment and exposure. Through the depth and breadth of the initial account setup phase, the data preparation flow is both visible and invisible on all three sites. By filling out the personality profile, providing information on lifestyle and tastes, or answering questions, users experience the data preparation, but the sites reveal little about how the information will be used. In the matching flow, EH continues to rely on concealment, and this flow is primarily unavailable to its members. Matching flows on Match and OKC are concealed to varying degrees. Users are allowed to engage in self-matching procedures, a process that appears to be transparent, but users only have data available to them that is gathered by the site. In addition, Match publicizes very little about its matching procedures, and OKC's complexity may limit the understanding of many participants. The EH communications flow is visible and deliberate, while Match's and OKC's remain invisible through their similarity to other, more familiar electronic communication technologies, such as e-mail and social-networking messaging.

It is in situatedness that the three sites converge; through examination of this condition, we will gain a better understanding of both the cultural assumptions that undergird each of the sites and their relationship to processes of subjectification. While all three sites rely on components of the automation as defined in chapter 1 (e.g., modulation), each site draws on specific, divergent ideologies that govern the kinds of user information they collect and store and how they connect users to one another. In the next chapter, I focus on the cultural logics and processes of subjectification that inform (and are informed by) each of the three sites and online dating generally.

NOTES

1. Alexander R. Galloway, *Protocol: How Control Exists after Decentralization.* Cambridge (MA: MIT Press, 2004), 52.
2. Ibid., 53.
3. Jeremy Packer and Stephen B. Crofts Wiley, *Communication Matters: Materialist Approaches to Media, Mobility, and Networks* (Hoboken: Taylor and Francis, 2011), 4.
4. Galloway, *Protocol*, 30.
5. Ibid., 31.
6. Ibid., 37–38.
7. Ibid., 81.
8. Ibid., 138–40.
9. Ibid., 87 (italics original).
10. Ibid., 55.
11. eHarmony.com, accessed on July 4, 2015, www.eharmony.com.
12. "Who We Are," eHarmony.com, accessed on July 4, 2015, www.eharmony.com/about/eharmony/.
13. "29 Dimensions of Compatibility," eHarmony.com, accessed on July 4, 2015, www.eharmony.com/why/dating-relationship-compatibility/.
14. Neil Clark Warren, *Falling in Love for All the Right Reasons: How to Find Your Soul Mate*, with the assistance of Ken Abraham (New York: Time Warner Book Group, 2005), 34.
15. Ibid., 56.
16. Ibid., 161–98.
17. Ibid., 194.
18. After completing the initial account setup phase without uploading a photograph and subscribing, my profile completion meter indicated 22 percent completion.
19. J. Galen Buckwalter et al., Method and System for Identifying People Who Are Likely to Have a Successful Relationship, US Patent 6,735,568, filed on August 10, 2000, and issued May 11, 2004, 4.

20. Ibid.
21. Ibid., 4.
22. If both users agree, they may choose to proceed directly to the final stage of Guided Communication, but that is not the site's preferred method. For an additional fee, users may also make voice calls through the site. Non-subscribing members may send icebreakers, one-sentence messages selected from a drop-down list (e.g., "Your profile brought a smile to my face!").
23. Warren, *Falling in Love*, 99.
24. Julia Angwin, "Love's Labor Lost: Online Matchmaker Still Seeks Love, Money," SFGate.com, February 12, 1998, www.sfgate.com/business/article/LOVE-S-LABOR-LOST-Online-matchmaker-still-seeks-3013097.php.
25. "Match Inc CEO Sam Yagan Reveals How Math and Data Predict Love Online," interview by Alyson Shontell, BusinessInsider.com, May 17, 2013, www.businessinsider.com/how-online-dating-sites-work-2013-5.
26. Ibid.
27. Emily Gould, "True Match," Technology Review, January/February 2010, 73–75, 75.
28. Nick Paumgarten, "Looking for Someone: Sex, Love, and Loneliness on the Internet" (includes an interview with Amarnath Thombre), *The New Yorker*, July 4, 2011, www.newyorker.com/ magazine/ 2011/07/04/looking-for-someone.
29. Ibid.
30. Ibid.
31. Netflix asks users to rank videos on a five-star scale. The site then uses its Cinematch system for matching users with movies, television shows, and so on that they might enjoy by comparing their rankings to aggregated data from all of its users. For more on Cinematch, see the discussion of the Netflix Prize in chapter 6.
32. "New Profile Design: Profile Display," Match.com, accessed on March 21, 2011 (page discontinued).
33. Match, "Try Something New :30--Match.com Official Commercial," YouTube video, :32, posted October 1, 2010, www.youtube.com/watch?v=CV5GLZTYDUs&sns=em.
34. danah boyd, "Guilt through Algorithmic Association," Apophenia (blog), September 12, 2011, www.zephoria.org/thoughts/archives/2011/09/12/guilt-through-algorithmic-association.html; John Cheney-Lippold, "A New Algorithmic Identity: Soft Biopolitics and the Modulation of Control, *Theory, Culture, & Society* 28, no. 6 (2011), 164–81; Alexis Madrigal, "Take the Data out of Dating," *Atlantic Monthly*, December 2010, 37.
35. boyd, "Guilt."
36. Mark Poster, *Information Please: Culture and Politics in the Age of Digital Machines*, (Durham, NC: Duke University Press, 2006), 113–14.
37. Ibid., 113.
38. See Heino, Rebecca D. Nichole B. Ellison, and Jennifer Gibbs, "Are We a Match? Choosing Partners in the Online Dating Market" (paper presentation, International Communication Association Annual Conference, New York City, New York, May 2005); Dawn Shepherd, "Match.com, Genre, and the Construction of Self" (paper presentation, Biennial Conference of the Rhetoric Society of America, Seattle, Washington, May 2008).
39. OkCupid does offer A-List, an add-on service that includes access to additional search filtering options and special discussion forums, the ability to add attachments to messages and to change usernames, and an advertisement-free view of the site.
40. "About Us," OkCupid.com, accessed on May 6, 2010, www.okcupid.com/about (page discontinued); "IAC's Match.Com Buys Rival OkCupid for $50m," *Bloomberg Business*, February 2, 2011, www.businessweek.com/ap/financialnews/D9L4QP681.htm.
41. Dan Avery, "OkCupid Adds Loads of New Sexuality, Gender Options for Members," NewNowNext.com, November 16, 2015, www.newnownext.com/okcupid-adds-loads-of-new-sexuality-gender-options-for-members/11/2014/.
42. Of the sites discussed, both OKC and Match use the same entry point for opposite- and same-sex matching. Although, at least according to its patent document, eHarmony's *Relationship Questionnaire* would be similar for heterosexual and gay and lesbian couples, eHarmony uses a different site (CompatiblePartners.net) for same-sex dating. In addition, only OKC

allows for bisexual matching and openly facilitates non-monogamous relationship formations (e.g., casual sexual, polyamorous). Only OKC has the same entry point for all matching, including international. In 2014, OKC began piloting new gender and sexuality options for a portion of its members. Those users may choose from twenty-two gender categories (Woman, Man, Agender, Androgynous, Bigender, Cis Man, Cis Woman, Genderfluid, Genderqueer, Gender Nonconforming, Hijra, Intersex, Non-Binary, Other, Pangender, Transfeminine, Transgender, Transmasculine, Transsexual, Trans Man, Trans Woman, Two Spirit) and select up to five sexualities from twelve options (Straight, Gay, Asexual, Bisexual, Demisexual, Heteroflexible, Homoflexible, Lesbian, Pansexual, Queer, Questioning, Sapiosexual). Avery, "OkCupid Adds."

43. A captcha requires users to enter a series of characters from an image into a text-box. This serves as a test to ensure the account is being requested by a live human being. Since OKC is the only site examined that does not require a subscription in order to send messages to other members, presumably this step is included to prevent infiltration of the site by spammers.

44. "Match Percentage," OkCupid.com, accessed on March 12, 2011, www.okcupid.com/support.

45. Users may choose to block other individual users. If User A blocks User B, B will not be able to view or contact A.

46. "Any Questions?" OkCupid.com, accessed on March 12, 2011, www.okcupid.com/support.

47. "Match Percentage," OkCupid.com.

48. OKC uses a margin of error based on the number of questions a pair of users has both answered (S). If S is 1, the margin of error is 1.00, and the highest possible true match percentage is 0. If S is 1000, the margin of error is .001, and the highest true match percentage is 99.9 percent. The full margin of error chart is available at www.okcupid.com/help/topics.

49. "Match Percentage," OkCupid.com.

50. Match's *Daily 5* appears to be based on OKC's *Quickmatch*.

51. Nick Paumgarten. "Looking for Someone" (includes interview with Chris Coyne).

52. Sam Yagan, interview by Mark Brooks, *Online Personals Watch*, September 21, 2009, www.onlinepersonalswatch.com/news/2009/09/sam-yagan-ceo-of-okcupid.html.

53. Paumgarten, "Looking for Someone."

54. *The Dating Persona* is a personality test. It includes thirty-two personality types, sixteen each for men (Backrubber, Bachelor, Slow Dancer, Gentleman, Billy Goat, 5-Night Stand, Mixed Messenger, False Messiah, Poolboy, Playboy, Boy Next Door, Lover Boy, Last Man on Earth, Hornivore, Man Child, Vapor Trail) and for women (Intern, Dirty Little Secret, Sonnet, Maid of Honor, Nymph, Stiletto, Priss, Battle Axe, Nurse, Playstation, Window Shopper, Peach, Half-Cocked, Genghis Khunt, Wild Rose, Sudden Departure). Like the Myers-Briggs, personalities are measured on four matrices (Deliberate/Random, Brutal/Gentle, Love/Sex, Dreamer/Master). For descriptions of all thirty-two types, see www.okcupid.com/personality?type=DGSD&g=1&o=1.

55. Paumgarten, "Looking for Someone."

56. The fourteen techniques of continuity identified by Galloway are as follows: conceal the source, eliminate dead links, eliminate no links, green means go, true identity, remove barriers, continuity between media types, prohibition against low resolution, highest speed possible, prohibition on crashes, prohibition on dead media, eliminate mediation, feedback loops, and anonymous but descriptive. Galloway, *Protocol*, 64–69; Ibid., 69 (italics original).

57. Galloway, *Protocol*.

58. Ibid.

Chapter Three

Online Dating and the Construction of Subjects

In 2014, Facebook directed the news feeds of nearly 700,000 site members to include items with a majority of either positively or negatively worded items in order to test the effects of "emotional contagion."[1] When the results of the study were published in *Proceedings of the National Academy of Sciences*, many observers questioned the ethics of the study, which included site members without their consent. Indeed, US Senator Mark Warner asked the Federal Trade Commission to investigate the company's privacy practices.[2] Shortly after the public outcry, OkCupid.com revealed that it had conducted three experiments on users in recent years. First, to commemorate the release of its mobile application for arranging blind dates, the company observed "Love Is Blind Day" by eliminating all pictures from user profiles. Second, in order to test the relative importance of users' physical attractiveness (profile pictures) and personality (profile text), the site engineered profile displays of a random sampling of users to show their full profile half the time and to present it without text the other half. Third, they manipulated the *Match Percentage* for a group of users, depicting bad matches (30 percent) as excellent ones (90 percent) with the aim of measuring the matching formula's capacity for predicting compatibility.[3]

OkCupid.com's manipulation of user information, especially in the third case, highlights the abstruseness of online dating procedures. In addition, it underscores the importance of understanding the relationship between the user's experience of a site and how the site works, and in this chapter I further the investigation of online dating sites as articulators by considering them as procedural rhetorics.[4] Just as procedurality defines permission-denial and transparency-opacity, so is it situated. Procedures may seem "flexible and porous, but they are crafted from a multitude of protracted, intersecting cultural processes."[5] Bogost's procedural rhetoric offers a opportunity to consider articulators in terms of "the coupling of a cultural process and its computational representation":[6]

> This sentiment [wondering how a process works] probably conjures images of mechanical devices like wristwatches, where procedural understanding implies taking a set of gears apart to see how they mesh. But procedurality can also entail the operation of cultural, social, and historical systems. In these cases, *how does this work?* requires taking a set of cultural systems apart to see what logics motivate their human actors.[7]

In the previous chapter, we disassembled online-dating procedures. Here we will parse the cultural systems they articulate.

While discussing *Freaky Flakes*, a videogame that teaches children advertising awareness by allowing them to design a cereal box, Bogost offers ideas for improving the game's procedural argument. He suggests that implementing a rule set that mirrors advertisers' techniques for motivating purchases would introduce failure into the activity, enabling players to understand better the actual tactics employed by marketers. He identifies the recommended process as *procedural enthymeme*. Enthymeme is commonly understood as a syllogism that requires the audience to supply its minor premise,[8] and with procedural enthymene "the player *literally* fills in the missing portion of the syllogism by interacting with the application [through action] constrained by the rules."[9] Brock argues that enthymeme more generally operates as a rhetorical algorithm, in that it contains the logic that limits the range of conclusions an audience can draw.[10] Brock and Shepherd suggest that through procedural enthymematic interactions we engage in processes of self-persuasion in which "we accept options made apparently available to us from a set of constrained possibilities."[11] In other words, we may feel that we are choosing to engage with a site in a particular way, but the site defines (and frequently conceals) the conditions for engagement.

Online dating sites depend on self-persuasion through both enthymematic procedures and thought processes. For example, when a Match.com user is matched with a man who shares her love of pet birds, she may presume that they are paired based on their responses to the question on pet ownership (*Which pets do you have (or you like but don't have)?*). It would be reasonable for her to expect *compatibility* with her counterpart—and to believe that Match.com somehow identified an affinity between them. To put it another way, she has supplied the minor premise of the following syllogism:

1. Full participation on Match.com requires answering a set of questions. (major premise)
2. Match.com employs Users A's and User B's responses to determine compatibility. (minor premise)

3. User A and User B make a match. (conclusion)

Based on her experience with Match.com, she applies enthymematic thought processes to reason that Match uses the information she explicitly provided to determine her and her match's compatibility. At the same time, all of her activities on the site require enthymematic procedures built on that same premise. In other words, participating in online dating works under the assumption that a user's participation—answering questions, reviewing matches, searching or browsing profiles, composing and responding to messages—are governed by a logic of compatibility. This chapter examines online dating as computational and cultural systems. In particular, it considers the persuasive processes described in the previous chapter in terms of enthymematic procedures and thought processes for eHarmony.com, Match.com, and OkCupid.com as well as online dating overall.

ONLINE DATING AND CULTURAL SYSTEMS

As discussed in chapter 2, online dating procedures can be divided into three flows (data preparation, matching, and communications). These flows are initially sequentially, but the latter two recur throughout the relationship between the user and the site, both in terms of the former's experience and the latter's functionality.

eHarmony.com (EH) was granted a patent in 2004 for its matchmaking method and system. During the data preparation flow, the system assigns numerical values to user responses and creates a user database. It then uses correlation matrices to identify and classify the user's potential for relationship satisfaction.[12] This stage lines up with the initial account setup phase, in which users complete the Relationship Questionnaire, a personality test addressing twenty-nine characteristics.[13] In the first-round matching flow, the site identifies other users of the appropriate gender (male for female, in the case of opposite-sex matching) with similar potential for relationship satisfaction.[14] Users can reinitiate a truncated data preparation flow by changing any of the small subset of data that can be altered (e.g., location). This stage is invisible to users, and matching does not become visible to them until the site shares the results during the communications flow. Finally, in the communications flow, EH presents the user with a list of matches and the choice of whether to communicate with them individually.[15] At this point, the user experience realigns with site processes. EH presents matches to users both on and off the site through the site's *My Matches* page and e-mail messages.

Match.com (Match) users play a role in all three flows and are even capable of triggering them. They initiate the data preparation flow procedures in the initial account setup phase and through any status changes (e.g., marital status) or profile updates (e.g., adding or changing "Last Read"). As soon as the original data preparation flow is complete, the site generates users' initial match set—group of potential romantic partners—and presents it to them. At that time, and at anytime during membership in the site, users may initiate matching flows for themselves by using the site's search functionality. They may create and save their own searches using variables stored by the site. Subscribers may contact users through the communications flow at any point as well.

During the OkCupid.com (OKC) initial data preparation flow, users complete a brief initial account setup phase, and the site converts user responses to numerical values and establishes a user database. Users must complete a secondary question-answering phase prior to the site's initializing the matching flow. Once a user has met a fifty-question benchmark, the site engages in a second round of data preparation, this time converting and weighting user responses for matching. In the matching flow, the site analyzes users' answers and preferences in order to match them with candidates. These matches are shared with users, both on and off site, as part of the communications flow. As is the case with Match, OKC users may begin self-matching and communicating as soon as they have completed the first-round data preparation flow. They can add or make changes to profile information and engage in self-matching at any point. In addition, they start communications flows by contacting other users throughout the process.

eHarmony.com

EH's functionality is rooted in a social-scientific, modernist understanding of subjectivity. Within this logic, individual identity is fixed and quantifiable, and relationship satisfaction is predictable. And under the tight control exerted through all flows, there is little likelihood that users could diverge from that construction. In other words, if EH determines that certain characteristics govern the potential for relationship satisfaction, and then weeds out all people who do not possess those characteristics, all EH users will possess them—and all EH successes will have relationships based on those principles. In *Falling in Love for All the Right Reasons*, cofounder Neil Clark Warren paints a pretty bleak picture of the prospect of marriage success—"'And they lived happily ever after' for many people is a colossal hoax!"[16] —before offering a pitch for EH's ideological perspective. He predicates his argument for EH's matchmaking method on the scientific makeup of the system:

> After countless hours of meticulous, empirical research, a team of psychologists and I have discovered the twenty-nine dimensions that happily married people have taken seriously, and unhappily married people have matched poorly. Simply put, if you learn to recognize these twenty-nine dimensions and

> discover how to determine whether you are well matched with a particular person, you will have a nearly foolproof litmus test to determine your potential for enjoying a lifetime of great marital success—or not—with that person.[17]

Interestingly, this mechanism for sorting the desirable from the undesirable is reminiscent of the historical construction of madness.

Foucault notes that until the mid-seventeenth century, the world was "strangely hospitable, in all senses, to madness," and insanity, along with unemployment and prostitution, hovered at the margins of society.[18] By 1656—he marks the founding of Paris's Hôpital Général as a significant milestone—madness was characterized as a moral shortcoming that should be restricted and scrutinized. For EH, individuals identified as deficient of character, who appear to have difficulty controlling their anger, who are not self-actualized, who are stubborn or defiant, who have been "raised in a dysfunctional family," who are neurotic, are denied access to their services.[19]

In a way, this denial works much like the exclusion of lepers in the Middle Ages and then madmen in the Renaissance. As late as the sixteenth century the salvation for madness was its isolation, though not necessarily its confinement. By the mid-seventeenth century, madness was reconfigured as something to be confined, studied, and treated. In the case under discussion, EH seems interested in isolating the "abnormality" without confining it. Indeed, EH makes an argument, both through the site's matching procedures and in Warren's self-help texts, that these maladies are resistant to treatment. As featured in Chemistry.com's "Rejected by eHarmony" advertising campaign, EH does not provide services to all users who complete its Relationship Questionnaire. First, the company did not offer same-sex matching until 2009, after settling a class-action lawsuit in California. Second, as discussed in the patent document, users with a low likelihood for relationship satisfaction may not be accepted as members. Finally, users whose Relationship Questionnaire responses do not satisfy the screening dimensions, such as good character or anger management, are liable to be excluded.

This practice of labeling certain behaviors or proclivities as risky is akin to the use of actuarial science by automobile insurance companies to classify unsafe drivers. As Packer notes, certain populations (e.g., motorcyclists, young people) are constructed as dangerous based on assessments of their potential for loss.[20] EH defines potential for success in terms of a particular kind of relationship: marriage of a heterosexual, usually religious, couple. Interestingly, the site does not address what might well be day-to-day safety concerns of their users; there is no criminal background check or sexually transmitted disease screening, for example. Rather, EH eliminates those individuals with certain personality characteristics or family backgrounds, presumably closing off, or at least limiting, their opportunities for procreative relationships.

Match.com

Individual identity is quantifiable—and essentially fixed—on Match, though not in the same way as on EH. Although users may edit their responses to nearly all questions that they answer during the data preparation flow, it is hard to imagine that many users would choose to review the answers from those seventeen pages. Despite this similarity to EH, the ideological perspectives of the two sites differ significantly. Rather than matching people based on personality profiles, Match's processes are based on patterns of consumption. Much of the information acquired during the data preparation flow is related to lifestyle, leisure activities, and tastes. In addition to being asked about their enjoyment of nearly fifty sports- and exercise-related activities and about experiences they would like to share with their dates, Match users must describe their favorite pastimes and places ("local hot spots and travel destinations") and reveal their tastes in music, television, film, food, and books. Much like the record-store owner protagonist of the 2000 film *High Fidelity*, Match conceptualizes compatibility as "what you like, not what you *are* like."[21] Clearly this envisages a particular kind of user—one with the money and free time to travel, dine out, participate in organized sports or exercise classes, and relax with a good book. Not only will the site's matching procedures be based, at least in part, on patterns of consumption, but also people's self-matching practices will generally be anchored in tastes, leisure activities, and the like.

As we saw in our discussion of the site's matching flow, Match relies on a) a comparison of whom users say they want to meet (stated preferences) and what their behavior says about whom they want to meet (revealed preferences) and b) the stated and revealed preferences, as well as behaviors, of similar users. *Behavior* on the site includes exchanging messages and "winks" with other users, to be sure, but much on-site activity is made up of viewing and evaluating user profiles. In that way, matching procedures are based on "consuming" other users as well. Indeed, as Match engineer Amarnath Thombre noted, this process mirrors Cinematch, Netflix's system for recommending media consumption to its clients.[22] As users' stated and revealed preferences are aggregated and rearticulated, new understandings of attractiveness and compatibility emerge. In his critique of categorization online, Cheney-Lippold calls this recursive process *soft biopolitics*, and he uses the example of gender identity as an illustrative case. He argues that a user's viewing of certain kinds of information online (e.g., visiting a news site) might be coded as male. As that user engages in other activities online, the viewing habits and activities of other users are combined through algorithmic calculation with the original user's. Then as new recommendations are made to the original user based on those combinations and calculations, *maleness* is constructed, reified, and re-presented throughout the process. For example,

Google's Ad Preferences develops individualized age and gender profiles to target advertising to user accounts,[23] and my frequent searches related to productivity software, legal affairs, and politics have marked me as a male, age 35 to 44.

Constructing gender identity through computational processing becomes a kind of "statistical stereotyping":[24]

> Rather than maintaining a particular and concrete relationship to maleness in the offline world, a cybernetic conception of gender would position maleness as a modulating category, based on a vector and composed of statistical assessments that define male group membership. . . . Instead of standards of maleness defining and disciplining bodies according to an ideal type of maleness, standards of maleness can be suggested to users based on one's presumed digital identity, from which the success of identification can be measured according to ad click-through rates, page views, and other assorted feedback mechanisms. The regulation of gender as a category then becomes wholly embedded within the logic of consumption, where categorical behaviors are statistically defined through a cybernetics of purchasing and research that marketers have deemed valuable for identification and categorization.[25]

Cheney-Lippold notes that these categorizations have become aligned with consumption. For example, advertising has long been a means for communicating information related to biopolitical endeavors for mitigating risky behavior, such as public-health campaigns for preventing sexually transmitted diseases. If based on my interests in mobile computing and education, Google characterizes me as a 34- to 44-year-old male, I may be presented with advertisements for condoms rather than, say, the Human Papillomavirus (HPV) vaccine. Of course, these same targeted advertising techniques are used in all media; however, when I watch a football game, I am aware that advertising content is generally geared toward a male audience, and I am familiar with, as most of us no doubt are, the uneasiness of being misconstrued by television advertising. But online ad tailoring is invisible, and these invisible technologies are permitting or denying access to information based on gender (or, as Pariser points out, political affiliation[26]). On Match, desirability is predicated on patterns of consumption—both in terms of users' individual tastes and lifestyles and their viewing of others' profiles. These patterns of consumption are reproduced through the site's matching procedures (i.e., soft biopower) and through the procreative relationships established on the site.

OkCupid.com

In some ways, OKC's conception of subjectivity is the most fluid. Unlike EH and Match, OKC does not presume a fixity of identity. OKC offers the most diverse array of relationship formations (e.g., bisexual, polyamorous, casual sexual, friendship), and it allows users to submit questions to be used during its matching flow. Additionally, OKC began piloting new gender and sexuality options for a portion of its users.[27] Like EH and Match, OKC requires a conception of identity as quantifiable. Still, OKC does not rely on traditional indicators of compatibility, such as particular personality traits or patterns of consumption. Instead, users are matched based on the comparison between their opinions, attitudes, values, and self-perceptions; the opinions, attitudes, values, and self-perceptions of others; and their relative importance to the individuals being compared. The questions designed to address compatibility range from the practical (Which best describes your method for paying off debt?) to the explicit (Have you ever tried sexual roleplaying?) to the mundane (Watching cartoons as an adult is . . . ?). Although the site doesn't seem invested in determining universal factors that predict compatibility, that is not to say that OKC is not rooted in a particular ideological perspective. First, although OKC does not screen users in the manner of EH, the site's flexibility, the openness to non-heteronormative sexualities and non-cisgender identities, and the tolerance of sexually explicit content may create an uncomfortable environment for some potential users. In addition, the carefree and humorous tone of much of the site, as well as its playful color scheme and graphics, may convey an air of flippancy that undermines credibility with other prospective members. Finally, similar to EH and its social-scientific approach to matching users, OKC relies on mathematical calculation, and OKC's mathematical approach, as opposed to personality profiles or consumption patterns, *is* the underlying ideology.

More than another site, OKC privileges quantification, not just as a tool for measuring compatibility but as the governing logic for its system. Conventionally, Matchmaking functions under a logic of compatibility aligned with human values, and we saw this with earlier discussions of EH and Match. According to cofounder Christian Rudder, "[OKC's founders] were all mathematically minded, and the site succeeded in large part because we applied that mind-set to dating: we brought some analysis and rigor to what has historically been the domain of love 'experts' and grinning warlocks like Dr. Phil. How the site works isn't all that sophisticated—it turns out the only math you need to model the process of two people getting to know each other is some sober arithmetic."[28] Just as Neil Clark Warren distinguishes EH's system from conventional notions of compatibility, Rudder delineates his "analysis and rigor" from "love 'experts'" and "grinning warlocks." This privileging of mathematics fits with a larger trend of quantification.

It is reasonable to expect that traditional matchmaking would involve some process of recommendation, epideictic rhetoric extolling the quality of a proposed match. Likewise, both EH and Match make arguments not only about the effectiveness of their computational systems but also the human logics (e.g., relationship satisfaction research) that govern them. OKC makes no such claims. In rhetorical terms, this is a difference in stasis. Stasis concerns what is at the heart of a matter, and includes four levels of questions: fact, definition, quality, and jurisdiction. Each level of stasis builds upon the next, so that it must first be established whether something exists or occurred (fact) before we can determine how to categorize it (definition), whether it is good or bad (quality), and how it should be handled (jurisdiction). Not all of these occur explicitly in every argument, but determining the level of stasis on which a debate occurs is a powerful opportunity. As Miller points out in her discussion of expertise in risk analysis, the tendency toward quantification (favoring of numbers over expertise) had serious implications on the nuclear-power safety debate. One rhetorical implication she identifies is the "technicizing" of the issue, or a stasis shift from evaluative to technical.[29] She notes that the third and fourth stasis levels are evaluative and call for decisions based on values. Conversely, the first and second stasis levels rely on knowledge or expertise. While online dating may not seem as important as determining the safety of nuclear-power facilities, it is noteworthy that the OKC system addresses not an evaluative question (is this a good match?) but rather a technical one (what is the square root of two users' aggregated match question scoring?).

ENTHYMEMATIC THOUGHT PROCESSES IN ONLINE DATING

Despite their differences, all three sites rely on similar enthymematic thought processes. As we saw in the introduction, online dating's combined computational systems require users to draw on cultural understandings to make sense of their participation. In the Match example presented earlier, a hypothetical user drew on what she understood about Match's notion of compatibility to make assumptions about how she and another were matched. Her assumptions might be right. Match, in fact, does rely on users' stated preferences (e.g., owning or liking birds as pets) when matching two users. But it also utilizes revealed preferences (e.g., profiles viewed by a user or similar users) and presents its premium subscribers as matches to all new members who fit within the subscribers' parameters. The user, then, has persuaded herself of the viability of this potential new romance based on connections she has made between her experience of the site and its actual workings with partial information that may or may not be correct. Rudder describes this enthymematic processing as a "tension between the continuity of the human condition and the fracture of the database."[30]

> The challenge facing sites and apps is thus to chop and jam the continuum of human experience into little buckets 1, 2, 3, without anyone noticing: to divide some vast ineffable process—for Facebook, friendship, for Reddit, commu-

> nity, for dating sites, love—into pieces a server can handle. At the same time you have to retain as much of the je ne sais quoi of the thing as you can, so the users believe what you're offering represents real life. It's a delicate illusion, the Internet.[31]

Online dating's particular delicate illusion both relies on contemporary systems indicative of the Internet and is tied to affect, to physical desire, to cultural narratives about romantic love and marriage. As our hypothetical user makes reasonable assumptions about her and her match's compatibility, she may also reasonably be thrilled about her potential new lover, anticipate making contact with him, and daydream about their future encounters.

Online dating sites exhibit characteristics of what Rettberg calls *filtering*. Rettberg chooses *filter* as a critical term in order to

> emphasise [sic] the similarities between the visual filters we apply to our photographs, the technological filters we apply to our blogs and other social media feeds and the cultural filters (norms, expectations, normative discursive strategies) that teach us, for instance, to mimic photo models in fashion magazines or Instagram selfie stars when we photograph ourselves.[32]

In addition to determining what goes in or stays out, filters define realm of possibility and quality of experience. Rettberg offers the example of the coffee filter, which not only prevents grounds from making their way into the brew but also allows the grounds to steep as the water passes through them.[33]

Similarly, Packer and Oswald label this process *screening*. Extending Williams's mobile privatization beyond the era of broadcast television, Packer and Oswald define screen technologies as "the technologies and practices that act as both personal viewing technologies (e.g., cell phones, Global Positioning System [GPS], netbooks) with practices and technologies of socially screening persons (e.g., airport security, biometrics, profiling) to manage appropriate conduct," and they intend *screen* in a double sense—as both the viewing screen and the sifting screen.[34] Digital iterations of these screening technologies act much like modular "sieves [that] can be altered one individual at a time, acting differently in one place rather than another and according to differing schedules."[35] They note that these screens not only capture viewers' attention but also gather resources and information related to them. The sieve adjusts itself based on that information, and in turn it can be adjusted as it is integrated into new spatial, social, and technical networks by viewers.

Packer and Oswald's sieve harkens back to Deleuze's discussion of control societies. Indeed, Deleuze contrasts the mold of disciplinary societies with control's modulation, "like a self-forming cast that will continuously change from one moment to the other, or like a sieve whose mesh will transmute from point to point."[36] In many ways, online dating exemplifies this conception of control technology. These technologies are, on the one hand, highly flexible, modular screening mechanisms for identifying potential relationship partners, and they work in concert with users' own practices (running on-site searches, sorting potential partners recommended by the site, interacting with and evaluating other singles both on- and offline). The way matches are made generally and the kinds of matches delivered to specific users can change based on any number of factors (updates to algorithms or computer processing, new social-scientific research, larger pools of user data, prolonged exposure to an individual user's behavior) as well.

This is not wholly dissimilar from the recursive human matchmaking processes, but there are some significant differences. When humans act as matchmakers for a friend, this procedure begins by first establishing personal characteristics of interest to that friend, cycling through acquaintances that possess those characteristics, and making recommendations based on knowledge of both. The process engaged in by online dating sites is like receiving a dating recommendation from the passenger seated next to us on a plane or a professional associate or anyone with whom we have a finite, situated relationship. There is a qualitative difference between a close friend's matchmaking and that of someone we've just met. No matter how much information users provide or how much time they spend on the site, all matching of users in online dating is based exclusively on what happens there. This requires identification with a particular role, the single subject. This context only butts up against other user identities—a single parent, a videogame enthusiast, or a devout Catholic, for example—but the site does not adequately account for them. It can match users who have children living at home with other users who are interested in (or not opposed to) meeting parents, but it cannot account for all the joys, anxieties, and energies that come with the role of single parent. I am reminded of a scene from the 1992 Cameron Crowe film *Singles*. After joining a video-dating service, a woman arranges a date with an avid cyclist. She embarks on a journey to meet him, on her new bike and decked out in new biking gear. After it becomes clear that there has been a miscommunication about the meeting location, she learns that he has received her address from the service and will meet her at her apartment. She arrives home, wearing bright pink spandex, to find her potential love interest dressed casually in jeans and a worn leather jacket.[37] Her expectations for this man, and the service's restrictive framing of him, are the result of the reliance on a narrow data set that is limited to information made available in the dating space.

In addition, the relationship between the site and its users is isolated in time, both in terms of how long they spend there and where in their histories their participation with the site falls. While a user's past experiences will no doubt influence how he interacts with the site, they are not present on the site per se. No matter how forthcoming we are with the stranger seated next to us, she cannot know our tastes, interests, and history as well as our oldest friend. Likewise, no matter how much information an online dating site gathers, it cannot account for a user's entire romantic history. Even sites that rely on personality profiling do so based on specific ideologies that only capture a particular part of user identity. Just as a skin biopsy may be used to check for skin but not breast cancer, a personality test can only be used to match based on the dimensions it accounts for. This is the paradox of online dating. These technologies offer great flexibility, but they operate under a limited, relatively fixed, quantifiable understanding of both identity generally and individual users specifically.

However, in the end, the effectiveness/ability of online dating to predict "real" compatibility is irrelevant. Indeed the privileging of compatibility in and of itself is more important. Compatibility functions as a free-floating legitimator, and in online dating it draws on affective dimensions—desires for companionship, soul mates, romantic fulfillment, physical connection—to motivate individual participation in computational systems that work to buttress "traditional" relationship formations and reform dangerous subjects. Identity is modulated into categorizations that are used for matching procedures. Singles are transformed into users. They are objectified, converted into units stored in databases, analyzed, and reconfigured as a *match*.

NOTES

1. Robinson Meyer, "Everything We Know about Facebook's Secret Mood Manipulation Experiment," *The Atlantic*, June 28, 2014, last modified September 8, 2014, www.theatlantic.com/technology/archive/2014/06/everything-we-know-about-facebooks-secret-mood-manipulation-experiment/373648/.
2. Dara Kerr, "Senator Asks FTC to Investigate Facebook's Mood Study," CNET.com, July 9, 2014, www.cnet.com/news/senator-asks-ftc-to-investigate-facebooks-mood-study/.
3. Christian Rudder, "We Experiment on Human Beings!" *OkTrends* (blog), July 28, 2014, blog.okcupid.com/index.php/we-experiment-on-human-beings/.
4. According to Packer, *articulators* provide "the mechanisms and processes that maintain the necessary movement and flows to keep an apparatus working smoothly." Jeremy Packer, "What Is an Archive? An Apparatus Model for Communications and Media History," *The Communication Review* 13, no. 1 (2010), 88–104, 102.
5. Ian Bogost, *Persuasive Games: The Expressive Power of Videogames* (Cambridge, MA: MIT Press, 2007), 7.
6. Ibid., 8.
7. Ibid.
8. The classic syllogism example deals with Socrates's mortality:

    1. All humans are mortal. (major premise)
    2. Socrates is human. (minor premise)
    3. Therefore, Socrates is mortal. (conclusion)

    An enthymeme would leave one of the premises unstated. For example, *Socrates, like all humans, is mortal.*
9. Ian Bogost, *Persuasive Games*, 34 (italics original).

10. Kevin Brock, "Enthymeme as Rhetorical Algorithm," *Present Tense: A Journal of Rhetoric in Society* 4, no. 1 (2014), www.presenttensejournal.org/volume-4/enthymeme-as-rhetorical-algorithm/.

11. Kevin Brock and Dawn Shepherd. "Understanding How Algorithms Work Persuasively through the Procedural Enthymeme," *Computers and Composition*, forthcoming.

12. J. Galen Buckwalter et al., *Method and System for Identifying People Who Are Likely to Have a Successful Relationship*. US Patent 6,735,568, filed on August 10, 2000, and issued May 11, 2004.

13. The Relationship Questionnaire consists of twenty-one screens, and responses allow eHarmony.com to develop the personality profile it uses during the matching process. Seventeen of those screens are broken into ten subsections (*About You*, *Self Descriptions*, *Personality Characteristics*, *About Your Feelings*, *Relationship Orientation and Values*, *Important Qualities*, *About Your Personality*, *Your Personal Interests*, *Living Skills*, and *Communication Style*), and I provide example instructions and responses for each in chapter 2.

14. Buckwalter et al., *Method and System*.

15. Ibid.

16. Neil Clark Warren, *Falling in Love for All the Right Reasons: How to Find Your Soul Mate*, with the assistance of Ken Abraham (New York: Time Warner Book Group, 2005), 41.

17. Ibid., 42.

18. Michel Foucault, *Madness and Civilization: A History of Insanity in the Age of Reason*, translated by Richard Howard (New York: Vintage Books, 1965), 37.

19. Warren, *Falling in Love*, 90.

20. Jeremy Packer, "Driving Mobility: Governing and Safety," in *Foucault, Cultural Studies, and Governmentality*, edited by Jack Z. Bratich, Jeremy Packer, and Cameron McCarthy (Albany, NY: State University of New York Press, 2003), 135–61.

21. *High Fidelity*, directed by Stephen Frears (2000; Burbank, CA: Touchstone Home Entertainment, 2000), DVD.

22. Nick Paumgarten, "Looking for Someone: Sex, Love, and Loneliness on the Internet," the *New Yorker*, July 4, 2011, www.newyorker.com/magazine/2011/07/04/looking-for-someone.

23. Casey Johnston, "Google Already Knows You're a 24-Year Old Woman Who Loves Wombats," *ArsTechnica*, January 26, 2012, arstechnica.com/gadgets/2012/01/google-already-knows-youre-a-24-year-old-woman-who-loves-wombats/.

24. John Cheney-Lippold, "A New Algorithmic Identity: Soft Biopolitics and the Modulation of Control," *Theory, Culture, & Society* 28, no. 6 (2011), 164–81, 171.

25. Ibid., 171.

26. Eli Pariser, *The Filter Bubble: What the Internet Is Hiding from You* (New York: Penguin Press, 2011).

27. In 2014, OKC began piloting new gender and sexuality options for a portion of its members. Those users may choose from up to five sexualities (Straight, Gay, Asexual, Bisexual, Demisexual, Heteroflexible, Homoflexible, Lesbian, Pansexual, Queer, Questioning, Sapiosexual) in addition to identifying their preferred relationship type(s). Dan Avery, "OkCupid Adds Loads of New Sexuality, Gender Options for Members," NewNowNext.com, November 16, 2014, www.newnownext.com/okcupid-adds-loads-of-new-sexuality-gender-options-for-members/11/2014/.

28. Christian Rudder, *Dataclysm: Who We Are (When We Think No One's Looking)* (New York: Crown Publishing Group, 2014), 9.

29. Carolyn R. Miller, "The Presumptions of Expertise: The Role of Ethos in Risk Analysis," *Configurations* 11, no. 2 (2003), 197.

30. Rudder, *Dataclysm*, 13.

31. Ibid.

32. Jill Rettberg, *Seeing Ourselves through Technology: How We Use Selfies, Blogs, and Wearable Devices to See and Shape Ourselves* (London: Palgrave Macmillan, 2014), Kindle edition.

33. Ibid.

34. Jeremy Packer and Kathleen F. Oswald, "From Windscreen to Widescreen: Screening Technologies and Mobile Communication," *The Communication Review* 13, no. 4 (2010), 309–39, 313.

35. Ibid., 314.

36. Gilles Deleuze, "Postscript on the Societies of Control," *October* 59 (Winter, 1992) 3–7, 4. For discussion of marriage and family in contemporary control societies, see chapter 5.

37. *Singles*, directed by Cameron Crowe (1992; Burbank, CA: Warner Bros. Entertainment, 1999). DVD.

Chapter Four

Online Dating, Biopower, and Discourses of Success

We are at an important moment for marriage in the United States. In the past two decades, public opinion on same-sex marriage has shifted dramatically. Even before the historic US Supreme Court ruling effectively legalizing same-sex marriage across the country, the Pew Research Center found that the majority of Americans in 2015 favored extending marriage rights to gay and lesbian couples. In 1996, less than a third supported it. Although the majority opposed marriage equality as recently as 2009, the population has been fairly evenly split in the past few years.[1] At the same time, heterosexual marriage is tenuous. While same-sex couples fought for access to legal matrimony, opposite-sex couples married both at a slower rate and later in life. For instance, the Pew Research Center reports that in 2010 just over half of Americans over the age of eighteen were married, a decline of 20 percent since 1960. In addition, the median age of first marriage has risen to nearly twenty-seven for women and twenty-nine for men.[2] During this transitional time, new techniques have emerged to help keep "traditional" marriage relevant. For example, founded in 1997, the National Marriage Project's mission is "to provide research and analysis on the health of marriage in America, to analyze the social and cultural forces shaping contemporary marriage, and to identify strategies to increase marital quality and stability."[3] Under such conditions, it is important to understand efforts at reforming risky single subjects.

In the previous chapter, we delved into the logics of compatibility and persuasive processes that compel participation in online dating. This chapter explores popular truth constructed through online dating success stories. It demonstrates how these narratives perform a double process of subjectification by serving both as confessional discourses through which the success couple is transformed and as instructional accounts that offer singles a subject position with which to identify. It goes on to connect online dating success stories and the privileging of marriage with other techniques of biopower, such as legislative efforts by the Tea Party Movement to exert control of reproduction.

DISCOURSE, POWER, AND KNOWLEDGE

This sort of concern for power over life is what Foucault calls *biopower*, "the set of mechanisms through which the basic biological gestures of the human species became the object of a political strategy, of a general strategy of power, or, in other words, how, starting from the eighteenth century, modern Western societies took on board the fundamental biological fact that human beings are a species."[4] He marks the advent of sexuality as a significant component of the development of biopower, and he describes the point of his project as concerned with three domains related to the "deployment of sexuality" (discourse, power, and knowledge).[5] He suggests that, despite conventional wisdom, the "*putting into discourse* of sex," actually increased at the end of the sixteenth century. In addition, "*techniques of power*" exerted on sex have followed a code of "dissemination and implantation of polymorphous sexualities." Finally, the "*will to knowledge*" constituted "a science of sexuality."[6] These three domains coalesce into the construction of *sexuality*: "a great network in which the stimulation of bodies, the intensification of pleasures, the incitement to discourse, the formation of special knowledges, the strengthening of controls and resistances, are linked to one another, in accordance with a few major strategies of knowledge and power."[7]

The spread of discourse about sex marked a shift in how it had been conceived of for centuries. Foucault notes that for the ancient Greeks, sex served as a mechanism for teaching and learning through the spread of knowledge between bodies through sex. Since that point, sex has shifted in the West from an educational technique to a "problem of truth."[8]

> Toward the beginning of the eighteenth century, there emerged a political, economic, and technical incitement to talk about sex. And not so much in the form of a general theory of sexuality as in the form of analysis, stocktaking, classification, and specification, of quantitative or causal studies. . . . [O]ne had to speak of it as a thing to be not simply condemned or tolerated but managed, inserted into systems of utility, regulated for the greater good of all, made to function according to an optimum. Sex was not something one simply judged; it was a thing one administered.[9]

It is during this time that the population emerges as "an economic and political problem," and the state's concern expands from a management of subjects as individual people to include the management of population "with its specific phenomena and variables: birth and death rates, life expectancy, fertility, state of health, frequency of illnesses, patterns of diet and habitation."[10]

As Coontz notes, it is also in the eighteenth century that marrying for love, what she calls the *love match*, became the ideal: "Until the late eighteenth century, most societies around the world saw marriage as too vital an economic and political institution to be left up to the free choice of the two individuals involved, especially if they were going to base their decision on something as unreasoning and transitory as love."[11] By the Victorian era, though, "marriage harbored all the hopes of romantic love, intimacy, personal fulfillment, and mutual happiness."[12] The love match had taken hold, and the character of marriage changed.

For Foucault, marriage historically has been synonymous with the *deployment of alliance*. This logic relies on an understanding of marriage concerned with establishing and maintaining blood relations that stabilize the social body."[13] By and large, marriage's purpose has been consolidating resources, facilitating cooperation, and managing procreative relationships. In the eighteenth century, alliance lost its prominence with the political and economic shifts we discussed in chapter 2, and the *deployment of sexuality* emerged. Alliance, on the one hand, is concerned with rules that distinguish the normal from the taboo. It produces familial, political, and social relations—especially edicts that address the married couple—while preserving the rules that govern them, and it is connected to the economy through the transmission of wealth.[14] Sexuality, on the other hand, is concerned with "proliferating, innovating, annexing, creating, and penetrating bodies in an increasingly detailed way, and in controlling populations in an increasingly comprehensive way."[15] It relies on multifaceted power mechanisms that extend to managing the couple in new ways, particularly through "the sensations of the body [and] the quality of pleasures."[16] It is connected to the economy primarily through the bodily production and consumption.[17]

The deployment of sexuality grows out of and builds on the deployment of alliance. For instance, under the deployment of alliance, the sacrament of the confession was interested in sex as it related to marriage and kinship. As a result, the content of the confession centered on transgressions such as adultery or incest. As alliance's importance waned, attention progressively shifted. While discussion of sex had long been an expected component of confession, the degree of detail and scope of content shifted during the Counter-Reformation. Parishioners had long been advised to discuss sins in great detail, including comprehensive description of the sex act. Within the movement's increased attention to self-examination and penance, the focus on confessions associated with sex shifted from detailed description of acts to thorough cataloging all sex-related transgressions, including dreams and impure thoughts. Moreover, talk of sex became a concern of public interest.[18]

Foucault indicates four significant components of the confession. First, confession includes a power dynamic that assumes an audience, whether present or not, that compels the confession and acts as its evaluator. In addition, the truth that develops through the confession is confirmed by that which it must overcome in order to be expressed. Next, it is through this very expression that the confessor is transformed. In this ritual, those who confess take on the double position of the subject, as both the confessing subject (the *I* of the confession) and the subject of power who is constituted by it.[19] Finally, confession comes from the bottom up through "an obligatory speech act."[20] With the confession, agency lies not with the confessor but with the one who listens, and the "discourse of truth finally takes effect" in the confessor.[21]

Foucault argues that the confession is significant not only to the church but also to broader knowledge production in the West. He notes that through most of history, people were understood in terms of their associations, a phenomenon we discussed at length in our examination of marriage and romantic matchmaking in chapter 2. Under the deployment of alliance, they were first judged in terms of their bloodlines, political relations, and kinship. As conditions changed, they were validated as individuals, and the "truthful confession was inscribed at the heart of procedures of individualization by power."[22] The confession has become "one of the West's most highly valued techniques for producing truth" and continues to pervade to this day: "It plays a part in justice, medicine, education, family relationships, and love relations, in the most ordinary affairs of everyday life, and in the most solemn rites."[23] We see it in the high-profile admissions of indiscretions by political and religious leaders, the rise of the literary genre of the memoir, the prevalence of over-sharing on social networking sites, and the proliferation of reality television programming and documentary-style sitcoms in which characters break the fourth wall in talking-head segments.

To understand how online dating success stories work as confessional discourse, we must build an archive for examination. In an apparatus approach to media analysis, the archive "maps the linkages that bind together knowledge claims—their production, circulation, and validation—with their political, juridical, technical, and economic application (which is to say power/knowledge)."[24] It seeks to understand how a particular condition is made to appear "accommodating and natural" as well as what truth claims are used to justify it.[25] The archive includes documentation from legislative and regulatory arenas as well as scientific and technical specifications and corporate communication, but it is not limited to these sanctioned or dominant bodies. The archive also contains other, informal or mainstream discourses in order to account for "popular knowledge and expertise [used] to address newfound anxieties and pressures that [accompany] the implementation of new forms of living."[26] In this case, the online dating success story serves as a sort of "popular truth,"[27] produced by regular users of the site, that normalizes "traditional" marriage through confessional narratives.

DISCOURSES OF SUCCESS

The very act of participating in online dating is confessional in nature. Users share information with multiple audiences in an attempt at overcoming their risky single status and being transformed into one-half of a happy couple; the ultimate confessional and transformative act on the online dating site is the success story. These narratives describe the experiences of couples who have met through a site. In another context, these first-person accounts might be called customer testimonials. They are, after all, a kind of recommendation for a service provided by satisfied patrons.

Online dating success stories also include the elements of confession as defined by Foucault. They address a dual audience—the site and its representatives as well as singles seeking partners. Likewise, just as sinners reach absolution through cataloging their transgressions and repenting, couples reach the status of success by detailing their journey. The assumption of the label of "successful relationship" also transforms individuals into the success couple, regardless of the realities of their day-to-day lives. Lastly, the occasion of the success story is manufactured and managed by the site rather than its members, and the pair of former singles are instantiated into the couple formation—transformed into the happy couple, an act that reifies a particular definition of relationship satisfaction through that coupling.

These accounts are more closely aligned with audience expectations for other kinds of success stories, especially origin stories such as the garage myth, than testimonials. Many well-known American companies—Apple, Disney, Hewlett-Packard, Mattel, and Xerox, among others—trace their origins to a founder's (or, in the case of Apple, a founder's parents') garage.[28] These myths "[evoke] the image of the lone individual who relies primarily on his or her extraordinary efforts and talent to overcome the difficulties inherent in creating a new business."[29] Similarly, online dating success stories chronicle the couples' endeavors to overcome the odds and find their soul mates. As we will see in the discussion that follows, online dating success stories seem much more in line with the complex work of constructing subjects and stabilizing our understanding of relationship success than the endorsement of a matchmaking service.

Many online dating sites include these user-generated success stories, and I examine narratives from three of the most popular matchmakers, eHarmony.com (EH), Match.com (Match), and OkCupid.com (OKC). On

EH, they are located on the public portion of the site, and visitors may read them without registering with the service. This subsection of the site, labeled *Single*, also contains a site tour and information on EH's personality profiling, matching technique, and communication procedures that are designed to convert visitors to users. Match's success stories are located on a *success@match* page. Site visitors can browse stories there or read weekly "Featured Success Couple" posts on *Up to Date*, the Match blog. OKC's success stories are less prominently featured on a subpage of the site. I analyzed one-third of the success stories available on EH and OKC, ninety-seven entries for the former and forty-three for the latter. Since Match's stories are not easily surveyed by category, I reviewed all of the fifty-three "Featured Success Couple" posts on *Up to Date*.[30]

EH's user-submitted success stories are organized into thirteen categories, and a couple's narrative may appear in more than one. Most of the categories seem designed to anticipate objections a visitor might have to joining the site in order to offer them the opportunity to read about successful users who were in similar situations or once had similar concerns. So, for example, older users or parents might choose to click the *50+* or *Singles with Children* links. Discouraged singles might identify with the *Nearly Gave Up* accounts, and those living in areas where they are unlikely to encounter people they don't already know might be drawn to the *Long Distance* page. Although some of the content of stories might be different based on the section in which they are located—those in the *Singles with Children* section usually reference the challenges of balancing dating with parenting, and *Long Distance* tales often discuss the unexpected advantages of starting a relationship with someone who lives far away—the structure and content of EH success stories are remarkably similar across categories.

The length of the couple profiles varies from a short paragraph to approaching eight hundred words. Most, however, are similar in length to the example in figure 4.1, which is in the range of three hundred words. Since users submit the stories, they are usually written as first-person accounts. Rarely users will choose to tell their stories in third person, but that is not the expectation. In addition, most are written by one member of the couple, usually the woman. Page elements present the most traditional view of success. Each page includes a picture of the couple, their first names, their location, their status, and the date of their union.

Although there are thirteen categories of stories, there are only three potential statuses for EH success couples: married, engaged, and matched. The site invites other kinds of success stories, however. The call to share their experiences reads, "If you met someone wonderful through eHarmony, we'd love to hear your news," and the gallery of images of success couples is labeled *eHarmony Dating Success Stories*. While undoubtedly many EH matches culminate in cohabitation or pleasant, long-term relationships that

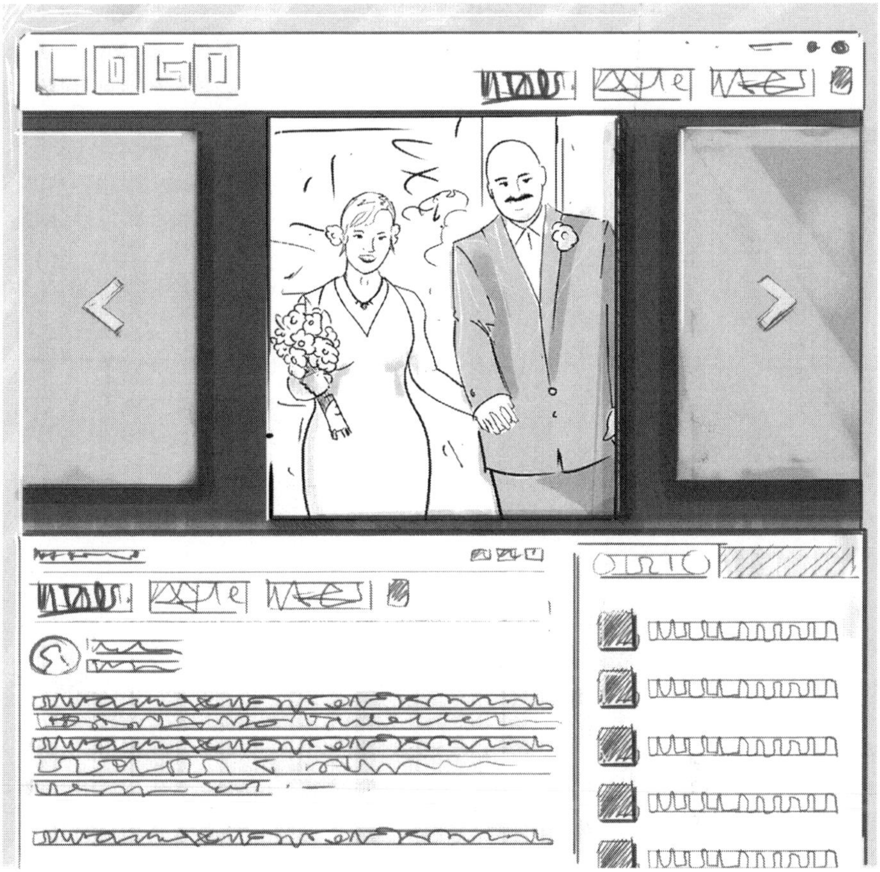

Figure 4.1. eHarmony.com Success Story

do not ultimately end in marriage, those relationship formations do not seem to be coded as successes by users or EH. Of the narratives reviewed, only one had a status of matched. Including a relationship status and date for every success story, much like EH's Guided Communication process,[31] conveys a sense of certainty. There is no ambiguity about the couple's relationship, and the photographs posted by users underscore this. Like the example, all but two of the photographs show the happy couple touching—either holding hands (as pictured), arms around waists, or faces cheek to cheek. Many of the entries feature professional-quality wedding or engagement photographs; few use candid snapshots. The couple is almost always smiling.

Match's success stories are displayed on both *success@match* and on the company's blog. Unlike EH, and despite the label used by the site, Match's success stories cannot be browsed easily. In other words, although visitors may click a link to *browse* stories, what they are actually doing would generally be described as searching them, a practice that models visitors' future experiences as Match members. Viewers may search success stories by location, keyword, and category. The ten categories include some used by EH (e.g., *50+*), but they also allow visitors to search for success stories of same-sex couples (*Gay*, *Lesbian*) and those who have chosen relationships other than marriage (*In a Relationship*). Like the couple profiles on EH, the profiles posted on the Match blog also vary in length, but they tend to be longer rather than shorter. They are also almost exclusively written in first person, and all are written by a member of the couple.

The success stories posted on the Match blog include the couple's first names and a photograph, but they are not formatted to include relationship status, though the posts do usually discuss proposals and marriages. Just as the kinds of relationships represented on Match are more varied than with EH, so are the photographs included in them. As with EH, nearly all photographs include the couple touching—holding hands, embracing or kissing, pressing faces together. Whereas EH users tend to share formal portraits or professional-quality photographs, Match blog posts are more likely to include snapshots, as in figure 4.2. Many show the couple in social environments—sports stadiums, restaurants, parties, cultural landmarks.

OKC success stories share a few things in common with EH and Match but diverge in many ways as well. Of the narratives examined, OKC's are the most difficult to find. They are located in the site's *About* section, but they are not linked either from the home page or the section's landing page.[32] The success stories are divided into six categories, and a story can be tagged with more than one classification. These categories are broad: *Long Distance Love*, *Second Chances*, *Almost Gave Up*, *Blinded with Science*, *Weddings*, and *OKBabies*. Page elements include the couple's first names, a title, and images of the couple. Couples may choose to include more than one photograph, and viewers may click the image currently displayed to move to the next one. Success stories on OKC are the least likely to include formal portraiture or professional-quality photography. Indeed, many stories include quirky snapshots, and the profile example in figure 4.3 contains both the image displayed with the couple in a carnival cutout and a photograph of them in Halloween costumes.

There are three important components that EH's, Match's, and OKC's confessional success stories generally share: discussion of the type of relationship, explanation of the reasons for engaging in online dating, and description of the first date. Additionally, success stories focus on other factors, such as describing the proposal or wedding, that are indicative of the particular relationship formations privileged by the site.

Figure 4.2. Match.com Success Story

Success stories across sites generally reveal the couple's relationship type. As discussed earlier, all EH success stories are labeled with the couple's status. The stories on this site portray the most "traditional" view of marriage. Of the stories I reviewed, only one couple was neither married nor engaged; their status was listed as *matched*. Marriage is the primary concern of all aspects of the site, of its marketing, and of the books authored by its co-founder Neil Clark Warren. In addition, EH focuses exclusively on opposite-sex matching,[33] and therefore all of its members have long had access to federally recognized marriage in the United States. OKC couples also discuss their relationship status; however, the variety of relationship types discussed in OKC success stories is unparalleled by the other two sites. The OKC sample included a number of gay and lesbian relationship successes, and one post from a bisexual woman who initially joined the site in hopes of meeting a woman but ended up in a relationship with a man. The site offers activities other than romantic matchmaking—such as quizzes, multi-player games, journals, and discussion forums. As a result, other relationship formations are possible, and the portrayals reviewed included a friendship success story. Of all the sites examined, Match's success stories are the least likely to explicitly state the relationship status of the couple, and over a quarter of the success

Figure 4.3. OkCupid.com Success Story

stories featured on the site's blog do not include it. This is consistent with the idea that Match lies somewhere between eHarmony and OKC. The relationship expectations are a bit more flexible than EH—Match, for example, offers same-sex dating—but are not as open as OKC, which welcomes users who are bisexual or polyamorous as well as those who are only interested in taking quizzes or playing games. Still, regardless of the relationship type, naming it as such is integral to the transformation into a successful union, and that is perhaps why other varieties of site success (e.g., finding numerous sexual partners) do not make their way onto the success story pages.

The second trait success stories shared across the sites is a reference to why the author either chose to join the specific site or to use online dating generally. On EH, these references are usually related either to a hectic schedule or distaste for conventional means of meeting potential partners (e.g., in bars, at parties). Specifically, many users are single parents who mention the difficulty of balancing those responsibilities with social activities. EH success couples also discuss how their friends or family encouraged them to participate. In fact, EH allows gift subscriptions and includes a category, *Gift of Love*, dedicated to success stories that resulted from someone else, usually a family member, purchasing the subscription for half of the couple. Like EH, many Match success story authors explain why they tried online dating. Usually they mention how busy they are ("my work schedule often kept me from going out and meeting people") or how a friend's success inspired them ("I owe my newfound happiness to my best friend and her involvement with Match.com"). When OKC users comment on why they joined, they usually refer to the site's non-dating components, such as "I'd only ever used the site to take tests and make a couple of long distance friends" or "I barely had a profile and was mainly on the site for the quizes [sic]." Providing the reasons for joining the site helps to contextualize the journey from singlehood to relationship success. Interestingly, however, success stories on none of the sites discuss the time commitment required to participate in online dating. When starting an account, EH members complete an extensive personality questionnaire. Match users rate hobbies, interests, and personality characteristics of themselves and their ideal partner. OKC participants answer, choose matches' acceptable answers for, and rate the importance of *Match Questions*, many of which deal with deeply personal issues. After that, sites require account maintenance, such as uploading photographs, writing profile narratives, and communicating with matches. While these confessional discourses necessitate detailing of the couple's dating journey, describing the arduousness of managing online dating profiles—and many users certainly register with more than one site—is antithetical to the goal of motivating others to take part.

The third trait shared by all three sites is a discussion of the first date. This detailing of the relationship journey is the ritual through which the site's vision of success is mostly clearly articulated. Most EH success stories discuss the first date, frequently in great detail. Here's an example from Emily and Sam, who were married in February 2007:

> Our first date was incredible. We spent that night talking over dinner from appetizers, to main course, to dessert, to coffee, and more coffee. The connection we had felt through emails, and on the phone, was now even more intense. [Sam] was attentive and engaging. I also remember him gently holding my hand on the way back to the car, and the sweet kiss on my cheek at the end of the night. We continued dating and both quickly agreed to withdraw our profiles from eHarmony.[34]

Accounts frequently end like this one, with a declaration of relationship status change that includes removing profiles or ending subscriptions.

Match success couples also discuss their first dates, though generally not in such detail. For example, Hannah and Dean's first date is recounted point by point in two sentences:

> A week later we had our first date at a local restaurant, followed by a movie, and a two hour conversation near a fountain. The agreed short dinner turned into a six hour date.[35]

Unlike EH, Match descriptions primarily recount the logistics of the date.

OKC first dates are depicted similarly, but sometimes include references to sexual contact, such as this excerpt from Spencer and Crowley's success story:

> Our first date was on Presidents' Day. We went to see Juno and he kissed me during the credits, out of no where, and when we went out side he had my cheap red lipstick all over his face. We laughed an awesome belly-laugh and went to Starbucks.[36]

The first-date descriptions from success stories are suggestive of each site's perspective on marriage and relationships. Sites' success stories also include at least one distinct characteristic that sets them apart from one another. For EH, it is a discussion of the proposal and/or wedding, for Match the communication process, for OKC sexual activity. In addition, sites' success stories share a tone that is representative of beliefs that undergird it.

Almost every EH profile examined talks about the proposal, the wedding, or both. Here's the description of Sam proposing to Emily (and a brief mention of their wedding):

> Through this all, I knew that [Sam] and I were kindred spirits. Then one evening in August 2006, [Sam] took me to the same restaurant he had taken me on our first date. [Sam] recreated that first night together, food and all. It was even a more magical moment then [sic] the first, ending with a marriage proposal! After a, "Yes of course I will marry you," we came home and celebrated with our children. We were married on February 15, 2007 in a small ceremony with our closet [sic] friends and family.[37]

These descriptions, coupled with the account of the first date, create a trajectory, a feeling of logical progression from match to first date to engagement to marriage, a kind of fairy-tale narrative with a happy ending. Although most online dating sites make claims about the relationships they have facilitated—EH, for example, claims that 4 percent of US marriages resulted from its matchmaking[38] —that may not convey the reality of most users' experiences. It does, however, create an ideal or model for unmarried people.

Likewise, the tone of EH's success stories conveys an energy and earnestness that is representative of the belief systems endorsed by the site. Some, such as Aria and Bobby's, even mention cofounder Neil Clark Warren by name:

> Thanks Dr. Warren for providing two people with a "harmony" of values and beliefs. We feel like we are well-positioned for a healthy, successful marriage![39]

Success couples focus on the value of the service offered by the site and their appreciation for it, and this gratitude is indicative of the tone of both EH's success stories and the site more generally.

Match, conversely, has an open and fluid communication process. Subscribers to the site may contact any users they find intriguing, and all users may send a "wink," an expression of interest similar to Facebook's "poke." Match users are far more likely to describe communicating on the site prior to meeting in person. For example, here's an excerpt from Mona and Cass's success story:

> He winked at me, and I being shy, winked back. Then he sent me a message. We talked through Match.com for a few weeks before we exchanged another method of contact.[40]

Often these stories even include days of the week or dates of communication, such as this example from Alison and Charlie:

> [Charlie] signed up for match.com on Sunday Nov. 29. I signed up on Monday Nov. 30, and he showed up on my daily 5 on Tuesday December 1. Friday Dec. 18 he emailed me . . . 3 times in an hour.[41]

Perhaps these descriptions are intended to be instructive for potential members. Since Match is less structured than EH, there is no presumption that all users proceed through the same communication process on the site. Authors may either be offering models for communicating with matches or reassurance that perseverance through multiple contacts can lead to a successful relationship, and by doing so they are acknowledging half of the dual audience addressed by their discourse. Whatever the case, this cataloging of communication activities, much like the point-by-point detailing of first-date activities, creates the impression of momentum seen in EH's proposal and wedding narratives.

Although Match may foreground dating in a way that EH does not, a recent Match advertising campaign features current subscribers on first dates, it still emphasizes its ability to lead to long-term relationships. Each spot ends with the following: "We've led to more dates, more relationships, and more marriages than any other site." There is a subtle difference between a site's framing itself as a place to find a marriage partner (EH) and positioning itself as a means for finding a relationship that could lead to marriage (Match). The former may seem like skipping the dating step, but in fact it replicates practices in prior configurations of the marriage apparatus, under the deployment of alliance when matchmaking was intermediated by parents or other authorities.[42] As the success stories above demonstrate, Match's successful couples take a tone that mirrors the site's current marketing campaign, emphasizing the process of dating. In addition, since Match is the online dating site most like ecommerce sites with which users are accustomed to interacting, it is not surprising that users would tell their stories in a more practical tone that focuses on site processes.

OKC users are more likely to mention sexual contact in their success stories. Portrayals on other sites rarely discuss physical contact, and when they do, it's a hug or, as we saw with Emily and Sam, a "sweet kiss on the cheek." OKC users, on the other hand, frequently describe sexual contact. Spencer and Crowley's success story not only includes an account of their first kiss but also a description of a prolonged one:

> On December 28th, 2008 he returned from visiting his parents up north and his father came back with him. There was no place for us to be alone. We searched college campuses and local parks, but everywhere was too crowded. So we settled for my parents' backyard porch-swing, and we kissed till we were out of breath.[43]

In addition, Meg and Toby's success story appears to allude to sexual intercourse on the first date:

> So we talked, email then phone, then met 2 weeks later. It was a hug, kiss & then back to my apt. I am not usually like that but it was right & we knew.[44]

These expressions of sexual contact set OKC apart from EH and Match. That is not to say that members of those sites do not use them to facilitate sexual encounters. Surely some, perhaps many, do. However, that is a hidden, unsanctioned use of the site. OKC is different. Users may select *casual sexual* as a preferred relationship type when they register. Likewise, much like Match's focus on dating, OKC's matching is concerned with connecting users who share enough commonalities that they will want to get to know each other, perhaps quickly scheduling a first date.[45] Still, the majority of success stories reviewed did portray marriages and engagements.

As with EH and Match, the tone of OKC success stories fits with the general tenor of the site. OKC posts largely maintain a casual, often sarcastic, tone. Here's an excerpt from Ashley and Benny's:

> Once upon a time . . . okay, let's cut that shit now. It's not how we roll. We are much more cool and crass than that and apparently you guys down at OkCupid already knew that about us and figured we'd be a really cute, callous and sarcastic couple. . . . It all started with a woo on the 26th of July, the summer of '07. It had been a really fun summer but I then decided I wanted to throw a

> cute boy into the mix which is why when this mohawked, dorkwad, cutiepie woo'ed me I messaged back.

Many successful OKC couples self-identify as outsiders. They describe themselves as nerds, jerks, dorks. Just as mention of sexual interaction in success stories resonates with the way the site presents itself, so does the tone of its success stories.

As we saw in chapter 2, online dating develops within a condition of mobile privatization. This paradoxically mobile and home-based configuration—in addition to the changing status of women, developments in birth control and fertility, and a growing concern for civil and human rights—necessitates new ways of managing unmarried adults. Some of that management has taken the form of voluntary participation in online dating and the perpetuation of established forms of relationship success (i.e., the heterosexual, white, middle-class married couple). Indeed, this need may also account for site use that does not fit the marriage-equals-success equation. Deemphasizing the marriage imperative, focusing on dating rather than marriage, allowing for alternate relationship formations, all serve to pull unmarried adults into a system that can both account for them and potentially motivate their establishing procreative relationships.

BIOPOWER AND MARRIAGE IN THE CONTEMPORARY UNITED STATES

As we discussed earlier, population emerged in the eighteenth century as a problem to be managed, and the accompanying tactics included a new drive to catalog bodies through tracking a number of statistics related to birth, fertility, health, death, diet, exercise, and so on. With falling birth and marriage rates in the United States,[46] recent legislative and juridical actions related to reproductive rights and marriage equality signal growing anxieties surrounding the management of population.

With developments such as the rise of the Tea Party Movement in the Republican Party, we have seen increasing attention paid to governmental intervention into citizens' everyday lives. One of the Tea Party's earliest stars was Michele Bachmann, who founded the US House of Representatives Tea Party Caucus in 2010, delivered the Tea Party response to the State of the Union address in 2011, and sought the Republican presidential nomination in 2012. While serving in the House from 2007 to 2015, Bachmann almost exclusively supported efforts to limit the powers of the federal government, including reducing federal taxes, eliminating the Department of Education, cutting energy regulation, and defunding the Affordable Care Act. There are two significant exceptions that relate directly to the management of life. First, Bachmann, a former member of the House's Pro-Life

Caucus, introduced the Heartbeat Informed Consent Act to the House, legislation requiring doctors to "make the heartbeat of an unborn child visible and audible to its mother" prior to receiving consent to perform an abortion.[47] Second, Bachmann was an original cosponsor of H. J. Res. 45, legislation supporting a Constitutional amendment defining marriage as the union of a man and a woman.[48] Bachmann's position on these two issues is consistent with views espoused by other Tea Party members of Congress, such as Idaho Representative Raul Labrador, and candidates for the 2016 Republican presidential nomination, such as Texas senator Ted Cruz and Florida senator Marco Rubio.

These paradoxical stances represent interventions by the state that are directly related to the management of life generally and control over reproduction in particular. Imposing limitations on how women maintain their reproductive health—the aforementioned proposed legislation but also resolutions that would require transvaginal ultrasounds prior to an abortion or that allow employers to deny prescription coverage of birth control or that defund Planned Parenthood—appear to be less about preserving the sanctity of life and more about privileging procreation as *the* sacred act. Even if we bracket issues related to preserving life from the rest of the political platform, the Tea Party seeks to defund the Affordable Care Act, which is connected to life expectancy and quality of life.

In the case of marriage equality, the inability to procreate is the only qualitative difference between same-sex matrimony and the heterosexual ideal. Denying gay and lesbian couples access to marriage—and to all of the legal rights that go along with it—seems to hinge, then, on the fact that they cannot create life. Oral arguments in *Obergefell v. Hodges*, the 2015 US Supreme Court case that granted marriage rights to same-sex couples, addressed whether the Constitution obliges states to grant same-sex marriage licenses and/or to recognize licenses issued in other states. Much of the discussion focused on procreation's centrality to marriage.[49] Making decisions about women's reproductive rights or same-sex couples' marital legitimacy are not issues of small government. Rather, these issues concern reforming dangerous subjects. For example, when arguing to limit access to abortion, women who choose not to carry a pregnancy to term are generally imagined to be sexually promiscuous and irresponsible. Rarely do we consider victims of spousal rape or couples who come to the difficult conclusion that they are not equipped to raise a special-needs child, for example. Without abortion and birth control, the argument goes, sex would be reformed as a procreative act between the married (heterosexual) couple. Efforts to ban same-sex marriage by amending the US Constitution have been unsuccessful. However, other techniques of power work to further entrench procreative relationships as the fundamental purpose of marriage. By synonymizing *marriage* and *success*, online dating success stories help do just that.

Examining online dating success stories as confessional, as a "ritual of discourse," illuminates the work they do in helping to keep marriage relevant. It should not be surprising that, although EH, Match, and OKC differ both in terms of how they interact with their users and how they position themselves in the larger culture, success couples on all three sites define success similarly. The tone and content of the sites and their success stories vary, but for the majority of stories examined marriage *is* success, and 44 percent of OKC's, sixty-four percent of Match's, and all but one of EH's success stories were for engaged or married couples. Matchmaking site success could be defined in a variety of ways—relationship satisfaction (regardless of marital status), length of relationships established, number of dates facilitated, percentage of dates that lead to sexual intercourse—but most online dating sites seem to mark success with the number of marriages they make possible.

Still, some online dating sites and participants resist the dominant marriage-equals-success discourse. For example, AshleyMadison.com, which bills itself as "the most recognized and reputable married dating site" and boasts nearly forty-million members,[50] facilitates extramarital affairs; other sites (e.g., AdultFriendFinder.com) are designed for arranging casual sexual relationships. At least one person, Jessica Sporty, admits to using Match to arrange dates for the purpose of receiving free food and drink, and she was able to garner twelve hundred dollars' worth of meals and cocktails monthly.[51] If we define success as meeting a goal or serving an intended purpose, then both Sporty and the taboo relationships arranged by online dating sites such as AshleyMadison.com meet that criterion; however, those successes do not fit with the larger narrative about romantic fulfillment and contemporary marriage.

Success stories help foster specific impressions about both online dating sites and contemporary marriage. Although these confessional narratives clearly do more than a standard customer testimonial, they do cast online dating in a positive light—at least in terms of a traditional notion of marriage, love, and romantic fulfillment. Couples describe their journey from single to relationship to (in most cases) marriage, documenting their resistances, their participation on the site, and their final destination. In doing so, they undergo a process of subjectification, and the "truth" of their conversion is corroborated by the detailed evidence of their experience. Most importantly, the act of sharing the success story, of confessing, "exonerates, redeems, and purifies"; it "unburdens [them] of [their] wrongs, liberates [them], and promises [their] salvation."[52] At the same time, success stories provide singles a subject position with which to identify and an outcome to desire. Risky singles can project themselves into these relatable narratives. *I, too, am unsure about online dating. I, too, don't have time to hang out in bars. I, too, want first kisses, romantic dinners, and a loving partner.* In aggregate, success stories appear to stabilize marriage. However, the paradoxical binary of online dating's voluntary status and the confession's compulsory nature makes strange bedfellows. In the next chapter, I examine the tenuous relationship of marriage, family, and online dating in contemporary societies of control.

NOTES

1. Carroll Dougherty, Jocelyn Kiley, and Rachel Weisel, *Behind Record Support for Same-Sex Marriage, Differences Endure* (Washington, DC: Pew Research Center, 2015), www.people-press.org /files/2015/06/6-8-15-Same-sex-marriage-release1.pdf; "Growing Public Support for Same-Sex Marriage," Pew Research Center, last modified on February 16, 2012, www.people-press.org/2012/02/07/growing-public-support-for-same-sex-marriage/
2. D'Vera Cohn, *Marriage Rate Declines and Marriage Age Rises* (Washington, DC: Pew Reasearch Center, 2011).
3. "About," National Marriage Project, accessed March 28, 2014, nationalmarriagepro ject.org/about/.
4. Michel Foucault, *Security, Territory, Population: Lectures at the College de France, 1977–78*, edited by Michel Senellart, translated by Graham Burchell (New York: Palgrave Macmillan, 2007), 1.
5. Michel Foucault, *The History of Sexuality, Vol. 1: An Introduction*, trans. Robert Hurley (New York: Pantheon Books, 1977), 12.
6. Ibid. (italics added).
7. Ibid., 106.
8. Ibid., 56.
9. Ibid., 24.
10. Ibid., 139.
11. Stephanie Coontz, *Marriage, a History: From Obedience to Intimacy, or How Love Conquered Marriage* (New York: Viking, 2005), 178.
12. Ibid.
13. Ibid., 107.
14. Ibid.
15. Ibid., 106–7.
16. Ibid., 106.
17. Ibid.
18. Ibid.
19. Ibid., 61–62.
20. Ibid., 62.
21. Ibid.
22. Ibid., 58–59
23. Ibid., 59.
24. Jeremy Packer, "What Is an Archive? An Apparatus Model for Communications and Media History," *The Communication Review* 13, no. 1 (2010): 88–104.
25. Ibid.
26. Ibid.
27. Ibid., 99.
28. Gail Sullivan, "Steve Wozniak Refutes Apple's Creation Story: 'The Garage Is a Bit of a Myth,'" *Washington Post*, December 5, 2014, www.washingtonpost.com/news/morning-mix/ wp/2014/12/05/steve-wozniak-refutes-apples-creation-story-the-garage-is-a-bit-of-a-myth/.
29. Pino G. Audia and Christopher I. Rider, "A Garage and an Idea: What More Does an Entrepreneur Need?" *California Management Review* 48, no. 1 (2005): 6–28.
30. I conducted all searches in early October 2011. Although all success stories are available on the public portion of the sites, I use pseudonyms for all couples and represent success story pages with drawings in order to protect couples' anonymity.

31. Unlike most dating sites, which allow members to determine how to communicate with others, EH's default contact sequence is its Guided Communication system. Guided Communication occurs in three sequential steps. First, users each answer five questions of the other's choosing. Next, they trade lists of ten relationship Makes and Breaks. Third, they each answer three open-ended questions selected by the other. Once a pair has completed these three exchanges, they proceed to EH Mail, which allows users to send messages through the site. Either user may terminate the connection at any point during the Guided Communication process.

32. The success stories are a relatively recent addition, probably a result of OKC's purchase by Match.

33. The company does provide same-sex matching services, but that matching is performed on a separate site, CompatiblePartners.net.

34. "Success Stories," eHarmony.com, accessed on November 11, 2011, www.eharmony.com/success/stories/single-with-children.

35. "Featured Success Couple," *Up to Date* (blog), accessed on October 24, 2011, blog.match.com/category/success-stories/.

36. "Success Stories," OkCupid.com, accessed on November 2, 2011, www.okcupid.com/success/.

37. "Success Stories," eHarmony.com.

38. Trent Gillies, "How to Surf the Web for a Mate: eHarmony Founder" (includes interview with Neil Clark Warren), CNBC, last modified May 9, 2015, www.cnbc.com/id/102662981.

39. "Success Stories," eHarmony.com, accessed on November 2, 2011, www.eharmony.com/success/stories/science-of-love/.

40. "Featured Success Couple, *Up to Date* (blog), accessed on October 24, 2011, blog.match.com/category/success-stories/.

41. "Success Stories," OkCupid.com.

42. For more on the marriage apparatus, see chapter 2.

43. "Success Stories," OkCupid.com, accessed November 2, 2011, www.okcupid.com/success/.

44. "Success Stories," OkCupid.com, accessed on November 2, 2011, www.okcupid.com/success/.

45. Nick Paumgarten, "Looking for Someone: Sex, Love, and Loneliness on the Internet," the *New Yorker*, July 4, 2011, www.newyorker.com/magazine/2011/07/04/looking-for-someone.

46. Birth rates are at the lowest since the late twentieth century, and marriage rates have declined by 20 percent since 1960. Cohn, *Marriage Rate*; Hamilton, Martin, and Ventura, *Births: Preliminary Data for 2010* (Atlanta, GA: Centers for Disease Control and Prevention, 2011).

47. "Life," Michele Bachmann's US House of Representatives page, accessed January 15, 2012, bachmann.house.gov/Issues/Issue/?IssueID=56263 (site discontinued).

48. "Family Values," Michele Bachmann's US House of Representatives page, accessed January 15, 2012, bachmann.house.gov/Issues/Issue/?IssueID=56264 (site discontinued). In the next chapter, I discuss same-sex marriage in relation to contemporary marriage and family in greater detail.

49. Amy Howe, "Same-Sex Marriage, in Plain English—Part II," *SCOTUSblog* (blog), April 27, 2015, www.scotusblog.com/2015/04/same-sex-marriage-in-plain-english-part-ii/.

50. AshleyMadison.com, accessed on June 25, 2015, https://ashleymadison.com.

51. Madeline Scinto, "This Young Woman Scored $1,200 a Month in Fancy Dinners Using Match.Com," *Business Insider*, November 29, 2011, www.businessinsider.com/confessions-how-she-made-1200-a-month-using-matchcom-2011-11. After numerous negative comments on the online article, BusinessInsider.com disabled commenting and Jessica Sporty was addressed by a pseudonym, Minerva McGonagall.

52. Foucault, *History of Sexuality*, 62.

Chapter Five

Online Dating, Marriage, and Family in Control Societies

As we saw in chapter 2, romantic matchmaking generally has been intended to facilitate marriages. First, leaders from nomadic collectives that traveled similar territories would arrange marriages between members of their groups in order to encourage resource sharing and cooperation. Once we settled down, marriage became a way to consolidate land ownership, wealth, and power. With religious, political, economic, and technological change, marriage changed as well.[1] In prior understandings, the terms of matching success were fairly obvious. A successful match was one that ended in an advantageous marriage for those on both sides of the aisle. The definition of advantageous has never been static, but it has generally meant that both families would profit from the union—again, by consolidating wealth, land, resources, power, access, etc. As people began to marry for different reasons, the activity of romantic matchmaking changed; however, its fundamental purpose did not. In other words, parents no longer held primary responsibility for initiating the matchmaking process. Marriageable individuals were allowed authority over their own matches and began establishing their own relationships with matchmakers. Technological developments also allowed for mediated matchmaking through the use of classified advertising, for example. Today, individuals maintain control over their own matchmaking, and romantic love has taken root as the expected reason for marrying. Rather than managing their own matchmaking through newspapers or other legacy media, singles may now rely on automated processes in online dating when searching for soul mates. At the same time, marriage has shifted from compulsory to optional and does not serve as gateway to adulthood. Indeed, the question today could be not only when or whom to marry but also whether to marry, a post-marriage condition that necessitates further explication.

This chapter seeks to understand efforts to maintain marriage's relevance under its current optional status by examining matrimony as a technique of power in contemporary control societies. I begin by positioning marriage and family in relation to institutions of power, especially Deleuze's conception of control. Next, I offer brief examples of the contemporary (heterosexual) marriage "crisis" and a discussion of oral arguments in *Obergefell v. Hodges*, the 2015 US Supreme Court case legalizing same-sex marriage, in order to demonstrate marriage's centrality to the institution of family. Finally, I argue that although it appears a fitting response to buttress marriage's boundaries, online dating's (and control's) synchronic logic is incompatible with the historically diachronic logic of marriage.

CONTROL SOCIETIES

In "Postscript on the Societies of Control," Deleuze contrasts Foucault's disciplinary societies and their reliance on confinement with control societies that use "free-floating control" to govern.[2] He notes that under control, we move from an analogical way of knowing with "independent variables" in which we are always beginning anew to a numerical one with "inseparable variations" in which we are never finished.[3] In disciplinary systems, we "went from school to barracks, from barracks to factory" and, although there may have been shared characteristics, we needed to learn the new particularities of each.[4] We can think here of how as children transition from home to school that they are subject to a new timetable (story time, recess) and new procedures from managing movement (lining up with classmates) that differ from those used at home, and these transitions are signaled by shifts in the environment, from the family home to the school building. In each stage of schooling, students move into new spatial enclosures, each distinct from the next, reminded at each juncture that this instantiation is different from the last. Middle school brings changing classes with bells and managing multiple subjects, for instance. High school layers in additional course variety, more distinct "tracks," greater scheduling flexibility. In systems of control, the factory gives way to business, and the school becomes "perpetual training."[5] Although school largely persists as a disciplinary apparatus, we are now caught in a loop of continuous education that extends beyond the walls of the classroom and long after our formal schooling ends, with corporate training programs, new certification and credentialing specifications, and so on. Deleuze offers the example of teacher merit pay, and teachers' required continuing education for licensure surely contributes to this phenomenon. Recent higher-education initiatives, such as massive open online courses (MOOCs) and digital "badges," help perpetuate this continuity as well.

According to Deleuze, disciplinary and control societies function under separate logics as well: "[Disciplinary] [e]nclosures are *molds*, distinct castings, but controls are a *modulation*, like a self-deforming cast that will continuously change from one moment to the other, or like a sieve whose mesh will transmute from point to point."[6] This is perhaps most clear in the individual-mass binary of disciplinarity, as "the disciplines never saw any incompatibility between [individual and mass], and because at the same time power individualizes and masses together, that is, constitutes those over whom it exercises power into a body and molds the individuality of each member of that body."[7] He notes that, rather than being caught between discipline's individual and mass, we are now *dividuals*, compilations of data that can be disassembled and reassembled as needed. In such a system, humans go from being the "discontinuous producer of energy" to "undulatory, in orbit, in a continuous network."[8] For example, the permissions encoded into the magnetic strip on my faculty ID allow me access to buildings on campus at times not available to students, monitoring my access and controlling my movement. At a change in status (promotion of student to instructor, my unfortunate dismissal) or situation (special event, heightened security concern), those magnetic strips can be reprogrammed to permit or deny access accordingly.

For Hardt, this eroding of disciplinary walls of enclosure is not a withdrawal of power structures from social space but rather "the generalization of the logics that previously functioned within these [institutional] domains across the entirety of society, spreading like a virus."[9] These logics work in "infinite undulations" and rely on "mobility and anonymity."[10] Identity shifts from its unchanging status within disciplinary systems to a flexible, modular "whatever identity" in control societies.[11] Similarly, Clough highlights control's attention to the subindividual, "a never-ending modulation of moods, capacities, affects, potentialities statistically assembled in genetic codes, identification numbers, ratings profiles, preference listings, risk statuses, that is, bodies of data and information (including human bodies as data and information)."[12] In his discussion of reality television and control, Bratich calls this "reality software," with the goal of "render[ing] reality visible, manageable, and programmable in accordance with the imperatives of a control society."[13] As I discussed in the introduction, this condition is what Galloway calls protocol. Galloway notes that control relies on information technologies to observe and manage movements of bodies, as demonstrated in the work of both Andrejevic and Packer discussed earlier, and "affects the functioning of bodies within social space and the creation of these bodies into forms of 'artificial life' that are *dividuated*, sampled and coded."[14]

In "Postscript," Deleuze highlights five important *environments of enclosure* that are threatened as control societies emerge.[15]

> We are in a generalized crisis in relation to all the environments of enclosure—prison, hospital, factory, school, family. The family is an "interior," in crisis like all other interiors—scholarly, professional, etc. The administrations in charge never cease announcing supposedly necessary reforms: to reform schools, to reform industries, hospitals, the armed forces, prisons. But everyone knows that these institutions are finished, whatever the length of their expiration periods. It's only a matter of administering their last rites and of keeping people employed until the installation of the new forces are knocking at the door. These are the societies of control, which are in the process of replacing disciplinary societies. . . . There is no need to fear or hope, but only to look for new weapons.[16]

He goes on to compare the logics that govern both disciplinary and control regimes and then to describe potential techniques of control.

At the end of "Postscript," Deleuze discusses some early techniques of confinement put into place in control societies. He notes that prison systems had begun to experiment with the use of electronic monitoring for non-violent offenders. School systems had entered a process of "continuous assessment" and had begun establishing relationships between schools and business interests across all levels. Hospital systems were using risk assessments of populations to determine treatment of individual patients. Business systems started to take advantage of "new ways of manipulating money, products, and men."[17] Deleuze's "Postscript" predicts many early tactics that have become pervasive in our contemporary culture. Although he explicitly names the family as an interior in crisis, Deleuze does not return to it in the end. Perhaps he does not address the threats to the family—and any attendant new weapons of control—because the family is a more problematic case than the prison, the hospital, the factory, or the school.

As the oldest social institution, both in terms of human history and individual lives, the family perseveres as a tool for bodily intervention, management of individuals at a distance, and population regulation. Unlike transitions into (and out of) other institutions in disciplinary societies, we perpetually exist in familial relations—as children, siblings, parents. In addition, even as the realities of those relations and how we define the term have changed across time, the concept of family has remained an important factor, both as an abstraction and as a component of everyday life. Even as economic and political systems ebb and flow, the family endures. However, the family's persistence does not ensure its sustainability as an environment of enclosure.

The first challenge to family's viability as an environment of enclosure is its very nature. First, the family is the one institution Deleuze mentions that isn't institutionalized. That is not to say that families are unregulated, as both same-sex marriage legislation and investments in social welfare programs related to mothers and children demonstrate; however, there is no spatial institutionalization of the family. The family—unlike the school, the hospital, the factory, and the prison—does not name an environment of enclosure. The spatial enclosure of the family is the family home, and even in disciplinary societies the family exists at a second arm's length from the surveilling body. In fact, even initiatives designed to protect children from neglect, abuse, and addiction intervene only in the most extreme cases. It is considered best to keep children with their families. Additionally, the contemporary American understanding of family is vast. Though the place of same-sex marriage is contested and children's well-being is of concern to government and social institutions, the variety of possible familial formations has expanded to include cohabiting couples, married couples without children, single parents and their children, divorced couples who co-parent, adult children caring for aged parents, urban tribes, and so on. Families can be blended, biological, adopted, constructed. While an expansive definition could indicate the sustainability of the family as a social institution, it also calls into question its continued efficacy as a disciplinary apparatus. After all, childless heterosexual couples, cohabitating homosexual couples, and latchkey children are all potentially "dangerous" identities that may resist traditional disciplinary techniques of confinement, heightening anxiety over these new familial formations. Likewise, the family's expansion limits the effectiveness of biopolitical control's techniques of power. When family can be defined in so many ways, traditional data tracking (e.g., divorce or birth rates) is less meaningful.

The second challenge to the family's viability as an environment of enclosure is the state of contemporary marriage. Just as our understanding of what it means to be a family has changed, so have the conditions of the married couple. As Coontz notes, the character of heterosexual marriage has "steadily become more fair, more fulfilling, and more effective in fostering the well-being of both adults and children than ever before in history."[18] Conventional discourse surrounding contemporary marriage is of an equitable, emotionally supportive relationship between two partners, and both access to and acceptance of divorce make those expectations real and urgent. In addition to political, economic, and technological changes that play a role in the improvements in marital relations, marriage's discretionary status permits greater flexibility.

Coontz also notes that marriage no longer serves as an obligatory part of a "credentialing process that people have to go through to gain adult responsibility and respectability."[19] Young adults do not need to marry in order to live outside the family home or seek employment and educational opportunities. She argues, "The historical record suggests that these two seemingly contradictory changes are inextricably intertwined. Even more than love and marriage, fulfilling and fragile seem to 'go together like a horse and carriage.'"[20] This fragility runs counter to marriage's position as the cornerstone of the institution of the family and may weaken the family's potential as a power mechanism.

Under disciplinary regimes of power, periodic marriage crises have been a source of anxiety. For instance, as we saw in chapter 1, shifting economic and labor conditions in the sixteenth century led to a class of young "'masterless' men and women" that caused concern for both church and state, and similar anxieties arose around increased attention to sex at the turn of the twentieth century.[21] In both instances, uneasiness is the by-product of the decreasing confinement of individuals within the family. In the case of the former, more young people were choosing wage labor over working on farms, as apprentices, or in the homes of others; they were therefore less likely to fall subject to parental or adult control. For the latter, teenagers and young adults were freed from parental scrutiny by automobility as well as working outside the home or attending college. Both instances produced increased religious, social, and governmental intervention into marriage in particular. Sermons began to address the couple directly, new laws were passed that prevented marriage by consent, and popular publications began including dating advice. Today, spatial and temporal distancing is intensified by mobility, immigration, and globalization, and the increasing acceptance of relationship formations that are not heterosexual matrimony has brought about apprehension in the face of a new marriage. More than any other of Deleuze's institutions in crisis, the family's ability to function as an environment of enclosure is threatened, and much of this threat currently comes from the fragile nature of contemporary marriage. Indeed, marriage is the linchpin around which the institution of the family coheres, and legislative edicts either permitting or prohibiting same-sex marriage are frequently cast as either saving or condemning the family.

MARRIAGE AND THE INSTITUTION OF FAMILY

Events surrounding recent actions on same-sex marriage serve as an example of marriage's importance to the institution of family. On the one hand, proponents point to same-sex marriage as important to the preservation of families. In his 2014 decision to overturn Oregon's ban on same-sex marriage, US District Judge Michael McShane wrote, "[T]he well-being of Oregon's children is not enhanced by destabilizing and limiting the rights and resources available to gay and lesbian *families*."[22] Likewise, Ed Murray, a Washington state senator, said of that body's January 2012 vote on the issue, "Those of us who support this legislation are not, and we should not be accused of, undermining *family* life or religious freedom. *Marriage* is how society says you are a *family*."[23] On the other hand, opponents see same-sex marriage as threatening the traditional family. When discussing Idaho's appeal of a federal court's decision to repeal his state's restriction on same-sex marriage, Governor C. L. "Butch" Otter explained that his decision goes "to the heart of Idaho's values and respect for the *family* unit as it's been embraced by society for millennia."[24] In a speech given to diplomats at the Vatican in 2012, Pope Benedict XVI noted the importance of family in children's development:

> Among these [settings], pride of place goes to the *family*, based on the marriage of a man and a woman. This is not a simple social convention, but rather the fundamental cell of every society. Consequently, policies which undermine the *family* threaten human dignity and the future of humanity itself.[25]

Marriage's centrality to the institution of family is one reason the stakes are so high in this issue.

Perhaps the clearest example of marriage's importance to the institution of family is the 2015 Supreme Court of the United States (SCOTUS) case that recognized the constitutionality of same-sex marriage. *Obergefell v. Hodges* addresses two issues, commonly referred to as the "marriage question" and the "recognition question":[26]

1. *Does the Fourteenth Amendment require a state to license a marriage between two people of the same sex?* and
2. *Does the Fourteenth Amendment require a state to recognize a marriage between two people of the same sex when their marriage was lawfully licensed and performed out-of-state?*[27]

Obergefell v. Hodges (Ohio) is linked to three other cases from the 6th Circuit US Court of Appeals—*Bourke v. Beshear* (Kentucky), *DeBoer v. Snyder* (Michigan), and *Tanco v. Haslam* (Tennessee); all but one of these focuses primarily on the recognition question. For example, James Obergefell and John Arthur had been together for over two decades when they married in Maryland in 2013. When Arthur died shortly thereafter, Obergefell filed suit to have their marriage recognized in their home state of Ohio in order for his name to appear on Arthur's death certificate, among other things. *DeBoer v. Synder* is related to the marriage question. April DeBoer and her partner, Jayne Rowse, have been a couple for ten years. They own property together and have three children—two adopted by Rowse, one by DeBoer. Since they cannot marry in their home state of Michigan, each does not have parental rights to the other's child(ren). As a result, they cannot share family health insurance coverage, and neither is assured custody of all of their children in the event of the other's death. They sought repeal of Michigan's law that precludes them from marrying.[28]

SCOTUS's 2013 decision in *United States v. Windsor* laid the groundwork for the 2015 case. In it, the Court struck down both California's Proposition 8, a ballot initiative that banned same-sex marriage, and the federal Defense of Marriage Act (DOMA), which defined marriage as "a legal union between one man and one woman as husband and wife."[29] For the former, the Justices found that proponents of Proposition 8 did not have legal "standing" to preserve the ban once the State of California chose not to support it upon appeal. For the latter, the Court decided that, since the states had historically held dominion over regulating marriage, the federal government did not have jurisdiction when defining marriage.[30] After the *United States v. Windsor* verdict in 2013, same-sex marriage bans were struck down in states across the country by the 4th, 7th, 9th, and 10th Circuit US Courts of Appeals.[31] When those states asked the Court to weigh in, it declined without providing justification.[32] Some reports suggest that Justice Ginsburg indicated that the rulings were in line with SCOTUS's decision and therefore did not require intervention.[33] However, the 6th Circuit ruled in favor of upholding bans in Kentucky, Michigan, Ohio, and Tennessee in November 2014, and that ruling became the basis for the recent case. Although *United States v. Windsor* does not assert explicitly that same-sex marriage bans are unconstitutional, most of the lower courts seemed to see it as a tacit endorsement of same-sex marriage. The 6th Circuit used the same decision as the basis for its opinion, suggesting that the final say on marriage equality should go to the citizenry, allowing people to "become the heroes of their own stories by meeting each other not as adversaries in a court system but as fellow citizens seeking to resolve a new social issue in a fair-minded way."[34]

The arguments of both the plaintiffs and the respondents in *Obergefell v. Hodges* hinge on *United States v. Windsor*. The respondents in the case stress the primary issue on both the marriage and the recognition questions in the precedent is that "the states have traditionally defined and regulated marriage."[35] Since DOMA was eliminated because it limited a state's (New York's) right to certify same-sex marriage by regulating couples' access to benefits, such as filing federal tax returns jointly, the four states contend that a requirement to license same-sex marriage—to recognize those marriages licensed in other states—would likewise nullify their citizens' choice to limit marriage to opposite-sex couples. They further assert that the issue is not same-sex marriage's quality but rather who possesses the privilege to make decisions about how to define marriage. Following the 6th Circuit's opinion, the respondents believed that the voters should decide. The plaintiffs point out that their position does not require the Court to imbue same-sex marriage with a new constitutionality. Rather, on the marriage question, they suggest that SCOTUS has long made clear that marriage is a fundamental right, and they are merely seeking equal access to that right. But even if that's not the case, they add, the Court's decision in the Windsor case means that states can't exclude same-sex couples from the dignity, rights, and benefits created by marriage just because they disapprove of them.[36] On the recognition question, they assert that a state's laws against accepting same-sex marriages licensed in other states disregards the decision of those states to certify marriage for gays and lesbians. How the respondents, plaintiffs, and SCOTUS Justices framed what is at issue in this case had striking implications for the understanding of marriage's position in the contemporary American institution of family.

When addressing the Court on the marriage question, the respondents defined the states' interests in marriage as maintaining families, specifically in sustaining the connection between parents and their biological offspring. Arguing for the states, Bursch suggested that allowing same-sex marriage could fundamentally change the purpose of that union for all couples from "believing that marriage is all about staying with their kids" to "believing that it's all about emotion and commitment."[37]

> I want you to think about two couples that are identically situated. They've been married for five years, and they each have a 3-year-old child. One grows up believing that marriage is about keeping that couple bound to that child forever. The other couple believes that that marriage is more about their emotional commitment to each other, and if that commitment fades, then they may not stay together.[38]

In adult-centered (the couple's emotional fulfillment is the reason for marrying)—as opposed to child-centered (procreating and providing for children is the reason for marrying)—marriage, the argument goes, the couple's attachment lasts only as long as their attraction, which may lead to growing divorce rates or increasing numbers of children born to unwed mothers. Under such a conception, marriage loses its value as a stronghold for the biological family, and children are in danger of being split from their (biological) parents. Bursch insists that extending marital access to same-sex couples necessarily weakens opposite-sex marriage because it disassociates marriage from childrearing.

The plaintiffs focused on marriage's importance to the establishment and protection of families when addressing the Court on the marriage question. When arguing for the plaintiffs in the case, Bonauto's opening statement began, "The intimate and committed relationships of same-sex couples, just like those of heterosexual couples, provide mutual support and are the foundation of family life in our society," and her argument is predicated on the notion that "a whole class of people who are denied the equal right to be able to join in this very extensive government institution that provides protection for families."[39] Likewise, Verrilli, also arguing for the plaintiffs, suggests that excluding same-sex couples from marriage not only debases them and their children but also denies them “the stabilizing structure that marriage affords.”[40]

POST-MARRIAGE CONDITION

In chapter 1, we discussed Foucault’s distinction between the deployment of alliance and that of sexuality. For Foucault, the transformation of marriage is closely associated with the emergence of sexuality, and the family “was the crystal in the deployment of sexuality.”[41]

> The family, in its contemporary form, must not be understood as a social, economic, and political structure of alliance that excludes or at least restrains sexuality, . . . its role is to anchor sexuality and provide it with a permanent support. It ensures the production of a sexuality that is not homogeneous with the privileges of alliance, while making it possible for the systems of alliance to be imbued with a new tactic of power which they otherwise would be impervious to. The family is the interchange of sexuality and alliance: it conveys the law and the juridical dimension in the deployment of sexuality; and it conveys the economy of pleasure and the intensity of sensations in the regime of alliance.[42]

For most of its history in the West, marriage was situated within a complex family past, influenced by the unions preceding it and influencing those that would follow. Under the deployment of alliance, marriage is concerned primarily with maintaining the social body, defining taboos, and governing the transmission of wealth. In such a context, matchmaking functions diachronically, taking into account prior marriages and current familial needs as well as anticipating future requirements and obligations.

When the deployment of sexuality is layered upon that of alliance, a system arises that relies on mobile and contingent relations, that expands its tactics to account for individual bodies in ever-increasing ways, that concerns itself with the population and economies of production and consumption. Currently, when the couple is the center of married life, when parents and in-laws have been pushed to the margins, when the expectation is that unmarried people make their own choices in marriage partners, matchmaking is isolated both spatially and temporally and functions primarily—as Galloway writes of collaborative filtering[43]—synchronically. Although singles undoubtedly consider their futures when making nuptial decisions (Will he be a good father? Will she take care of me when I grow old?), in most cases they hold the ultimate responsibility for deciding whom to marry. Couples are no longer expected to share a dwelling with, or even live in the same city as, their families, further destabilizing the family’s potential for enclosure. The synchronic logics of control, which presuppose a *dividual* subject, seem better equipped for reforming contemporary risky single subjects. Perhaps the most promising technique of power for the family in contemporary control societies is online dating, a system that serves to shore up marriage—and, by extension, the family—during this period of transformation and destabilization.

Much like online dating success stories work to normalize contemporary marriage, the practice of online dating bolsters it, as it *identifies* single adults who need to be reformed. Galloway and Thacker situate *identification* as the central problem of the twenty-first century: "The current century will be the era of universal standards of identification. In the same way that universals of communication were levied to solve crises in global command and control, the future's universals of identification will solve today's crises of locability and identification."[44] They offer examples. Profiling is replaced by DNA matching as the standard of evidence in criminal prosecution; logistics is replaced by RFID databases as the standard of efficiency in consumer-product distribution. This centrality of identification is indicative of the shift to biopolitical control.[45] The no-fly list is one instance of how control circulates. Online dating is another.

Online dating sites catalog and categorize vast amounts of user data. For instance, eHarmony.com uses twenty-nine traits to match its clients with one another, and new members answer over three hundred questions as part of personality testing and initial account setup. Match.com accounts for more than fifteen hundred variables, relying on users' stated and revealed preferences when making recommendations. OkCupid.com requires users to answer and rank at least fifty questions before beginning the matching process, the average user answers three hundred, and some report having answered more than four thousand.[46] In protocological terms, individuals are absorbed here by the (distributed) network of control societies, and this distinction can be likened to the difference between site functionality and user experience. Online dating sites *dividuate* users, collecting information on their behaviors, tastes, etc., converting it to data, and aggregating that data with information gathered from others. Regardless of how the online dating experience may feel for users, sites do not pair individuals one to another. This process creates a feeling of *individuation* for users. All the while, online dating helps marriage to remain relevant by pulling the contemporary unmarried men and women into the enclosure of marriage. Indeed, it is not insignificant that this contemporary romantic matchmaking takes place on *sites*, as this spatial metaphor underscores the importance of locability and identification of singles, regardless of their intentions. For even with the use of an online dating site for unsanctioned activities, such as experimenting with identity formation or arranging fleeting sexual encounters, participants are being constituted as marriageable individuals, quantified and managed.

In addition to serving as a valuable tool for identifying single adults, online dating uses persuasive mechanisms for reestablishing the marriage trajectory and preserving the marriage imperative. Site procedures are designed to create confidence in their users. eHarmony.com uses constrained data preparation and matching processes, as well as rigidly structured user interactions, to set an expectation of reliability and trustworthiness. In addition, the site features marriage successes in its marketing campaigns. The company's television commercials often show company cofounder Neil Clark Warren interacting with couples who met on the site to the soundtrack of "(This Will Be) An Everlasting Love" by Natalie Cole. Likewise, eHarmony.com claims to be responsible for 438 marriages daily,[47] and its user-generated success stories generally create a trajectory from match to date to relationship to marriage that contradicts both the actual experience of many, perhaps most, online dating users as well as the contemporary destabilization of marriage.

We earlier identified the historical purpose of matchmaking as facilitating advantageous unions, generally related to cooperation/consolidation and procreation across family units, broadly speaking. In other words, romantic matchmaking has been an act of family making. Enlarging nomadic groups' hunting capacity by aligning with each other, increasing clans' power by uniting neighboring farms, and augmenting merchants' holdings by consolidating wealth all (re)make a family via an advantageous match. Under these conditions, marriage functions as the yoke and brail line of the family's cast net. As an environment of enclosure, the netlike family has long been able to expand in order to envelop subjects within its domain; however, marriage has offered the constriction, the mechanism for tightening around subjects so that they may be managed. To put it another way, marriage has been the mold that allowed for the family's modulation. It has empowered the institution of the family to become increasingly flexible in response to economic, political, and social shifts. Even as the conditions within marriage have changed dramatically, its primary purpose has remained relatively stable.

Today, though, marriage finds itself in a precarious position. At the same time that heterosexual marriage has become less obligatory and more optional, the struggle for same-sex marriage rights reached SCOTUS. Marriage has undergone a fundamental change, and it can be traced back to the love match, or development of the couple's romantic fulfillment as the primary purpose for marriage.[48] The love match emerged during the Victorian era, and marriage has no doubt changed since that time, something that chapter 2 documents. The deployment of alliance and sexuality coalesce during the same period, and marriage and family expand into a technique of power with both legal intervention into sexuality and the insertion of desire and sensation into alliance. We can think here of the simultaneous appearance of seemingly contradictory phenomena: tracking birth, morality, and fertility rates (alliance into sexuality) and marrying for love (sexuality into alliance).

Although these tactics had taken hold in the eighteenth century, shifts in the twentieth and early twenty-first centuries—most notably the changing status of women, the development of birth control and fertility technologies, the availability of adoption, and the fight for marriage equality but also the conditions associated with mobile privatization—solidified matchmaking and marriage's ultimate goal: uniting the happy couple. The love match, and the accompanying importance of individual gratification, seems highly compatible with control's (and online dating's) synchronic logic. Online dating, then, may appear to be an ideal technique of power in control societies as a result of this compatibility. However, this modularity damages the brail lines, weakening their ability to tighten the cast net and undermining their capacity for enclosure. On the one hand, online dating's consumptive potential endures; marketing campaigns and public relations efforts, as well as sites' procedures and couple success narratives, offer persuasive means for compelling use and have propelled online dating to a billion-dollar industry.[49] On the other hand, its modularity undermines its (re)productive potential. Online dating's quantitative nature, its reliance on computational processing and big data, reinforces the integration of alliance into sexuality but negates the incorporation of sexuality into alliance. The affective dimension is largely unaccounted for in the compatibility systems that power online dating. To put it another way, there is no love in the match. And although online dating may hold great promise for identifying risky single subjects, it seems incompatible with the historically diachronic logics that have allowed marriage to play mold to the family's modulation. As a result, the search for the family's new weapon continues.

NOTES

1. Stephanie Coontz, *Marriage, a History: From Obedience to Intimacy, or How Love Conquered Marriage* (New York: Viking, 2005).
2. Gilles Deleuze, "Postscript on the Societies of Control," *October* 59 (Winter, 1992), 3–7, 4.
3. Ibid.
4. Ibid.
5. Ibid., 5.
6. Ibid., 4.
7. Ibid., 5.
8. Ibid.
9. Michael Hardt, "The Withering of Civil Society," *Social Text* 45 (Winter, 1995), 27–44, 36.
10. Ibid.
11. Ibid., 40–1.
12. Patricia Ticineto Clough, "Technoscience, Global Politics, and Cultural Criticism," *Social Text 80* 22, no. 3 (2004), 1–23, 14.
13. Jack Z. Bratich, "'Nothing Is Left Alone for Too Long': Reality Programming and Control Society Subjects," *Journal of Communication Inquiry* 30, no. 1 (2006), 65–83, 11.

14. Alexander R. Galloway, *Protocol: How Control Exists after Decentralization* (Cambridge, MA: MIT Press, 2004), 12. See also Mark Andrejevic, "Monitored Mobility in the Era of Mass Customization," *Space and Culture* 6, no. 2 (2003), 132–50, 142; Jeremy Packer, "Becoming Bombs: Mobilizing Mobility in the War on Terror," *Cultural Studies* 20, no. 4–5 (2006), 378–99, 392.

15. Deleuze, "Postscript," 3.

16. Ibid., 3–4.

17. Ibid., 7.

18. Stephanie Coontz, *Marriage, a History*, 301.

19. Ibid., 276.

20. Ibid.

21. Ibid.

22. Maria L. La Ganga, "Same-Sex Couples Marry in Oregon as Judge Strikes Down State Ban" (includes interview with Michael McShane), *Los Angeles Times*, May 19, 2014, www.latimes.com/nation/la-na-oregon-gay-marriage-20140520-story.html (italics added).

23. Rachel la Corte, "Washington State Approves Bill to Legalize Gay Marriage" (includes interview with Ed Murray), *Boston Globe*, February 2, 2012, www.bostonglobe.com/news/nation/2012/02/02/wash-state-approves-bill-legalize-gay-marriage/aW48oiNEpEKRD15wq Yjr8N/story.html (italics added).

24. Office of Governor C. L. "Butch" Otter, "Governor Otter Seeks Emergency Stay on Same-Sex Marriage Ruling," news release, May 14, 2014, gov.idaho.gov/mediacenter /press/pr2014/5.%20May/pr_030.html (italics added).

25. Benedict XVI, "To The Members of the Diplomatic Corps Accredited to the Holy See," January 9, 2012, w2.vatican.va/content/benedict-xvi/en/speeches/2012/january/documents/hf_ben-xvi_spe_20120109_diplomatic-corps.html (italics added).

26. Amy Howe, "Same-Sex Marriage, in Plain English—Part I," *SCOTUSblog* (blog), April 27, 2015, www.scotusblog.com/2015/04/same-sex-marriage-in-plain-english-part-i/.

27. "Obergefell v. Hodges," *SCOTUSblog*, accessed on June 23, 2015, www.scotusblog.com/case-files/cases/obergefell-v-hodges/.

28. Howe, "Part I."

29. Defense of Marriage Act, HR 3396, 104th Cong., 2nd sess. (1996).

30. Amy Howe, "Same-Sex Marriage, in Plain English—Part II," *SCOTUSblog* (blog), April 27, 2015, www.scotusblog.com/2015/04/same-sex-marriage-in-plain-english-part-ii/.

31. Robert Barnes, "Supreme Court Agrees to Hear Gay Marriage Issue," the *Washington Post*, January 16, 2015, www.washingtonpost.com/politics/courts_law/supreme-court-agrees-to-hear-gay-marriage-issue/2015/01/16/865149ec-9d96-11e4-a7ee-526210d665b4 _story.html.

32. Ibid.; Howe, "Part II."

33. Barnes, "Supreme Court."

34. Ibid.

35. Howe, "Part II."

36. Howe, "Part I."

37. *Obergefell v. Hodges* (oral arguments), April 28, 2015, 15, www.supremecourt.gov / oral_arguments/argument_transcripts/14-556q1_6k47.pdf.

38. Ibid.

39. Ibid., 5.

40. Ibid., 28.

41. Michel Foucault, *The History of Sexuality, Vol. 1: An Introduction*, translated by Robert Hurley (New York: Pantheon Books, 1977), 111.

42. Foucault, *History*, 108.

43. Galloway, *Protocol.*

44. Alexander R. Galloway and Eugene Thacker, *The Exploit: A Theory of Networks* (Minneapolis: University of Minnesota Press, 2007), 131.

45. Ibid.

46. Nick Paumgarten, "Looking for Someone: Sex, Love, and Loneliness on the Internet," *The New Yorker*, July 4, 2011, www.newyorker.com/magazine/2011/07/04/ looking-for-someone; Christian Rudder, *Dataclysm: Who We Are (When We Think No One's Looking)* (New

York: Crown Publishing Group, 2014), 90); Ali (AlisAddiction), Twitter post, December 14, 2012, 12:41pm, https://twitter.com /AlisAddiction/status/276063750874877952.

47. "Who We Are," eHarmony.com, accessed June 11, 2015, www.eharmony.com /about/ eharmony/.

48. Coontz, *Marriage, a History*, 178.

49. Nick Paumgarten, "Looking for Someone."

Chapter Six

Postscript on Technologies of Matching

The January 2011 issue of *Wired* magazine featured a special report on artificial intelligence. The cover showcased a toppled windup toy robot with the headline, "Artificial Intelligence is here. But it's nothing like we expected." In the introductory essay, veteran technology reporter Steven Levy described an artificial intelligence (AI) revolution that is very different from the one imagined by science fiction, one based in algorithmic models rather than robots and replicated human thinking. The report included accounts of algorithmic processing that affects our lives, from its current role in our financial system to the future of the driverless car. As we have seen, AI has permeated the corners of everyday life through seemingly innocuous technologies that rely on intricate processes of calculation and categorization to make us into matches. We saw that romantic matchmaking exists in a set of contingent social, cultural, familial, economic, and technological relationships within the marriage apparatus, and that those relationships can be described in terms of how they mediate interactions of the matcher and the matched. In the condition of automation, the primary automator of matching for romantic purposes is online dating, and a complex of computational processes, big data, and cultural assumptions enable it. Online dating sites rely on persuasive procedures to build connections with users, even as they are converting them into data objects. Singles emerge on the other side, or so the success story goes, as (re)productive subjects. This process of modulation is a significant feature of control societies, and online dating's computational and cultural systems help keep contemporary marriage relevant. Now we will take a step back from online dating to think about contemporary AI more generally.

COMPUTATIONAL PROCESSING AND HUMAN DECISION-MAKING

These complex computational systems rely on the interrelated procedures of categorization and matching. According to Lakoff, classifying is foundational to human understanding:

> Categorization is not a matter to be taken lightly. There is nothing more basic than categorization to our thought, perception, action, and speech. Every time we see something as a kind of thing, for example, a tree, we are categorizing. . . . Without the ability to categorize, we could not function at all, either in the physical world or in our social and intellectual lives. An understanding of how we categorize is central to any understanding of what makes us human.[1]

Although both human and machine reason rely on categorization, computation uses a classical understanding of categorizing that cannot sufficiently account for the human activity. This classical notion is "represented by sets, which in turn are defined by the properties of their members."[2] Such understanding is closely related to considering human reason as a form of deduction, and positioning human thought thusly creates the impression that computation can replicate it. While many human categorizations follow this logic, mere similarity cannot account for all human thought processes. By way of illustration, Lakoff offers an example from a study of Dyribal, an aboriginal Australian language that has four basic categories for all things. Although some Dyribal classifications may seem anomalous, they are instead based on the "domain of experience," which develops from the significance of certain components, cultural mythos, and personal experience.[3]

In order to understand the human processes of matching and categorization, the case of making a reading recommendation to a friend serves as an example. We must sort through what we know about our friend's interests and reading preferences before we can make a suggestion. For example, I might reflect on conversations he and I have had about books he has read and sort those books into categories based on any number of criteria (e.g., his enjoyment level, the plot's complexity, the book's genre). I can eliminate titles I know he won't enjoy, based on his distaste of predictable plots and detective fiction, and select works he might enjoy, based on his preferences for fantasy and intricate storylines. Using these categorizations, I can then recommend to him (*match* him *with*) a number of options I think he would like.

These processes of matching and categorization are recursive. In fact, categorizing is a special case of matching. Matching, in its primary, definitional sense, is a method of making associations or connections between X and Y. While these connections may be based on any number of qualities—equality, similarity, compatibility, complementarity, opposition—they are connections nonetheless. Categorization, more specifically, would link X with Y based primarily on similarity. When I am categorizing books based on genre or plot complexity, then, I am engaging in a particular kind of matching, classifying like items. When I use that information (e.g., genre categorization) to determine whether my friend would enjoy a particular book, I am engaging in a different kind of matching, putting together complementary components. The other significant point that arises from this example is that matching *precedes* recommending. I must categorize books and match some of them to my friend before I can advocate one over another. Much like the Dyribal, I rely on the domain of experience when making recommendations. These realms that inform human decision-making cannot be outsourced to machine processing, at least not yet.

According to Dreyfus, the largest challenge for machine processing's attempt to replicate human thinking is determining "relevance and significance," or "being-in-a-situation."[4] Machine processing requires context for decision-making, but artificial intelligence can only interpret context as a rule, which requires a hierarchy of contexts.

> Computers can only deal with facts, but man—the source of facts—is not a fact or set of facts, but a being who creates himself and the world of facts in the process of living in the world. This human world with its recognizable objects is organized by human beings using their embodied capacities to satisfy their embodied needs.[5]

This is similar to the domain of experience as discussed by Lakoff and captures the challenge that led the IBM supercomputer Watson to stumble when it competed against former *Jeopardy!* champions Ken Jennings and Brad Rutter. When asked to identify the US city in which the two largest airports are named after a hero and a battle, both from World War II, Watson answered Toronto. This misstep likely resulted from how the system categorized information. In this instance, responses in the show's category *US Cities* could include not only city names but also the names of rivers, sports teams, airports, or anything else closely aligned with a major city. As a result, Watson could not limit its responses only to cities in the United States. Again, this points to a problem of relevance and significance, or an inability to account for context. Even state-of-the-art contemporary AI (e.g., robotics that interact with the physical world) cannot identify "what matters when."[6]

At least part of the controversy surrounding the recent Turing test victory is related to a similar issue. The winning program convinced a third of jurors that it was Eugene Goostman, a thirteen-year-old boy from Odessa, Ukraine.[7] Some have called Eugene "psychological smoke and mirrors," since it used vagaries and deflection in earlier instantiations and relied on its supposed youth and lack of English fluency to fool judges.[8] Perhaps more significantly, some have argued that the problem is with the test itself, since it does not measure the ability to replicate human thinking, a process that relies on relationships with the environment around us.[9] According to Shanahan, "It overemphasises [sic] language at the expense of the issue of embodiment."[10] The question of embodiment here is akin to Dreyfus's *being-in-a-situation*. In other words, the difficulty for the Turing Test, and the challenge for developing machine processing that replicates human thought, is a rhetorical problem. Human decision-making is situational and responds to a complex of contingent material, cultural, economic, and social conditions, but such computational systems are decontextualized. Although they instantiate logics circulating within the culture in which they develop, they are arhetorical.

MASS CUSTOMIZATION AND FILTER BUBBLES

One mainstay of this new AI, and a process that most of us have encountered online, is mass customization. A concept adopted from manufacturing and engineering, mass customization is "the ability to provide customised [sic] products and services to individual customers or niches on a large scale, without losing the benefits of mass production."[11] For example, some computer manufacturers allow customer's to personalize certain laptop components—hard-drive specifications, software preloads, shell color—at the time of purchase, enabling the computer manufacturer to customize the purchase while still producing its laptops on a large scale. With the flexibility that Internet technologies afford, it is no wonder that many of the most-visited sites on the Web offer some degree of mass customization. During the height of the dotcom boom, high-profile sites such as Yahoo! allowed users to customize content (e.g., movie listings, weather, horoscopes) displayed on a Web portal page designed to serve as the user's primary access point for the Internet. And while some companies had already begun to employ aggregated data to enhance the user experience, Amazon was able to patent its BookMatcher recommendation service in 2000 after years of use.[12] Today, Amazon uses information on customers' past buying and viewing behaviors to display individualized pages for its users. More recently, Google Instant updates search results onscreen as users type.

Andrejevic describes mass customization as functioning on both a symbolic and a practical level:

> It advertises itself as a means of overcoming the stultifying uniformity of mass culture while simultaneously allowing for intensified forms of consumer surveillance and exploitation. . . . The attempt to both manage and anticipate increasingly volatile patterns of consumption (patterns that correspond to a need to sell the goods made available by enhanced productive capacity) leads

> to an intensive reliance on marketing research. The development of information and communications technologies (ICTs) has allowed for the proliferation of strategies for rationalizing such research, and, perhaps more significantly, for shifting much of the burden to the consumer. One of the cannier strategies of mass customization is its claim that surveillance works to the advantage of consumers by allowing producers to more closely meet their wants and needs.[13]

He also notes that, by providing the information marketers rely on for producing these customizations, at minimum consumers are taking on the responsibility of providing market research for the companies they patronize. Beyond that, when participating in the design process (e.g., choosing a shell color or selecting software when ordering a new laptop), customers are not only not paid for the work they are doing but also "buy[ing] back the added value generated by the additional labor."[14] As Terranova points out, this "free labor" is a pervasive component of the postindustrial economy that destabilizes the boundary between production and consumption by turning consumers into producers of that which they consume.[15] Online dating exemplifies this phenomenon. Members experience mass customization through both site- and self-matching procedures, and through their participation (completing personality questionnaires, answering match questions, rating profiles) they generate value by adding to big data for internal (site matching, company research) and external (marketable data sets for scholarly research) use.

In addition to collapsing production and consumption, this personalization creates what Pariser calls *filter bubbles*:

> The new generation of Internet filters looks at the things you seem to like—the actual things you've done, or the things people like you like—and tries to extrapolate. They are prediction engines, constantly creating and refining a theory of who you are and what you'll do and want next. Together, these engines create a unique universe of information for each of us—what I've come to call a filter bubble—which fundamentally alters the way we encounter ideas and information.[16]

Pariser identifies three important characteristics of filter bubbles. First, they are isolating. While we may choose the information we consume through other media—such as picking which television programming to watch, which cable news sources to trust—we are still experiencing the same programming as millions of other viewers. With personalization, we are solitary; filter bubbles act as "a centrifugal force, pulling us apart."[17] Second, they are concealed. Search engines do not tell us how they produce their results. We don't know how—or in many cases, whether—they are making assumptions about us. As we saw with Match.com's use of revealed preferences and OkCupid.com's experiments, sites may intentionally mislead us.[18] Third, filter bubbles are compulsory. We elect to watch Fox News or to read the *New York Times*. And while we may not always take the opportunity, we can choose to engage with these media critically. We can examine, evaluate, and interrogate their credibility. We "don't make the same kind of choice with personalized filters,"[19] primarily because of their concealment. As co-founder Chris Hughes noted early in the site's history, Facebook is closely tied to users' status and identity, making it difficult to opt out:

> If you don't have a Facebook profile, you don't have an online identity. . . . It doesn't mean that you are antisocial, or you are a bad person, but where are the traces of your existence in this college community? You don't exist—online, at least. That's why we get so many people to join up. You need to be on it.[20]

Facebook's proliferation both on and off campus in the past decade has made it even more difficult to avoid. And although we may choose not to use sites such as Facebook or Google that tailor information to our interests and tastes, there are still traces of us there—in photographs or status updates of our friends and family and in the information about us on other sites. These mass-customized technologies utilize complex, interrelated methods of categorization and matching to determine what (and sometimes how) information is presented to us, and these protocols rely on problematic computational systems to stand in for human decision-making.

MATCHING TECHNOLOGIES

We engage with matching and categorization processes online, but they are not labeled as such. They are known by a variety of names (e.g., search engines, recommender systems, collaborative filtering), but none sufficiently describes the computational processes at work. In fact, they belie both a) the tension between these processes and how users experience them and b) their relationship to other technologies (e.g., those used for military and intelligence or surveillance purposes). In order to shed light on the tension between user experiences on the Web and the underlying processes that enable them, I propose a new term, *matching technologies*, that describes the combination of big data, computational processing, and the values and cultural assumptions built into informationalization. Indeed, matching is a more appropriate descriptor than the three processes most often named in their current labels: searching, collaborating, and recommending.

Search, in general use, has an exploratory connotation—organizing a search party, searching for lost keys or the meaning of life. In that sense, it is appropriate to describe the user experience in many Web-enabled transactions (e.g., using Google to find statistics on global broadband penetration or Match.com to find a date). However, search does not adequately describe the technological processes that enable users to locate the information (or product or person) they seek. In fact, Amit Singhal, a Google Fellow instrumental in the search engine's 2001 rewrite, notes that "The holy grail of search is to understand what the user wants. Then you are not *matching words*; you are actually trying to *match* meaning."[21] Singhal's comments also underscore the importance of distinguishing between what the user experiences (search) and the technical processes that enable that experience (match), and we saw clear distinctions in our examinations of online dating in chapters 3 and 4.

Recommending and collaborating are distinct processes, but the technologies named after them (i.e., recommender systems and collaborative filtering) often work in concert. In other words, recommender systems generally rely on collaborative filtering to match aggregated user data and make suggestions. Collaborative filtering makes use of data on user choices, preferences, and behaviors, and the most popular sites on the Internet integrate these processes into the user experience. For example, as we have seen, Match.com aggregates the behavior of similar users in order to make matches. Similarly, Facebook makes friend suggestions based on a combination of factors, including friends two users have in common; similarly, the "like" feature enables the site to determine which items to include in a user's news feed, which ads to show him, and what to list in his *Recommended Pages* section (e.g., musical artists he might enjoy, businesses he might patronize). Likewise, Amazon uses data on the items purchased, viewed, and wished for by consumers to make recommendations to individual users, and Netflix's Cinematch uses millions of movie ratings to predict the viewing choices of its members.

Although thinking of these technologies as collaborative filtering or recommender systems is convenient, it is not sufficiently descriptive. First, collaboration is generally associated with cooperation. In the classroom, instructors implement collaborative-learning projects that require students to work together in order to facilitate their learning and to hone their ability to perform as members of a team. When musicians team up for a performance or organizations pool resources, we refer to their efforts as a collaboration. There is also a presumption of intent whenever we collaborate. To put it differently, we know when we are involved in a collaboration. The technologies that allow Facebook and Amazon to suggest friends and books do not require users' intentional participation in the process, or even their knowledge of it. Although users choose to participate in online dating, surely few understand how information about them is used in combination with others' data.

Similarly, recommending has a specific rhetorical connotation. We are accustomed to receiving recommendations from people—friends and librarians, salespeople and critics. The act of recommending, then, could be—and generally would be—more than sharing a list of names or titles. A recommendation is an argument, specifically an engagement in epideictic rhetoric in which the recommender praises that which is recommended. The intricacy of the argument varies based on any number of factors. For example, when recommending a book to a friend, I might simply hand it to him and say, "Here, read this." Based on our previous discussions of books we've both enjoyed and his knowledge of my attitudes and tastes, he could fill in the assumptions that underlie my imperative. However, while working summers in a bookstore, I found few customers who would take my recommendations at face value. It might require offering the similarities between my pick and other books the customer liked, identifying the genre to which it belonged, or providing a synopsis of the plot. Other recommendation processes work similarly. Only the closest of friends could arrange a blind date without providing some discussion of the qualities that make the potential suitor attractive. And the most formal of these processes, such as recommending someone for employment, often require documentation (i.e., the letter of recommendation). These technologies jettison the act of praising, offering instead a list of potential books to read (or people to date or employees to hire) with little reasoning as to why it is being offered. When reasoning is provided, it is categorical (e.g., others with similar purchasing patterns also purchased these products) rather than celebratory.

Whereas recommending involves demonstrating the value of the recommendation, matching precedes the recommendation, discerning whether two (or more) things fit together. We might match our shoes to our belt, an interlocutor's intensity in an argument, or cards in a children's game. As we saw in chapter 2, this is a difference in stasis. One rhetorical implication of the trend of quantification is "technicizing" of an issue, or a stasis shift from evaluative to technical.[22] Miller notes that the third and fourth stasis levels are evaluative and call for decisions based on values. Conversely, the first and second stasis levels rely on knowledge or expertise. Much like OkCupid.com's matching system, the process of recommending as practiced by recommendation engines is technical rather than evaluative. Indeed, unifying these technologies as matching technologies highlights the underlying similarities among these processes, including their reliance on similar protocols.

AREAS FOR FURTHER STUDY

Other matching technologies address different problems of power and establish different relationships to subjects. As we have seen, for example, making reading or viewing recommendations is similar to online dating. Sites like Amazon and Netflix use comparable processes. For example, just as Match.com makes matches based on the preferences users state explicitly in their account profiles and the preferences they reveal in their behavior on the site, Netflix make suggestions to users based on both how they rate specific films or television shows and the films and television shows they choose to watch. In both cases, the sites employ big data aggregated from millions of users in order to make mass-customized recommendations for individuals, and both provide seemingly unique user experiences that contradict the dividualizing procedures that enable them.

There are two subtle yet significant differences between online dating and most commercial applications of matching technologies. The first distinction is that online dating engages in a process of double dividuation. As we saw in chapter 1, online dating users take on two roles on the site: presenter (of self) and evaluator (of others). The initial account setup procedure and ongoing account management are processes of self-presentation in which users portray themselves as desirable individuals whom their matches should get to know better. Viewing profiles of site- and self-selected matches are processes of other-evaluation in which users decide whether they would like to get to know their matches better. In other words, on both sides of these processes are dividuals, people who have been translated into discrete data units. In contrast, most commercial applications of matching technologies engage in one-sided dividuation. These technologies use member data to match them with objects (products, media content, travel destinations, etc.) rather than other members.

The second distinction is that online dating sites and other matching technologies create different subjectivities for their users. As we have seen, online dating requires users to assume the subject position of the *single subject*. Both online dating and consumer matching rely on the quantification of components of identity for their ability to reconstitute individuals into new subjectivities, but the other matching technologies call for a different identity, the consumer. These systems are able to deconstruct components in such specialized ways that some of the most closely held identifiers (e.g., race, gender) can be filtered out entirely. While romantic matchmaking still relies heavily on categories such as race, class, gender, age, and sexual orientation for its pairings, sites like Netflix don't consider them at all. The old joke that nobody knows you're a dog online is not only outdated but also potentially, depending on whom "nobody" refers to, wrong. Netflix knows you're a dog; it just doesn't care. The site stores identifying information, but that is not what the site uses for its matching. For Netflix, we are consumers, and it is much more interested in consumption than anything else.

Although some online dating sites, especially Match.com, rely heavily on consumption and consumer logics to make connections between users, that reliance is concealed. While much of the information gathered relates to leisure, lifestyle, and popular culture, the site matches based on compatibility logics rather than the consumption that stands front and center. And while all online dating participants are necessarily both consumer and product—both in terms of evaluating others/being evaluated and serving simultaneously as site customers and inventory—it is qualitatively different from being a consumer and a product in traditional ecommerce contexts. Simply, when shopping on Amazon or choosing what to watch on Netflix, as Match.com engineer Amarnath Thombre puts it, "The movie doesn't have to like you back."[23] Acts of evaluation in online dating are tempered by this notion, and acts of presentation are complicated by it. There are no visible ratings or reviews in online dating, though OkCupid.com does allow users to give others (positive) badges such as *Eye Candy*. If there were rating or review systems, it is difficult to imagine that they would be anything other than positive, for fear of retribution or a poor reflection on the evaluator. In addition, users' self-presentation seems to function more as a process of identity construction than as product advertisement.[24]

In order to develop an understanding of matching technologies more broadly, we must investigate other examples as procedural rhetorics. One potential case study is Netflix, its Cinematch system, and the Netflix Prize. Launched in 1999, Netflix initially allowed users unlimited DVD rentals by mail for a monthly subscription fee, and within three years it had six hundred thousand customers. In 2007, Netflix enhanced its service with "Watch Instantly," which allowed members to watch streaming video; today it has over twenty million subscribers."[25] Much like brick-and-mortar video stores of the past relied on knowledgeable staff to set them apart from others, Netflix depends on the suggestions it makes to attract and keep its clients. The expectation is that by understanding users' viewing preferences, the site is able to make more accurate recommendations for future viewing. The happier users are with what they are watching, the more likely they are to maintain their subscriptions.[26] An increasingly crowded streaming video market that includes Hulu, iTunes, Amazon, cable service providers, and even Wal-Mart necessitates maintenance of existing memberships. The company uses Cinematch, a matching technology that compares user viewing habits and ratings of content with the viewing habits and ratings of others, to match users with viewing options.

> The Cinematch system automatically analyzes the accumulated movie ratings weekly using a variant of Pearson's correlation with all other movies to determine a list of "similar" movies that are predictive of enjoyment for the movie. Then, as the user provides ratings, an on-line, real-time portion of the system computes a multivariate regression based on these correlations to determine a unique, personalized prediction for each predictable movie based on those ratings. If no personalized prediction is available, the average rating based on all ratings for the film is used. These predictions are displayed on the Website as red stars.[27]

By 2007, Netflix had gathered nearly two billion ratings of more than eighty-five thousand videos by nearly twelve million users at a pace of two million ratings a day.[28]

Rather than make human resources investments to ensure continued advancements in its Cinematch system, Netflix introduced the Netflix Prize on October 6, 2006.[29] The international competition was open to anyone, except residents of six countries (Cuba, Iran, Myanmar, North Korea, Sudan, Syria), and the goal of the contest was to produce a predictive algorithm that would net at least a 10 percent increase in the accuracy of the company's current matching technology; winning came with a million-dollar paycheck.[30] The Netflix Prize was scheduled to run through October 2, 2011; however, two teams submitted solutions that met the contest criteria on July 26, 2009.[31] During its nearly three-year run, the Netflix Prize received attempts from over forty thousand teams representing nearly two hundred countries. The winning team, BellKor's Pragmatic Chaos, was a collaboration of three teams, and their solutions have all been published and are easily available for download on the Netflix site. This offers access to logics that govern these particular matching technologies, logics that are usually concealed with the "black box."[32]

This brief discussion of Netflix, Cinematch, and the Netflix Prize has given us one potential case study, but there is much more to be done. First, this work has focused primarily on understanding Web applications in terms of how they work and how they establish and maintain relationships with users. Although chapter 3 did include the "voices" of users in terms of how they define success, there has been little discussion of how users actually engage with the sites. One potential avenue, then, is ethnographic research that gets at actual user practices, enhancing our understanding of how users conform to and subvert site expectations for subjectification online. Do user motivations persist or shift over time, and how do those motivations impact user practices? For example, we have engaged with site-sanctioned online dating success stories, but do they reflect the experiences of a broad constituency of users? Are the relationship formations presented in success stories the kinds established, or even sought, by users? What difference does it make, then, that Match.com is perfecting its ability to match users with one another, or Netflix is spending millions of dollars on research and development, if users don't actually pay attention to the recommendations?

Second, it will be important to understand matching technologies and the attendant user practices in circulation with users' offline practices and other media as well. We have positioned online dating historically, but an in-depth examination of popular-culture artifacts could enhance our understanding. For example, how do women's magazines, reality television, and self-help books contribute to biopower and control in distributed networks? How do matching technologies remediate old media within those distributed networks?

Finally, in addition to subversive use of online dating, it is important to look beyond marriage and procreative relationships. This study maintained a heteronormative perspective both in order to understand better the relationship between online dating and the institutions of family and marriage and because most of the dating industry, and the coverage thereof, does the same. One potential case study could compare eHarmony's heterosexual matching procedures with those on the company's same-sex dating site, CompatiblePartners.net. By beginning to explore these areas, we can lay a foundation for an informed, historical understanding of matching technologies and their relationship to subject formation and regimes of power.

NOTES

1. George Lakoff, *Women, Fire, and Dangerous Things: What Categories Reveal about the Mind* (Chicago: University of Chicago Press, 1987), 6–7.
2. Ibid., 8.
3. Ibid., 95.
4. Hubert L. Dreyfus, *What Computers Still Can't Do: A Critique of Artificial Reason* (Cambridge, MA: MIT Press, 1992), 215.
5. Ibid.
6. Sean Dorrance Kelly and Hubert L. Dreyfus, "Watson Still Can't Think," *Opinionator* (blog), *New York Times*, February 28, 2011, opinionator.blogs. nytimes.com/2011/02/28/watson-still-cant-think/.
7. Douglas McCormick, "Virtual Teen Passes Turing Test," *IEEE Spectrum*, June 10, 2014, spectrum.ieee.org/tech-talk/robotics/artificial-intelligence/virtual-tween-passes-turing-test.
8. David Auerbach, "A Computer Program Finally Passed the Turing Test? Not so fast," *Decoding the Tech World* (blog), *Slate*, June 10, 2014, www.slate.com /articles/technology/bitwise/2014/06/turing_test_reading_university_did_eugene_goostman_finally_make_the_grade.html.
9. Kelly Oakes, "No, A Computer Did Not Just Pass the Turing Test" (includes interview with Murray Shanahan), *BuzzFeed*, June 9, 2014, www.buzzfeed.com /kellyoakes/no-a-computer-did-not-just-pass-the-turing-test#.euvEgBaxD.
10. Ibid.
11. Bart MacCarthy, "Weapon of Choice," *IEEE News*, April 2003, 34, 34.
12. Troy Wolverton, "Amazon Snags Patent for Recommendation Service," CNet.com, May 31, 2000, news.cnet.com/2100-1017-241267.html.
13. Mark Andrejevic, "Kinder, Gentler Gaze of Big Brother: Reality TV in the Era of Digital Capitalism," *New Media & Society* 4, no. 2 (2002), 251–70, 256.
14. Ibid., 257.
15. Tiziana Terranova, "Free Labor: Producing Culture for the Digital Economy," *Social Text* 18, no. 2 (2000), 33–58.
16. Eli Pariser, *The Filter Bubble: What the Internet Is Hiding from You* (New York: Penguin Press, 2011), 9.
17. Ibid.
18. Nick Paumgarten, "Looking for Someone: Sex, Love, and Loneliness on the Internet," *The New Yorker*. July 4, 2011. www.newyorker.com/magazine /2011/07/04/looking-for-someone; Christian Rudder, "We Experiment on Human Beings!" *OKTrends* (blog), July 28, 2014, blog.okcupid.com/index.php/we-experiment-on-human-beings/.

19. Ibid., 10.

20. John Cassidy, "Me Media" (includes interview with Chris Hughes), *The New Yorker*, May 15, 2006, 50-9.

21. Steven Levy, "Inside the Box" (includes interview with Amit Singhal), *Wired* magazine, February 22, 2010, www.wired.com/2010/02/ff_google_algorithm/ (italics added).

22. Carolyn R. Miller, "The Presumptions of Expertise: The Role of Ethos in Risk Analysis," *Configurations* 11, no. 2 (2003), 197.

23. Paumgarten, "Looking for Love" (includes interview with Amarnath Thombre).

24. Dawn Shepherd, "Match.com, Genre, and the Conception of Self." Paper presentation, Biennial Conference of the Rhetoric Society of America, Seattle, Washington, May 2008.

25. "Overview," *Netflix.com*, accessed on November 21, 2011, ir.netflix.com/.

26. Robert M. Bell et al., "The Million Dollar Programming Prize," *IEEE Spectrum*, May 1, 2009, spectrum.ieee.org/computing/software/the-million-dollar-programming-prize.

27. James Bennett and Stan Lanning, "The Netflix Prize," *Proceedings of the KDD Cup and Workshop* (2007), www.netflixprize.com/assets/NetflixPrizeKDD _to_appear.pdf.

28. Ibid.

29. Bell et al., "The Million Dollar Programming Prize."

30. Bell et al., "The Million Dollar Programming Prize"; Bennett and Lanning, "The Netflix Prize"; "The Netflix Prize Rules," *The Netflix Prize*, accessed on November 22, 2011, www.netflixprize.com/rules.

31. "The Netflix Prize Rules"; "Congratulations!" *The Netflix Prize*, accessed on November 22, 2011, www. netflixprize.com/index.

32. "Congratulations!"

Bibliography

Adams, Paul C. *The Boundless Self: Communication in Physical and Virtual Spaces*. Syracuse: Syracuse University Press, 1998.

Agamben, Giorgio. *What Is an Apparatus? and Other Essays*. Stanford: Stanford University Press, 2009.

Ahuvia, Aaron C., and Mara B. Adelman. "Formal Intermediaries in the Marriage Market: A Typology and Review." *Journal of Marriage and Family* 54, no. 2 (1992), 452–63.

Alis (AlisAddiction). Twitter post. December 4, 2012, 12:41 p.m. twitter.com/AlisAddiction/status/276063750874877952.

Andrejevic, Mark. "Kinder, Gentler Gaze of Big Brother: Reality TV in the Era of Digital Capitalism." *New Media & Society* 4, no. 2 (2002), 251–70.

———. "Monitored Mobility in the Era of Mass Customization." *Space and Culture* 6, no. 2 (2003), 132–50.

Angwin, Julia. "Love's Labor Lost: Online Matchmaker Still Seeks Love, Money." SFGate.com, February 12, 1998, www.sfgate.com/business/article/LOVE-S-LABOR-LOST-Online-matchmaker-still-seeks-3013097.php.

Arvidsson, Adam. "'Quality Singles': Internet Dating and the World of Fantasy." *New Media & Society* 8, no. 4 (2006), 671–90.

Ashley Madison. Home page. Accessed on June 25, 2015. ashleymadison.com.

Audia, Pino G., and Christopher I. Rider. "A Garage and an Idea: What More Does an Entrepreneur Need?" *California Management Review* 48, no. 1 (2005), 6–28.

Auerbach, David. "A Computer Program Finally Passed the Turing Test? Not so fast." *Decoding the Tech World* (blog), *Slate*. June 10, 2014. www.slate.com/articles/technology/bitwise/2014/06/turing_test_reading_university_did_eugene_goostman_finally_make_the_grade.html.

Avery, Dan. "OkCupid Adds Loads of New Sexuality, Gender Options for Members." NewNowNext.com. November 16, 2014. www.newnownext.com/ okcupid-adds-loads-of-new-sexuality-gender-options-for-members/11/2014/.

Bachmann, Michele. "Family Values." US House of Representatives. Accessed January 15, 2012. bachmann.house.gov/Issues/Issue/?IssueID=56264 (page discontinued).

———. "Life." US House of Representatives. Accessed January 15, 2012. bachmann.house.gov/Issues/Issue/?IssueID=56263 (page discontinued).

Bain, Sandy Kyle. "Singles: Standing by Their Clubs." The *Washington Post*. June 4, 1982. www.washingtonpost.com/archive/lifestyle/1982/06/04/singles-standing-by-their-clubs/6959b80e-05f8-4d36-93c8-d0a3da33a8f0/

Barad, Karen. *Meeting the Universe Halfway: Quantum Physics and the Entanglement of Matter and Meaning*. Durham, NC: Duke University Press, 2007.

Barnes, Robert. "Supreme Court Agrees to Hear Gay Marriage Issue." The *Washington Post*. January 16, 2015. www.washingtonpost.com/politics/courts_law /supreme-court-agrees-to-hear-gay-marriage-issue/2015/01/16/865149ec-9d96-11e4-a7ee-526210d665b4_story.html.

Barnett, Scot. "Toward an object-oriented rhetoric: A review of Tool-being: Heidegger and the metaphysics of objects and Guerrilla metaphysics: Phenomenology and the carpentry of things by Graham Harman." *Enculturation* 7 (2010). enculturation.net/toward-an-object-oriented-rhetoric.

Beer, David. "Power through the Algorithm? Participatory Web Cultures and the Technological Unconscious." *New Media & Society* 11, no. 6 (2009), 985–1002.

Bell, Robert M., et al. "The Million Dollar Programming Prize." *IEEE Spectrum*. May 1, 2009. spectrum.ieee.org/computing/software/the-million-dollar-programming-prize.

Benedict XVI. "To The Members of the Diplomatic Corps Accredited to the Holy See." January 9, 2012. w2.vatican.va/content/benedict-xvi/en/speeches /2012/january /documents/hf_ben-xvi_spe_20120109_diplomatic-corps.html.

Bennett, James, and Stan Lanning. "The Netflix Prize." *Proceedings of the KDD Cup and Workshop* (2007). www.netflixprize.com/assets/NetflixPrizeKDD _to_appear.pdf.

Bogost, Ian. *Persuasive Games: The Expressive Power of Videogames*. Cambridge, MA: MIT Press, 2007.

boyd, danah. "Guilt through Algorithmic Association." *Apophenia* (blog). September 12, 2011. www.zephoria.org/thoughts/archives/2011/09/12/guilt-through-algorithmic-association .html.

boyd, danah, and Kate Crawford. "Critical Questions for Big Data." *Information, Communication, & Society* 15, no. 5 (2012), 662–79.

Bratich, Jack Z. "'Nothing Is Left Alone for Too Long': Reality Programming and Control Society Subjects." *Journal of Communication Inquiry* 30, no. 1 (2006), 65–83.

Brock, Kevin. "Enthymeme as Rhetorical Algorithm." *Present Tense: A Journal of Rhetoric in Society* 4, no. 1 (2014). www.presenttensejournal.org/volume-4/enthymeme-as-rhetorical-algorithm/.

Brock, Kevin, and Dawn Shepherd. "Understanding How Algorithms Work Persuasively through the Procedural Enthymeme." *Computers and Composition*, forthcoming.

Brown Jr., James J. "The Decorum of Objects." *Clinamen*. September 12, 2011. clinamen.jamesjbrownjr.net /2011/09/12/the-decorum-of-objects/.

———. "Crossing State Lines: Rhetoric and Software Studies." In *Rhetoric and Digital Humanities*. Edited by Jim Ridolfo and William Hart-Davidson. Chicago: University of Chicago Press, 2015, 20–32.

Buckwalter, J. Galen, et al. Method and System for Identifying People Who Are Likely to Have a Successful Relationship. US Patent 6,735,568, filed on August 10, 2000, and issued May 11, 2004.

Bugeja, Michael J. "Facing the Facebook." *Chronicle of Higher Education* January 23, 2006. chronicle.com/article/Facing-the-Facebook/46904.

Burrow, Roger, and Nick Ellison. "Sorting out Places? Towards a Social Politics of Neighbourhood Information." *Information, Communication & Society* 7, no. 3 (2004), 321–36.

Caspi, Tamar. "Make Me a Match." *The Jewish Advocate*. December 11, 2009, 31.

Cassidy, John. "Me Media." *The New Yorker*. May 15, 2006, 50–59.

Castells, Manuel, *Rise of the Network Society (The Information Age: Economy, Society, and Culture, Volume I)*. Oxford: Blackwell Publishers, 1996.

Castells, Manuel, Shujiro Yazawa, and Emma Kiselyova. "Insurgents Against the Global Order: A Comparative Analysis of the Zapatistas in Mexico, the American Militia and Japan's AUM Shinrikyo." *Berkeley Journal of Sociology* 40 (1995–1996), 21–59.

Cheney-Lippold, John. "A New Algorithmic Identity: Soft Biopolitics and the Modulation of Control." *Theory, Culture, & Society* 28, no. 6 (2011), 164–81.

Clough, Patricia Ticineto. "Technoscience, Global Politics, and Cultural Criticism." *Social Text 80* 22, no. 3 (2004), 1–23.

Cocks, H. G. *Classified: The Secret History of the Personal Column*. London: Random House Books, 2009.

Cohn, D'Vera. *Marriage Rate Declines and Marriage Age Rises*. Washington, DC: Pew Research Center, 2011.

Coontz, Stephanie. *Marriage, a History: From Obedience to Intimacy, or How Love Conquered Marriage*. New York: Viking, 2005.

Daly, Willie. *The Last Matchmaker: The Heartwarming True Story of the Man Who Brought Love to Ireland*. London: Hachette, 2010.

Del Sesto, Christine. "Looking for Love at First Byte." The *Washington Post*. June 4, 1987. www.washingtonpost.com/archive/lifestyle/1987/06/04/looking-for-love-at-first-byte/98a8c6d9-96e8-4376-9847-2b685c6739b7/.

Deleuze, Gilles. "Postscript on the Societies of Control." *October* 59 (Winter, 1992), 3–7.

Deziel, Shanda. "Love Inc." *MacLean's* February 14, 2005, 28–34.

Dougherty, Carroll Jocelyn Kiley, and Rachel Weisel. *Behind Record Support for Same-Sex Marriage, Differences Endure*. Washington, DC: Pew Research Center, 2015. www.people-press.org /files/2015/06/6-8-15-Same-sex-marriage-release1.pdf.

Dreyfus, Hubert L. *What Computers Still Can't Do: A Critique of Artificial Reason*. Cambridge, MA: MIT Press, 1992.

Duhigg, Charles. "How Companies Learn Your Secrets." The *New York Times Magazine*. February 16, 2012. www.nytimes.com/2012/02/19/magazine/shopping-habits.html.

eHarmony, Inc. "29 Dimensions of Compatibility." Accessed on July 4, 2015, www.eharmony.com/why/dating-relationship-compatibility/.

———. Home page. Accessed June 11, 2015. www. eharmony.com/.

———. "Who We Are." Accessed June 11, 2015. www. eharmony.com /about/eharmony/.

Ellison, Nichole B., Rebecca D. Heino, and Jennifer Gibbs. "Managing Impressions Online: Self-Presentation Processes in the Online Dating Environment." *Journal of Computer-Mediated Communication* 11 (2006), 415–41.

Foucault, Michel. *Discipline and Punish: The Birth of the Prison*. Translated by Alan Sheridan. New York: Vintage Books, 1977.

———. *The History of Sexuality, Vol. 1: An Introduction*. Translated by Robert Hurley. New York: Pantheon Books, 1977.

———. *Madness and Civilization: A History of Insanity in the Age of Reason*. Translated by Richard Howard. New York: Vintage Books, 1965.

———. *Security, Territory, Population: Lectures at the College de France, 1977–78*. Edited by Michel Senellart. Translated by Graham Burchell. New York: Palgrave Macmillan, 2007.

Galloway, Alexander R. "Protocol." *Theory, Culture & Society* 23, nos. 2–3 (2006), 317–20.

———. *Protocol: How Control Exists after Decentralization*. Cambridge, MA: MIT Press, 2004.

Galloway, Alexander R., and Eugene Thacker. *The Exploit: A Theory of Networks*. Minneapolis: University of Minnesota Press, 2007.

Geraci, Ron. "A Headhunter for My Heart," *Men's Health*, June 2004: 166–71.

Gibbs, Jennifer, Nichole B. Ellison, and Rebecca D. Heino. "Self-Presentation in Online Personals: The Role of Anticipated Future Interaction, Self-Disclosure, and Perceived Success in Internet Dating." *Communication Research* 33 (2006), 152–77.

Gillies, Trent. "How to Surf the Web for a Mate: eHarmony Founder." CNBC. Last modified May 9, 2015. www.cnbc.com/id/102662981.

Gofen, Charlie. "Lonely Hearts: Dating Service Opens Its Doors in Inverness," *St. Petersburg Times*, April 25, 1988.

Gould, Emily. "True Match." *Technology Review*, January/February 2010, 73–75.

Graham, Stephen D. N. "Software-Sorted Geographies." *Progress in Human Geography* 29 (2005), 562–80.

Graham, Stephen D. N., and David Wood. "Digitizing Surveillance: Categorization, Space, and Inequality." *Critical Social Policy* 23 (2005), 227–48.

Hamilton, Brady E., Joyce A. Martin, and Stephanie J. Ventura. *Births: Preliminary Data for 2010*. Atlanta, GA: Centers for Disease Control and Prevention, 2011.

Hardt, Michael. "The Withering of Civil Society." *Social Text* 45 (Winter, 1995), 27–44.

Hayles, N. Katherine. *My Mother Was a Computer: Digital Subjects and Literary Texts*. Chicago: University of Chicago Press, 2005.

Heino, Rebecca D., Nichole B. Ellison, and Jennifer Gibbs. "Are We a Match? Choosing Partners in the Online Dating Market." Paper presented at the International Communication Association Annual Conference, New York City, New York, May 2005.

High Fidelity. Directed by Stephen Frears. 2000. Burbank, CA: Touchstone Home Entertainment, 2000. DVD.

Hinsberg, Pat. "Romancing the Singles Scene: Video Dating Services Push Marriage in a Bid to Survive." *Adweek*, March 11, 1991.

Hitsch, Gunter J., Ali Hortaçsu, and Dan Ariely. "Matching and Sorting in Online Dating." *American Economic Review* 100, no. 1 (2010), 130–63.

Howe, Amy. *SCOTUSblog* (blog). April 27, 2015. www.scotusblog.com/author/amy-howe/.

"IAC's Match.Com Buys Rival OkCupid for $50m." *Bloomberg Business*. February 2, 2011. www.businessweek.com/ap/financialnews/D9L4QP681.htm.

Introna, Lucas D. and David Wood. "Picturing Algorithmic Surveillance: The Politics of Facial Recognition Software." *Surveillance and Society* 2, no. 2–3 (2004), 177–98.

Johnston, Casey. "Google Already Knows You're a 24-Year-Old Woman Who Loves Wombats." *ArsTechnica*. January 26, 2012. arstechnica.com/gadgets/2012/01/google-already-knows-youre-a-24-year-old-woman-who-loves-wombats/.

Kelly, Sean Dorrance, and Hubert L. Dreyfus. "Watson Still Can't Think." *Opinionator* (blog), *New York Times*. February 28, 2011. opinionator.blogs.nytimes.com /2011/02/28/watson-still-cant-think/.

Kerr, Dara. "Senator Asks FTC to Investigate Facebook's Mood Study." *CNET.com*. July 9, 2014. www.cnet.com/news/senator-asks-ftc-to-investigate-facebooks-mood-study/.

King, Matt. "Procedural Rhetorics /Rhetoric's Procedures: Rhetorical Peaks and What It Means to Win the Game." *Currents in Electronic Literacy* (2010). currents.dwrl.utexas.edu /2010/ king_procedural_rhetorics_rhetorics_procedures.

Krucoff, Carol. " . . . And a Matchmaker." The *Washington Post*, December 11, 1979.

la Corte, Rachel. "Washington State Approves Bill to Legalize Gay Marriage." *Boston Globe*. February 2, 2012. www.bostonglobe.com/news/nation/2012/02 /02 /wash-state-approves-bill-legalize-gay-marriage/aW48oiNEpEKRD15wqYjr8N/story.html.

La Ganga, Maria L. "Same-Sex Couples Marry in Oregon as Judge Strikes Down State Ban." *Los Angeles Times*, May 19, 2014. www.latimes.com/nation/la-na-oregon-gay-marriage-20140520-story.html.

Lakoff, George. *Women, Fire, and Dangerous Things: What Categories Reveal about the Mind*. Chicago: University of Chicago Press (1987).

Lessig, Lawrence. *Code: And Other Laws of Cyberspace*, Version 2.0. New York: Basic Books, 2006.

Lev-On, Azi, and Bernard Manin. "Happy Accidents: Deliberation and Online Exposure to Opposing Views." In *Online Deliberation: Design, Research, and Practice*, ed. Tom Davies and Seeta Pena Gangadharan. Stanford: CSLI Publications, 2009, 105–22.

Levey, Bob. "$15,000 Buys Computer Cold Shoulder." The *Washington Post*. April 14, 1988.

Levy, Steven. "Inside the Box." *Wired* magazine. February 22, 2010. www.wired.com /2010/ 02/ff_google_algorithm/.

Lyon, David. "Technology vs. 'Terrorism': Circuits of City Surveillance since September 11th." *International Journal of Urban and Regional Planning Research* 27, no. 3 (2003), 666–78.

Madrigal, Alexis. "Take the Data out of Dating." *Atlantic Monthly*. December 2010, 37.

Match.com, LLC. "New Profile Design: Profile Display." Accessed on March 21, 2011 (page discontinued).

———. "Try Something New :30--Match.com Official Commercial." YouTube video, :32. Posted October 1, 2010. www.youtube.com/watch?v=CV5GLZTYDUs&sns=em.

"Match Inc. CEO Sam Yagan Reveals How Math and Data Predict Love Online." Interview by Alyson Shontell. *BusinessInsder.com*. May 17, 2013. www.businessinsider.com /how-online-dating-sites-work-2013-5.

MacCarthy, Bart. "Weapon of Choice." *IEEE News*. April 2003, 34.

McCormick, Douglas. "Virtual Teen Passes Turing Test," *IEEE Spectrum*. June 10, 2014. spectrum.ieee.org/tech-talk/robotics/artificial-intelligence/virtual-tween-passes-turing-test.

McManus, Kevin. "Strike a Match." The *Washington Post*. November 17, 1989. www.washingtonpost.com/archive/lifestyle/1989/11/17/strike-a-match/fe6d37b2-643e-4773-a2f7-f2adbd1850ef/.

Meyer, Robinson. "Everything We Know about Facebook's Secret Mood Manipulation Experiment," *The Atlantic*, June 28, 2014. Last modified September 8, 2014. www.theatlantic.com/technology/archive/2014/06/everything-we-know-about-facebooks-secret-mood-manipulation-experiment/373648/.

Miller, Carolyn R. "The Presumptions of Expertise: The Role of Ethos in Risk Analysis." *Configurations* 11, no. 2 (2003), 163–202.

National Marriage Project. "About." Accessed March 28, 2014. nationalmarriageproject.org/about/.

Netflix. "Overview." Accessed on November 21, 2011. ir.netflix.com/.

Netflix Prize. "Congratulations!" Accessed on November 22, 2011. www.netflixprize.com/index.

Netflix Prize. "The Netflix Prize Rules." Accessed on November 22, 2011. www.netflixprize.com/rules.

Nguyen, Hanh. "'Millionaire Matchmaker' Offers Tough Love." Zap2It.com. January 22, 2008. www.zap2it.com/blogs/zap-story-pattistanger-millionairematchmaker.

Norman, Don. *Adaptive Path*. Interview by Peter Merholz. December 13, 2007. www. adaptivepath.com/ideas/e000862/.

Oakes, Kelly. "No, A Computer Did Not Just Pass the Turing Test." BuzzFeed. June 9, 2014. www.buzzfeed.com/kellyoakes/no-a-computer-did-not-just-pass-the-turing-test#.euvEgBaxD.

Obergefell v. Hodges (oral arguments). April 28, 2015. www.supremecourt.gov/oral_arguments/argument_transcripts/14-556q1_6k47.pdf.

OkCupid. "Any Questions?" Accessed on June 29, 2015. www.okcupid.com/support.

———. "About Us." Accessed on May 6, 2010. www.okcupid.com/about (page discontinued).

———."Match Percentage." Accessed on March 12, 2011. www.okcupid.com/support.

Office of Governor C. L. "Butch" Otter. "Governor Otter Seeks Emergency Stay on Same-Sex Marriage Ruling." News release. May 14, 2014. gov.idaho.gov/mediacenter /press/pr2014/5.%20May/pr_030.html.

Packer, Jeremy. "Becoming Bombs: Mobilizing Mobility in the War on Terror." *Cultural Studies* 20, no. 4–5 (2006), 378–99.

———. "Driving Mobility: Governing and Safety." In *Foucault, Cultural Studies, and Governmentality*. Edited by Jack Z. Bratich, Jeremy Packer, and Cameron McCarthy. Albany, NY: State University of New York Press, 2003, 135–61.

———. "What Is an Archive? An Apparatus Model for Communications and Media History." *The Communication Review* 13, no. 1 (2010), 88–104.

Packer, Jeremy, and Kathleen F. Oswald. "From Windscreen to Widescreen: Screening Technologies and Mobile Communication." *Communication Review* 13, no. 4 (2010), 309–39.

Packer, Jeremy, and Stephen B. Crofts Wiley. *Communication Matters: Materialist Approaches to Media, Mobility, and Networks*. Hoboken: Taylor and Francis, 2011.

Pariser, Eli. *The Filter Bubble: What the Internet Is Hiding from You*. New York: Penguin Press, 2011.

Pauley, Perry M., and Tara M. Emmers-Sommer. "The Impact of Internet Technologies on Primary and Secondary Romantic Relationship Development." *Communication Studies* 58, no. 4 (2007), 411–27.

Paumgarten, Nick. "Looking for Someone: Sex, Love, and Loneliness on the Internet" (includes interviews with Chris Coyne and Amarnath Thombre). *The New Yorker*. July 4, 2011. www.newyorker.com/magazine/2011/07/04/looking-for-someone.

Pew Research Center. "Growing Public Support for Same-Sex Marriage." Last modified on February 16, 2012. www.people-press.org/2012/02/07/growing-public-support-for-same-sex-marriage/.

Poster, Mark. *Information Please: Culture and Politics in the Age of Digital Machines*. Durham, NC: Duke University Press, 2006.

Rapp, David. "The Love Machine." *American Heritage*. November/December 2005, 13–4.

Rettberg, Jill. *Seeing Ourselves through Technology: How We Use Selfies, Blogs, and Wearable Devices to See and Shape Ourselves*. London: Palgrave Macmillan, 2014. Kindle edition.

Rudder, Christian. *Dataclysm: Who We Are (When We Think No One's Looking)*. New York: Crown Publishing Group, 2014.

———. "We Experiment on Human Beings!" *OKTrends* (blog). July 28, 2014. blog.okcupid.com/index.php/we-experiment-on-human-beings/.

Schaefer, Laura. "Looking for Love Online, or on Paper." The *New York Times*. February 14, 2003. www.nytimes.com/2003/02/14/opinion/looking-for-love-online-or-on-paper.html.

Scinto, Madeline. "This Young Woman Scored $1,200 a Month in Fancy Dinners Using Match.Com." *Business Insider*. November 29, 2011. www.businessinsider.com/confessions-how-she-made-1200-a-month-using-matchcom-2011-11.

SCOTUSblog. Accessed on June 23, 2015. www.scotusblog.com/case-files/cases/obergefell-v-hodges/.

Shepherd, Dawn. "Match.com, Genre, and the Conception of Self." Paper presented at the Biennial Conference of the Rhetoric Society of America, Seattle, Washington, May 2008.

Sidelinger, Robert J. "Couples Go Online: Relational Maintenance Behaviors and Relational Characteristics Use in Dating Relationships." *Human Communication* 11, no. 3 (2006), 341–56.

Singles. Directed by Cameron Crowe. 1992. Burbank, CA: Warner Bros. Entertainment, 1999. DVD.

Smith, Aaron, and Maeve Duggan. *Online Dating and Relationships*. Washington, DC: Pew Research Center's Internet and American Life Project, 2013.

Spano, Susan. "Her World: Rural Ireland Is in the Green with Bachelors at Matchmaking Festival." The *Los Angeles Times*. June 17, 2001.

Storment, Molly. "Using Games to Teach Procedural Rhetoric." *HASTAC: Humanities, Arts, Science, and Technology Alliance and Collaboratory*. October 22, 2011. www.hastac.org/blogs/mhstorment/2011/10/22/using-games-teach-procedural-rhetoric.

Sullivan, Gail. "Steve Wozniak Refutes Apple's Creation Story: 'The Garage Is a Bit of a Myth.'" The *Washington Post*. December 5, 2014. www.washingtonpost.com/news/morning-mix/wp/2014/12/05/steve-wozniak-refutes-apples-creation-story-the-garage-is-a-bit-of-a-myth/.

Terranova, Tiziana. "Free Labor: Producing Culture for the Digital Economy." *Social Text* 18, no. 2 (2000), 33–58.

Toma, Catalina L., Jeffrey T. Hancock, and Nichole B. Ellison. "Separating Fact from Fiction: An Examination of Self-Presentation in Online Dating Profiles." *Personality Social Psychology Bulletin* 38, no. 8 (2008), 1023–36.

Turkle, Sherry. *Life on the Screen: Identity in the Age of the Internet*. New York: Touchstone, 1995.

Vee, Annette. "Understanding Computer Programming as Literacy." *Literacy in Composition Studies* 1, no. 2 (2013). 42–64.

Warren, Neil Clark. *Falling in Love for All the Right Reasons: How to Find Your Soul Mate*. With the assistance of Ken Abraham. New York: Time Warner Book Group, 2005.

Wildermuth, Susan M., and Sally Vogl-Bauer. "We Met on the Net: Exploring the Perceptions of Online Romantic Relationship Participation." *Southern Communication Journal* 72, no. 3 (2007), 211–27.

Williams, Raymond. *Television: Technology and Cultural Form*. New York: Schocken Books, 1975.

Winn, Jasper. "Mix 'N' Match," *Geographical*. February 2003, 32–36.

Wolverton, Troy. "Amazon Snags Patent for Recommendation Service," CNet.com. May 31, 2000, news.cnet.com/2100-1017-241267.html.

Yagan, Sam. OnlinePersonalsWatch.com. Interview by Mark Brooks. September 21, 2009. www.onlinepersonalswatch.com/news/2009/09/sam-yagan-ceo-of-okcupid.html.

Yurchisin, Jennifer, Kittichai Watchravesringkan, and Deborah Brown McCabe. "An Exploration of Identity Re-Creation in the Context of Internet Dating." *Social Behavior and Personality* 33, no. 8 (2005), 735–50.

Index

About the Author

Dawn Shepherd is an assistant professor of English and associate director of the First-Year Writing Program at Boise State University. Her work has been published in edited collections as well as *The Norton Book of Composition Studies* and *WPA: Writing Program Administration*, and two pieces on genre and weblog, coauthored with Carolyn R. Miller, have been translated into Portuguese. Her research on romantic matchmaking and algorithmic culture has been featured in local and international media, including BBC World and the *London Times*. She teaches undergraduate courses in rhetoric and writing with a new media emphasis and graduate seminars on teaching writing, research methods, and digital rhetoric. She also regularly facilitates workshops on integrating mobile strategies into the classroom and teaching with technology.

ARCHITECTURE, ETHICS, AND TECHNOLOGY

Edited by
LOUISE PELLETIER and
ALBERTO PÉREZ-GÓMEZ

ARCHITECTURE, ÉTHIQUE ET TECHNOLOGIE

Textes réunis par
LOUISE PELLETIER et
ALBERTO PÉREZ-GÓMEZ

Institut de recherche en histoire de l'architecture
Canadian Centre for Architecture • Centre canadien d'architecture
McGill University
Université de Montréal

McGill-Queen's University Press
Montreal & Kingston • London • Buffalo

ISBN 0-7735-1148-2

Legal deposit first quarter 1994
Bibliothèque nationale du Québec

Printed in Canada on acid-free paper.

Données de catalogage avant publication (Canada)

Vedette principale au titre:
Architecture, ethics and technology – Architecture, éthique et technologie
Texte en anglais et en français.
Textes d'un symposium organisé par l'Institut de recherche en histoire de l'architecture et présenté au Centre canadien d'architecture, les 15–16 novembre 1991.
ISBN 0-7735-1148-2
1. Architecture – Innovations – Congrès. 2. Architecture – Aspect moral – Congrès. 3. Architecture – Art d'écrire – Congrès. I. Pelletier, Louise, 1963– . II. Pérez Gómez, Alberto, 1949– . III. Institut de recherche en histoire de l'architecture. IV. Titre: Architecture, éthique et technologie.
NA21.A73 1994 720'.1'05
C93-090630-6F

Canadian Cataloguing in Publication Data

Main entry under title:
Architecture, ethics and technology – Architecture, éthique et technologie
Text in English and French.
Texts from a symposium organized by the Institut de recherche en histoire de l'architecture and held in the Canadian Centre for Architecture, Novembre 15–16, 1991.
ISBN 0-7735-1148-2
1. Architecture – Technological innovations – Congresses. 2. Architecture – Moral and ethical aspects – Congresses. 3. Architectural writing – Congresses. I. Pelletier, Louise, 1963– . II. Pérez Gómez, Alberto, 1949– . III. Institut de recherche en histoire de l'architecture. IV. Title: Architecture, éthique et technologie.
NA21.A73 1994 720'.1'05
C93-090630-6E

This book was typeset by Typo Litho composition Inc. in Garamond and Futura.

CONTENTS / TABLE DES MATIÈRES

ACKNOWLEDGMENTS

In 1990, the Canadian Centre for Architecture, the Université de Montréal, and McGill University signed a protocol creating the *Institut de recherche en histoire de l'architecture.* The main purpose of this body is to promote interdisciplinary research in the field of architectural history, addressing issues of concern to architects from the point of view of both the creation of architecture and its relevance to the contemporary world. The symposium on "Architecture, Ethics, and Technology," which took place in Montreal at the Canadian Centre for Architecture on November 15th and 16th, 1991, was the first public event organized and funded by IRHA. The essays included in this volume were initially delivered as papers during that event.

A symposium such as this would not have taken place without the generous participation of the important architects and academics who unselfishly accepted our invitation and contributed their time. On behalf of IRHA, I would also like to thank many other people who contributed both to the symposium and to this book. Deserving of special mention is Louise Pelletier, executive coordinator, who worked tirelessly to see the project through from its inception to its culmination in the present publication. While I must accept responsibility for the initial idea, the faculty members of IRHA played a major role in clarifying the symposium's objectives and in determining its final structure. Members of the faculty also kindly volunteered their time to serve on the editorial board that made this publication possible. Invaluable throughout this process was the assistance of staff members of the Canadian Centre for Architecture, particularly Nicholas Olsberg.

Professor Steve Parcell of the Technical University of Nova Scotia read and edited the introduction. At McGill University, Susie Spurdens, secretary to the Saidye Rosner Bronfman Chair in the History of Architecture, provided devoted assistance. Indra McEwen provided the English translation of the abstracts for the papers written by Alain Findeli, Jean-Pierre Hardenne, Irena Latek, Robert Prost, and Jacques Rousseau; Alberto Pérez-Gómez was responsible for the abstracts of the contributions prepared by Philipe Madec and Lukas Sosoe. The French translations of the introduction and of the abstracts of English papers were prepared by Marie-Andrée Lamontagne. Michel Forand edited both French and English papers and summaries.

Dr Alberto Pérez-Gómez, Director, IRHA

REMERCIEMENTS

En 1990, le Centre canadien d'architecture, l'Université de Montréal et McGill University ont signé un protocole d'entente qui prévoyait la création de l'Institut de recherche en histoire de l'architecture. La principale tâche de cet organisme est de promouvoir la recherche interdisciplinaire dans le domaine de l'histoire de l'architecture et de débattre de questions qui intéressent les architectes, tant du point de vue de la création architecturale que de sa pertinence dans le monde contemporain. La première manifestation publique organisée et réalisée par l'Institut fut un symposium consacré aux rapports entre l'architecture, l'éthique et la technologie, qui s'est déroulé à Montréal au Centre canadien d'architecture les 15 et 16 novembre 1991. Les essais réunis dans le présent volume firent d'abord l'objet de communications présentées lors de ce symposium.

Un symposium de cette envergure n'aurait pu avoir lieu sans la participation généreuse d'architectes et de chercheurs réputés qui ont accepté notre invitation et donné généreusement de leur temps. Au nom de l'Institut, je voudrais également remercier les nombreuses personnes qui ont à la fois collaboré au symposium et à ce recueil. Je tiens notamment à souligner le travail de la coordonnatrice générale, Louise Pelletier, qui a travaillé sans relâche pour mener le projet à terme, depuis sa conception jusqu'à son aboutissement dans cette publication. Bien qu'il faille me créditer l'idée de départ, on se saurait négliger l'apport important des membres du conseil de l'Institut, qui ont contribué à expliciter les objectifs du symposium et à lui donner une forme définitive. De plus, ils ont bien voulu siéger bénévolement au comité éditorial qui a rendu possible ce recueil d'essais. L'aide du personnel du Centre canadien d'architecture tout au long de ce processus a été inestimable, en particulier celle de Nicholas Olsberg. Je remercie également le professeur Steve Parcell, de la Technical University of Nova Scotia, qui a bien voulu lire et corriger l'introduction. L'aide de Susie Spurdens, secrétaire de la chaire Saidye Rosner Bronfman en histoire de l'architecture, a également été précieuse. Indra McEwen a traduit en anglais les résumés des textes rédigés par Alain Findeli, Jean-Pierre Hardenne, Irena Latek, Robert Prost et Jacques Rousseau, tandis que les résumés des textes de Philippe Madec et Lukas Sosoe ont été préparés par moi-même. Marie-Andrée Lamontagne a traduit en français l'introduction et les résumés des autres textes rédigés en anglais. La révision de tous les essais et résumés français et anglais a été réalisée par Michel Forand.

Alberto Pérez-Gómez, directeur de l'IRHA

ARCHITECTURE, ETHICS, AND TECHNOLOGY

ARCHITECTURE, ÉTHIQUE ET TECHNOLOGIE

INTRODUCTION

ALBERTO PÉREZ-GÓMEZ

Why is it that, in the last decade of our millennium, the design professions have suddenly decided to focus on the question of ethics and to devote numerous conferences to this topic? To state that the common good must be a primary concern in architecture may sound hollow in the context of a contemporary practice determined by economic or political forces. It may even be deemed downright problematic by some of the poststructuralist philosophers who claim that the old humanist values must be deconstructed – a position that has had a significant impact on North American architecture.

Yet, I would argue that the common good has always been a primary concern in architecture. This was evident in the writings of Vitruvius and others over the centuries who have attempted to elucidate the meaning of architectural praxis. If this ethical concern is difficult to recognize, it is because it is internal to the practice of architecture rather than an external adjunct to some formalistic or technical activity. Because of their irrepressible desire to disclose a symbolic order, architects have traditionally sought to provide individual existence with a built world that reflects the purpose of social institutions and of life in general. Human thoughts and actions, ranging from myth and religion to science and art, have been responses in different times and different places to fundamental questions that have remained essentially unchanged. Far from being merely an aesthetic or technical concern, architecture seeks to set human action within an "appropriate" frame, fulfilling this reconciliatory role despite man's tendency to control and dominate fellow human beings and the environment.

Today, we may finally realize that the reduction of the fine arts to a morally inconsequential aesthetic formalism is not an absolute paradigm but rather a historical event related to the glorification of scientific reason during the eighteenth century. Believing that positive science is capable of disclosing absolute truth – a belief whose roots were indeed theological – the rationalists implicitly relegated art (and "non-scientific" architecture) to a marginal, illegitimate zone. Architectural practice then began to be defined by two assumptions: architecture is differentiated from other building practices by its interest in formal manipulations; and architecture must adequately resolve the internal conditions generated by a "program" that is external to the architect's discipline. Imbued with the same scientific logic, ethics was assumed to be autonomous, external to those formal concerns of the "ornamental builders"; it was deemed to be practically subsumed by the paradigmatic technological values of efficiency and economy. Is there anything else for a builder to be concerned with than to design the most efficient building possible and to spend as little as possible in achieving that goal?

The polarization of ethics and aesthetics has been exacerbated by the confusing nihilism of our post-Nietzschean culture, the Babel of specialized disciplines and interests, and the overwhelming power of technology and simulation. Often, ethics is still regarded as external to architecture (and to the other design disciplines) – as ultimately subjective and therefore inconsequential in our pluralist world. And yet, can it not be argued that the cultural relevance of architecture is ultimately related to its capacity to embody ethical issues?

If architects are to play a role in the complex world of the twenty-first century – a world more conscious of environmental limits and cultural differences while civilization continues to embrace the goal of technological globalization – they must ponder strategies to disclose their discipline's potential for embodying an ethical intentionality. The inherent ambivalence and opacity of architecture as a poetic universe of discourse that speaks directly to humanity's perceptual and imaginative (rather than intellectual) understanding, must be embraced as an asset rather than lamented as a limitation. While affirming the necessary conditions of modern/postmodern architecture as a discipline that is both functional and conventional, grounded in use-value but also able to acknowledge the crucial role of syntax in the potential generation of poetic metaphor, is it possible to see beyond mere formal (stylistic) relativism and pragmatic professionalism? Can we contemplate architecture as a non-discursive embodiment of knowledge born from a concern with the *other*?

Well-meaning cultural critics, seeking to deconstruct oppressive power structures from our tradition and to empower the free individual, have at the same time proclaimed the self-referentiality of art – the postmodern labyrinth of mirrors and relative values – implying that the personal imagination, and with it the very possibility of ethical action, may be an illusion. Declaring the fallacy of an absolute, ontological ground, however, need not lead to the absolute self-referentiality of language and art. Critics with diverse pursuits (Paul Ricœur, Jean-François Lyotard, Richard Kearney, and Julia Kristeva, among others) have emphasized, each in his or her own way, an alternative to the traditional mimetic imagination – the Aristotelian concept presupposing an intersubjective cosmos that stands for all and need only be "imitated" – and to the Romantic egocentric imagination presupposing the possibility of creation *ex nihilo.* While acknowledging the inevitability of "intertextuality" and convention (taking seriously Rimbaud's discovery of the ultimate autonomy of language), these critics posit the possibility of an embodied imagination that still speaks about something other than itself, about a possibly different future founded on values gleaned from the traces of history. The potential integration of ethic and aesthetic concerns has profound consequences for architectural practice. Indeed, it confronts the issue of form generation at its inception.

Furthermore, to consider the generation of architecture with all its implications, fully acknowledging the public dimension of its meaning, a careful understanding of the dominant technological character of our world culture is paramount. It is within that world that the architect must work. Technology is far more than a question of machines; it is more than just one of many determinants of our culture. It has become clear that there is an intimate link between the nature of technology and a number of cultural traits that define both modernity and postmodernity. Technology, especially after the Industrial Revolution, opened up the possibility of a human world that is fully constructed. However, a technological world of artifacts would act more as a wall between man and nature than as a bridge between our consciousness and an external reality that we have not created. There is clearly a chasm between modern/postmodern technology and the traditional techniques of pre-industrial cultures. Technology substitutes a "picture" for the world of our primary experience. However mutable and historically determined this world of primary experience may be, phenomenology discloses it as one where the universal and the specific are given *simultaneously* in the mystery of perception, in the space *between* Being and Becoming. The technological substitution of a picture creates the condition that Martin Heidegger has called "enframing" (*Ge-stell*). While seeking to answer the same inveterate questions about the measure of man and his place amidst the immeasurable and alien, technology opts for control and domination, for arrogant and efficient action upon the reduced picture. Success within this instrumental mode of action has led to an affirmation of man's quasi-divine ambitions in the face of a universe that, he strongly suspects, may not have been made for him but may nevertheless be exploited for all its resources.

This all-encompassing technology has been intimately bound up with architectural ideation and practice since the early nineteenth century. Despite considerable resistance, architectural theory was quick to adopt the productive models of applied science. Architecture, the art of mediation *par excellence*, has tended to become a tool of domination. After many failures and dead ends, it is not surprising that today we seek a reformulation of technology. This new understanding would have an immediate impact on architectural representation, the production of architecture, the role of scientific models, and the use of language to articulate design intentions.

These aspects of architectural ideation are major concerns for theoreticians of architecture as well as for its practitioners. The cultural and political relevance of the architectural profession in the late twentieth century hinges upon these questions. Issues arising from the relationship between architecture and the technological world are the crucial theoretical problems for an ethical practice of that activity. The implications of that relationship are far more important than the superficial stylistic distinctions that consume such a great proportion of the contemporary architectural debate.

Technology cannot be simply dismissed or criticized from the point of view of traditional art, metaphysics, or humanism. It would be difficult to deny that the thrust behind its accomplishments and failures is indeed the human thirst for transcendence, especially after the epistemological turning point signalled by the works of Copernicus and Galileo. As it became clear that neither God nor man was at the centre of the world, technology became a morally justifiable response to the inveterate human *lack* in the face of the cosmos. As shown by Dalibor Vesely's incisive discussion of the historical origins of technology, its quest is related to the age-old intention of European man to inhabit a space of ontological continuity, to dwell poetically upon the earth. Was this quest abandoned in favour of a "false" alternative that would appear to be true only in the absence of God? Obsessed by the accomplishments of instrumentality and its own assertive will to power, technology has led postmodern humanity into a gnostic trap in which the pre-existing "world out there" is forgotten. Today, we are often controlled by forces that we thought we were capable of mastering. The environmental crisis and the dilemmas created by genetic engineering are two prime examples.

However, as Heidegger pointed out in his essay on technology, there are dangers present in the technological world that are more subtle but yet more serious than humanity's potential for self-destruction and the threat of ecological disaster. The denial of death as a positive limit and essential qualifier of life may indeed end up jeopardizing the possibility that human existence can remain open to meaning and thus perpetuate civilization through its cultural institutions and symbols. A nihilism of despair may become a powerfully destructive force, impossible to control and overriding the necessary affirmative nihilism that we must embrace with Nietzsche. While questioning the absolute truth of western metaphysics as something outside experience (the Platonic "truth as correspondence"), this form of affirmative nihilism still enables humanity to remain open to the powerful silence of art and architecture, to the dark radiance of light, and it maintains its sensitivity to the vibrations of the quiet footsteps of the gods that may come dancing by. The problem is, finally, to diagnose the power of the technological enframing, to understand that while we may be "enframed," technology itself may be "weak," that while God and man may be irretrievable, the planet Earth (and the sky) are present as a central ground for our full embodied experience, one through which we are now capable of questioning the hegemony of abstract, theoretical constructs.... Paradoxically, this realization becomes stronger as we can now observe Earth as a unique oasis of life in the desert of outer space....

In architecture, the dangers of technology are directly associated with reductive representation, cultural homogenization, and the death of the city as a place for public interaction. The architect must become aware of these threats and accept responsibility. A first step towards an enlightened discussion would be to understand that the premise of the postmodern polemic against the modern movement, with its nostalgic escape from the discourse of technology, is overly simplistic. The way out may then be sought by acknowledging the mysterious origins and historical transformations of technology itself. We may be called to question our often arrogant, instrumental relationship towards the external world and to explore other options, including a strategy to destructure the languages of science and technology through a poetic vision. Eventually, a different relationship may emerge between practice and the cultural artifacts that make up our architectural tradition.

The philosophy of postmodernity, as exemplified by the writings of Gianni Vattimo and his interpretation of Nietzsche and Heidegger, seems intent on erasing the inherited difference between technological and aesthetic cultures, and with it, the difference between rationalism and irrationalism that has sustained modernity since at least the early nineteenth century. The very idea of historical progress, intimately bound up with the technological assimilation of architecture and the unquestionable value of the new, is no longer operative. In this context, an enlightened discussion would question commonly held assumptions about the nature of architectural history, theory, representation, and ideation, as well as professional ethics and the production of buildings in the postindustrial city.

THAT WAS THE CHALLENGE offered to the distinguished professionals and academics who contributed their expertise to the symposium and to this collection of essays. The complexity of the issues suggested many approaches from a variety of disciplines. As might be expected, the results and positions are diverse, and there are probably more questions raised than answers given. But because of that complexity and because of the novelty of the problems in the postmodern discourse about architecture, diversity is the strength of the compendium. The essays are provocative invitations to focus on crucial ethical concerns that have largely been absent from the practice of architecture over the past two centuries. Their authors are aware that our concepts of architecture are in crisis, and they seek to enhance its relevance in the late twentieth century.

The book is divided into four parts. The introductory section includes three major presentations – a keynote address by Dalibor Vesely that focuses on the epistemological origins of technology in the early modern European context, viewed within the framework of continental hermeneutic philosophy; and two alternative visions of ethics and its potential relevance for architecture, offered by ethicists Lukas Sosoe and Margaret Somerville. In the tradition of analytic philosophy, Lukas Sosoe cautions against a potential incompatibility between the categories of ethics and the creation of architecture. He suggests that in architectural practice, following common sense and a professional code may be all that is needed. In sharp contrast, Margaret Somerville advocates applied ethics, based on her experience in medicine and biotechnology. In order to avoid disastrous consequences, ethics must be brought to bear, by specialists making decisions within their disciplines, on seemingly straightforward technical or rational issues. Architecture, a highly technical and public discipline, must also take account of ethical concerns at its inception: the world of images and forms is not value-free; and the architect, in control not only of systems and forms but also, sometimes, of means of production and program, is partly responsible for the psychic and somatic health of human culture.

The three main sections of the book broadly follow the structure of the symposium. The topics of the sessions were related mainly to the interests of the practising architects who presented their work and provided a tone for each session.

Part I, "Ethics and Architectural Representation in the Age of Simulation," asks important questions about how we represent buildings and about the ethical values involved in representation. In the context of instrumental theory, it has long been assumed that plans, sections, and elevations are neutral projections – that they are value-free carriers of information necessary for the construction of a building. One could argue that this assumption has been questioned ever since Piranesi produced his *Carceri* etchings. More recently, however, this questioning has become more intense in works by architects such as Aldo Rossi, Daniel Libeskind, and John Hejduk.

Recall that Gothic cathedrals were never fully planned before construction and were rarely conceived as whole images. Long after the foundations had been laid, patrons and architects often debated the geometry that would allow their elevations to be raised. In the Renaissance, Filarete insisted that the architect was like a mother who had to carry the project like a baby during a nine-month period of gestation. After birth, the projected building had to be nourished into maturity through a series of transmutations from the initial mathematical idea to the line drawing, to the model, and finally to the construction. These transmutations were never automatic: the mathematical essence or symbolic order of the project was expected to prevail under radically different appearances, but changes were expected and welcome at every stage. A sensitivity to experienced reality always took precedence over any platonic ideal solution or a priori planning decision.

In a project for the Old Port of Montreal, presented by the architect Peter Rose, a sensitivity to the site and to the importance of historical traces demonstrates a different mode of architectural ideation that deals with a chronologically layered condition. In a world where the homogeneous space of electronic communications reigns supreme, a regard for marginal industrial sites as a legitimate part of our cultural heritage becomes paradigmatic; these ruins of industry are perhaps the only existing places where the mysterious origins of technology may be revealed. Their dark spaces – usually associated with the "edge" rather than the "centre" of cities – may even disclose a sense of immanent purpose for our urban culture. The act of revealing the traces (and the scars) of a site and the unmasking of cosmetic applications by urbanists suggest a heroic architecture of resistance.

Philippe Madec's poetic text may be read as a representation of architecture – one that nevertheless acknowledges a distance requiring a *translation* between representation and built work. His words evoke an architecture beyond formalism and technology. Madec's architecture is posited as an open work capable of accepting (rather than repressing) life's own predisposition to the poetic, declaring quietly a subtle polemic – that it is possible to marry poetic dwelling and the "style" of technology into a truly open reading. While the architect is now able to avoid the excesses of the avant-garde, having acknowledged the fallacies of romantic, egocentric "creation," what are the limits of his/her responsibility? How does he/she exercise imagination? What constitutes a truly open and democratic work of architecture in a world where panopticism and the values of technology are the dominant referents to be found at the surface of all productive work?

The representation of architecture embodies values that are carried through to its execution, although the relationship between what we conceive and what we build is not always unambiguous. That issue is addressed directly by Jean-Pierre Hardenne and Dan Hoffman, each expressing a different position and consequence for architectural practice. Hardenne enthusiastically advocates the use of computers in architectural design. Exploring the potential of binary logic in its more evolved, interactive forms, he emphasizes the importance of employing the most sophisticated imaging tools at our disposal. Hoffman, on the other hand, is implicitly critical of the progressive assumption that underlies Hardenne's position. In his project, Hoffman attempts to deal with technology from within, while taking full account of the importance of embodied making, celebrating the role of the architect as craftsman and builder. Using the process of construction and its "recording" as vehicles to disclose wonder in a potentially different architecture, Hoffman avoids the dangers of a reductive design that would simply propose pictures of recognizable buildings. Recalling the priority of architecture as a verb (rather than a noun), Hoffman constructs a wall that can never be "finished" and thus emphasizes the building process rather than the final product. The project is a recasting of the myth of Sisyphus in a contemporary context, suggesting a profound ethical position as paradigmatic for the architect's *métier* in a nihilist, technological era.

The second part of the collection, "Ethics and Poetics in the Context of Technological Production," focuses more specifically on the relationship between ethical values and architectural production. What is the role of philosophical ethics – i.e., a rational structure of categories – in architectural practice? Even if one could use scientific method and quantitative surveys to establish appropriate ethical values for a certain project (through a democratic consensus, for example), is it possible – or even ethically desirable – to incorporate these reflections into the generation of architectural form, expecting to render our buildings more human and meaningful?

The issue of ethics in architectural production touches upon the relationships between *thinking* and *acting* – not as mechanistic or instrumental functions but as interrelations of a much more complex nature. The aloofness or proximity of the architect to the building process, the distance between the generating design ideas and the constructed work, the scale of the operation, the embodied knowledge of the architect concerning materials and construction – all of these issues, so difficult to name and impossible to quantify, are indeed crucial for the development of an architecture capable of transcending the muteness of reductive technological processes. Recent philosophical speculation, particularly in the area of existential phenomenology, has provided architects with words to describe something of what has traditionally been the experience of true *builders* but has been forgotten or ignored in our technological age. This dimension was deemed subjective and irrelevant after traditional making and architectural apprenticeship were replaced by architecture schools and by purportedly more efficient processes of production. But these were predicated precisely on a mechanistic relationship between ideas and execution, making the architect somewhat analogous to the computer mind of a robot, and the builder, to its mechanical limbs. The modern polemic against this Cartesian mind/body split has been argued admirably by Maurice Merleau-Ponty, among others. While it is inappropriate even to sketch that argument here, it may be suggested that the intuition of a nondualistic understanding of reality, in which fully embodied consciousness is no longer conceived as a brain plus organs but as a unity, needs to be radicalized to acknowledge the mysterious "thinking hands" of the architect/builder – the effective medium of the architect's ethical imagination.

Traditional buildings crafted with care, even when otherwise derivative or ordinary, demonstrate a certain eloquence, as is still evident in the older sections of our cities. In view of this, what are the alternatives that we may contemplate in our technological world without simply being nostalgic? Accepting the intrinsic connection between ethics and poetics, the work of Jacques Rousseau asks us to reflect upon the importance of issues such as scale and the irreducible meaning of a site, as well as the qualities of a work that express simplicity, silence, and doubt, rather than absolute truth – qualities that may reveal the presence of the *other*, while inviting further speech. More than a purely formal aesthetic, this work poses a different relationship between the designer-architect and the builder. At the end of the twentieth century, we can still recognize the difference between the true art and craft of the architect and the inhuman buildings resulting from processes that are estranged from poetic construction.

The two critics whose work is also included in Part II explore diverse possibilities for scientific discourse in the discussion of ethics in architecture. Within an analytical framework, Alain Findeli asks us to begin by reconsidering the assumed autonomy of architecture in favour of a profound unity of aesthetic and ethical issues for an architecture in which practice has been radically modified by technology, while Robert Prost sketches a methodological model that may enable ethics to play a role in architecture as process or "work in progress."

"The Architectural Uses of History and Narrative in a Technocratic World" is the title of Part III. The presentations and discussions in this last section explore alternatives for articulating an ethical attitude in forms of discourse other than philosophy and science. Objectified values may be clear and easy to measure, following their own internal logic. However, there is always a danger that this structural logic may become a source of delusion, given the complex nature of architectural creation and the inherent ambiguity of architecture itself, whose intended symbolic order is read, re-read, and mis-read through time by mortal human beings whose own nature is always incomplete. Are the alternatives proposed to the architect by history and narrative fruitful? Do they suggest ways to deconstruct the negativity of technology while maintaining its desirable accomplishments? How might this affect the city? Distinguished philosophers and writers such as Hans-Georg Gadamer and Octavio Paz have insisted that the great themes of philosophy (and morality) have migrated to literature since the early nineteenth century. In addition, the works of Martin Heidegger and Paul Ricœur have led us to a better understanding of the relationship between action and language, and between language (metaphor, in particular) and imagination. These insights suggest narrative forms of discourse as an effective vehicle for ethical intentionality in architecture.

Cultural role and use traditionally constituted the temporal dimension of space. The appropriateness of a building to this "script" or "liturgy" – the Vitruvian *commoditas* – was culturally given until the end of the eighteenth century, together with the belief that transcendent meaning could be communicated through proportions and the classical orders. While *caractère* remained an important issue until the time of Claude-Nicolas Ledoux and Étienne-Louis Boullée, it was finally dismissed as a futile concern during the early nineteenth century by the first fully modern architect, Jean-Nicolas-Louis Durand, the recognized father of functionalism. Today, the program for an architectural project is usually summarized as a list of rooms and square footages. This, of course, greatly reduces the ambit of architecture and raises a crucial question concerning the ethical values embodied by a building. What is the role of the architect in drafting and implementing a different kind of program – one that may incorporate a plot for poetic dwelling in the context of the institutions of the late twentieth century?

Richard Henriquez addresses this and other related questions through his work, exploring the possibility of using contemporary rituals (a graduation ceremony at Trent University, in his example) to construe potential sites for architecture. Coupling the academic function with an evocation of nature's ecological "wholeness," he reminds young scientists of their own ethical responsibilities beyond the confines of specialized disciplines. Architecture is here intended to function as an event with its appropriate physical framework. Henriquez explores through his work opportunities for re-creating history and designing an architecture for invented rituals. Together with functional requirements, rituals such as these may become crucial for disclosing an enduring sense of order *in time*, one that is needed to preserve our humanity and cultural identity.

This section also includes a critical essay by Lily Chi, in which she discusses the use of narrative in work by architects such as Aldo Rossi and John Hejduk, and extracts valuable insights concerning the poetic embodiment of ethics and the concept of emplotment in architecture. In closing this compendium, Irena Latek discusses some pedagogical implications of an architecture based on narrative through her recent work with students of the Unité d'architecture urbaine at the Université de Montréal.

THERE IS, OF COURSE, NO FINAL WORD. These essays reveal the complexity of the challenge offered to practising architects at the conclusion of the present millennium. As both an expression of personal imagination and a political and social activity, architecture is profoundly affected by the crises and pathologies of our technological world. However, it is wonderfully positioned to take advantage of any possibilities that may emerge from a greater and more universal human compassion, from an increasing awareness of the limits that make human life possible, and from a new sense of human identity that respects cultural distinctiveness. Many of these essays already suggest that the role of the architect must be redefined to serve greater cultural interests, anticipating new openings for architecture in our postmodern epoch "at the end of history." Faced by the challenge of acknowledging the specificities of culture and site as well as the universalities of technology and space, the architect may still become a crucial figure in the making of an order for humanity beyond tyranny and anarchy – an order predicated on a wholly new relationship between thought and action.

Montréal, Canada, May 1992.

INTRODUCTION

ALBERTO PÉREZ-GÓMEZ

D'où vient qu'en cette dernière décennie de notre millénaire, les professions du design se préoccupent soudainement des questions d'éthique et y consacrent de nombreux colloques? Dans un contexte où la pratique contemporaine en architecture est dictée par des impératifs économiques ou politiques, dire que le bien commun doit être un souci fondamental paraîtra vide de sens ou tout simplement faux aux yeux de certains philosophes post-structuralistes, qui estiment qu'il faut « déconstruire » les vieilles valeurs humanistes – une position qui a eu sur l'architecture nord-américaine une influence considérable.

Et pourtant, il me semble que le bien commun a toujours été une préoccupation fondamentale en architecture. On le constate dans les écrits de Vitruve et d'autres qui, au cours des siècles, ont cherché à élucider le sens de la praxis architecturale. Si la dimension éthique ne se perçoit pas aisément, c'est qu'elle est inhérente à cette praxis plutôt qu'elle n'est un ajout extérieur à une quelconque activité purement technique ou formelle. Par leur volonté inlassable de dévoiler un ordre symbolique, les architectes s'efforcent depuis toujours de procurer à l'existence individuelle un environnement construit qui reflète la finalité des institutions sociales et de la vie en général. La pensée et l'action humaines, depuis les mythes et la religion jusqu'à la science et aux arts, représentent des réponses apportées, à différentes époques et en des lieux divers, à des questions fondamentales qui sont essentiellement les mêmes. Loin de répondre uniquement à un souci esthétique ou technique, l'architecture s'efforce se situer l'action humaine dans un cadre approprié, jouant ce rôle conciliateur malgré la propension de l'homme à contrôler et dominer ses semblables et son environnement.

Aujourd'hui, nous comprenons peut-être enfin que la réduction des beaux-arts à un formalisme esthétique dénué de considérations morales n'est pas un paradigme absolu mais plutôt un événement historique lié à la glorification de la raison scientifique au dix-huitième siècle. Les rationalistes, persuadés que les sciences exactes pouvaient révéler la vérité absolue – une croyance dont l'origine est du reste théologique –, ont implicitement relégué l'art (et l'architecture « non scientifique ») dans une zone marginale et illégitime. La définition de la pratique architecturale en vint alors à reposer sur deux postulats: d'abord, l'architecture se distingue des autres modes de construction par l'intérêt qu'elle montre pour les manipulations formelles; et ensuite, elle doit apporter une solution adéquate aux conditions internes engendrées par un « programme » étranger à la discipline de l'architecte. À partir de la même logique scientifique, on supposait que l'éthique était autonome, elle aussi étrangère aux préoccupations formelles résiduelles des « constructeurs ornementaux »; on estimait qu'elle se trouvait subsumée dans les valeurs technologiques paradigmatiques de l'efficacité et de l'économie. Un constructeur peut-il avoir d'autre souci que de concevoir le bâtiment le plus fonctionnel qui soit et de dépenser le moins possible pour réaliser cet objectif?

La polarisation de l'éthique et de l'esthétique a été exacerbée par le nihilisme confus qui imprègne notre culture post-nietzschéenne, le babel des disciplines et des intérêts spécialisés, l'omnipotence de la technologie et de la simulation. Encore aujourd'hui, on considère souvent l'éthique comme extérieure à l'architecture (et aux autres disciplines du design), comme un système essentiellement subjectif et donc sans conséquences dans notre univers pluraliste. Et pourtant, ne pourrait-on pas soutenir qu'en dernière analyse, c'est à sa capacité d'exprimer des valeurs éthiques qu'est liée la pertinence culturelle de l'architecture?

Si l'architecte doit jouer un rôle au vingt-et-unième siècle, dans un monde complexe et plus conscient des contraintes environnementales et des différences culturelles, un monde où la technique continuera néanmoins de s'étendre à l'échelle de la planète, il doit méditer sur des stratégies propres à révéler la capacité de sa discipline à concrétiser une intentionnalité éthique. Il faut célébrer comme des atouts l'ambiguïté et l'opacité inhérentes à l'architecture conçue comme un univers poétique de discours qui s'adresse directement à la compréhension perceptive et imaginative (plutôt qu'intellectuelle) de l'humanité, et non les déplorer comme si elles étaient des limites. Peut-on dépasser le relativisme purement formel (stylistique) et le professionnalisme pragmatique, tout en maintenant les conditions indispensables qui font de l'architecture moderne/post-moderne une discipline à la fois fonctionnelle et conventionnelle, reposant sur la valeur d'usage du bâtiment mais aussi capable de reconnaître le rôle central que joue la syntaxe dans la production éventuelle d'une métaphore poétique? Peut-on contempler l'architecture comme si elle était l'expression non discursive d'un savoir né d'un souci de l'*autre*?

Des critiques culturels bien intentionnés, cherchant à « déconstruire » les structures oppressives du pouvoir au sein de notre tradition et d'accroître celui de l'individu libre, ont en même temps proclamé l'autoréférentialité de l'art – le labyrinthe post-moderne des miroirs et des valeurs relatives –, ce qui supposait implicitement que l'imagination individuelle, et avec elle la possibilité même d'une action éthique, pourraient être des illusions. Mais le fait d'affirmer le caractère fallacieux d'un fondement absolu et ontologique ne conduit pas forcément à l'autoréférentialité absolue du langage et de l'art. Poursuivant des buts différents, certains critiques (Paul Ricœur, Jean-François Lyotard, Richard Kearney et Julia Kristeva, notamment) privilégient, chacun et chacune à sa manière, une voie qui se distingue à la fois de l'imagination mimétique traditionnelle – de la conception aristotélicienne qui présuppose l'existence d'un cosmos intersubjectif qui est accessible à tous et n'a besoin que d'être « imité » – et de l'imagination égocentrique des Romantiques, qui présuppose la possibilité d'une création *ex nihilo*. Tout en reconnaissant le caractère inévitable de « l'intertextualité » et des conventions – et, donc, en prenant au sérieux la découverte de l'autonomie ultime du langage faite par Rimbaud –, ces critiques admettent au départ la possibilité d'une manifestation de l'imagination qui parle encore de quelque chose qui lui est autre, d'un avenir éventuellement différent fondé sur des valeurs glanées parmi les traces de l'histoire. L'intégration éventuelle des préoccupations éthiques et esthétiques a des conséquences profondes pour la pratique architecturale, car elle pose la question de la production de la forme au moment même de la conception de cette dernière.

Il faut, par ailleurs, avoir une connaissance approfondie du technologisme qui domine notre univers culturel pour considérer la genèse de l'architecture sous toutes ses facettes et comprendre pleinement la dimension publique de sa signification, puisque c'est à l'intérieur de cet univers que l'architecte doit travailler. La technologie est bien plus qu'une simple question de machines; elle est bien plus qu'une caractéristique de notre culture parmi d'autres. Il est clair, désormais, qu'il y a un lien étroit entre la nature de la technologie et un certain nombre de traits culturels qui définissent à la fois la modernité et la post-modernité. Depuis la Révolution industrielle, notamment, la technologie a rendu possible la création d'un environnement entièrement construit. Mais un monde fait de telles constructions agirait bien plus comme un mur entre l'homme et la nature que comme un pont entre notre conscience et une réalité extérieure que nous n'avons pas créée. Un abîme sépare en effet la technologie moderne/post-moderne des techniques traditionnelles des cultures pré-industrielles. La technologie substitue une « image » à l'univers perçu par notre expérience première. Si mutable et historiquement déterminé que soit cet univers de l'expérience première, la phénoménologie révèle que l'universel et le spécifique s'y présentent de façon *simultanée* dans le mystère de la perception, dans l'espace situé *entre* l'Être et le Devenir. La substitution technologique d'une image crée une condition que Martin Heidegger a appelée l'« en-cadrement » (*Ge-stell*). La technologie, qui cherche également à répondre à la question sans cesse reposée de la mesure de l'homme et de la place qu'il occupe au sein de l'incommensurable et de l'inconnu, opte pour le contrôle et la domination, pour l'arrogance et l'efficacité agissant sur « l'image » réduite. Le succès remporté au sein de ce mode instrumental a amené l'homme à proclamer des ambitions quasi divines face à un univers dont il soupçonne fortement qu'il n'a pas été créé pour lui mais dont il peut néanmoins exploiter toutes les ressources.

Cette technologie englobante a été intimement liée à l'idéation et à la pratique architecturales depuis le début du dix-neuvième siècle. Malgré des résistances considérables, la théorie architecturale n'a pas tardé à adopter les modèles productifs des sciences appliquées. L'architecture, art de médiation par excellence, a eu tendance à devenir un outil de domination. Après avoir connu nombre d'échecs et d'impasses, il n'est donc pas étonnant que nous cherchions aujourd'hui à redéfinir la technologie. Cette compréhension nouvelle aurait un effet immédiat sur la représentation architecturale, sur la production de l'architecture, sur le rôle des modèles scientifiques et sur le recours au langage pour préciser les intentions des concepteurs.

Ces aspects de l'idéation architecturale préoccupent au premier chef aussi bien les théoriciens que les praticiens de l'architecture. En cette fin du vingtième siècle, la pertinence culturelle et politique de la profession dépend des réponses qu'on apportera à ces questions. Les rapports entre l'architecture et le monde technologique sont au cœur des problèmes théoriques que pose l'élaboration d'une pratique fondée sur l'éthique. Les incidences de ces rapports importent bien davantage que les distinctions stylistiques superficielles qui accaparent une large part du débat contemporain sur l'architecture.

Il ne suffit pas de rejeter ou critiquer la technologie du seul point de vue de l'art, de la métaphysique ou de l'humanisme traditionnels. Il serait difficile de nier que l'impulsion qui anime ses réalisations et ses échecs est effectivement la soif de transcendance de l'homme, surtout après le point tournant épistémologique qu'on représenté les travaux de Copernic et de Galilée. À mesure qu'il devenait évident que ni Dieu ni l'homme n'était au centre de l'univers, la technologie est devenue une réponse moralement justifiable à l'éternelle difficulté d'adaptation de l'homme au cosmos. Dalibor Vesely fait une analyse pénétrante de l'origine historique de la technologie et montre que cette quête est liée au désir plus que millénaire de l'homme européen de vivre dans un univers ontologique continu, d'habiter la terre dans un mode poétique. Cette quête a-t-elle été abandonnée au profit d'une « fausse » solution qui ne paraîtrait vraisemblable que si Dieu était absent? Obsédée par les réalisations de l'instrumentalité et par l'affirmation de sa propre volonté de pouvoir, la technologie a mené l'humanité post-moderne dans un piège gnostique, où le monde réel pré-existant est oublié. Aujourd'hui, nous sommes souvent victimes de forces que nous avons cru pouvoir maîtriser. La crise de l'environnement et les dilemmes que suscite la manipulation biogénétique en sont deux exemples probants.

Comme l'a montré Heidegger dans son essai sur la technologie, les dangers actuels qu'elle présente sont à la fois plus subtils et plus graves que la capacité d'autodestruction de l'humanité et la menace d'une catastrophe écologique. Le rejet de la mort comme limite positive et essentielle de la vie pourrait, en effet, finir par compromettre l'ouverture de l'existence humaine à la signification et, par là, la perpétuation de la civilisation par l'entremise des institutions culturelles et d'autres symboles. Le nihilisme du désespoir pourrait se révéler une force destructrice et impossible à contrôler, qui supplanterait le nihilisme affirmatif dont Nietzsche a montré la nécessité. Bien qu'il conteste la vérité absolue de la métaphysique occidentale, qu'il conçoit comme échappant à l'expérience (la conception platonicienne de la vérité comme « correspondance »), ce nihilisme affirmatif permet néanmoins à l'humanité de rester ouverte au puissant silence de l'art et de l'architecture, au sombre rayonnement de la lumière, tout en restant sensible aux échos des pas silencieux des dieux qui s'approchent en dansant... Au bout du compte, le problème consiste à comprendre le pouvoir de l'encadrement technologique, à comprendre que, même si nous sommes en-cadrés, la technologie en elle-même peut être « faible » – que, même si Dieu et l'homme ne peuvent être récupérés, la Terre (et le ciel) forment un lieu central où nous pouvons vivre notre pleine expérience matérielle, qui nous permet désormais de contester l'hégémonie des constructions abstraites et théoriques... Dernier paradoxe, cette prise de conscience est confortée par le fait que nous savons maintenant que la Terre est le seul oasis de vie dans le désert du cosmos...

En architecture, les dangers de la technologie sont directement liés à la représentation réductrice, à l'homogénéisation culturelle, à la mort de la ville en tant que lieu d'interaction publique. L'architecte doit reconnaître ces dangers et en assumer sa part de responsabilité. Un premier pas vers un débat éclairé serait de comprendre que la prémisse de la polémique qui oppose le post-modernisme au mouvement moderne – la fuite nostalgique en vue d'éluder le discours technologique – est par trop simpliste. Pour éviter ce recul, peut-être faut-il reconnaître les origines mystérieuses et les transformations de la technologie elle-même au cours de l'histoire. Il nous faudra peut-être remettre en question notre relation instrumentale, souvent arrogante, avec le monde extérieur et explorer d'autres voies, y compris l'élaboration d'une stratégie visant à « déconstruire » le langage de la science et de la technologie au moyen d'une vision poétique. Il pourrait éventuellement en naître une nouvelle relation entre la pratique et les objets culturels dont est constituée notre tradition architecturale.

La philosophie de la post-modernité, telle qu'exprimée dans les écrits de Gianni Vattimo et ses interprétations de Nietzsche et Heidegger, semble déterminée à faire disparaître l'écart qui s'est créé entre les cultures technologique et esthétique et, avec lui, l'écart entre le rationalisme et l'irrationalisme qui nourrit la modernité au moins depuis les débuts du dix-neuvième siècle. L'idée même du progrès historique, étroitement lié à l'assimilation de la technologie par l'architecture et à la valeur indiscutable de ce qui est nouveau, est devenue intenable. Dans ce contexte, une analyse éclairée devrait remettre en cause les idées reçues sur l'histoire de l'architecture, sur la théorie, la représentation et l'idéation architecturales, enfin sur l'éthique professionnelle et la production de bâtiments dans la ville post-industrielle.

C'EST LÀ LE DÉFI qui était proposé aux éminents praticiens et chercheurs qui ont prêté leurs connaissances et leur expérience au symposium et à ce recueil d'essais. La complexité même des questions abordées permettait d'emprunter une approche diversifiée auprès de plusieurs disciplines. Comme on pouvait s'y attendre, les résultats des recherches et les opinions exprimées reflétaient cette diversité, et les débats ont probablement soulevé plus de questions qu'ils n'en ont résolu. Étant donné la complexité des problèmes et leur nouveauté dans le discours post-moderne sur l'architecture, c'est cette diversité même qui donne au présent recueil toute sa valeur. Les essais qu'il renferme visent à provoquer un débat qui mettra l'accent sur des préoccupations éthiques fondamentales, restées dans l'ombre au cours des deux derniers siècles de pratique architecturale. Leurs auteurs savent bien que l'architecture est en période de crise et ils cherchent à en accroître la pertinence en cette fin du vingtième siècle.

Le recueil se divise en quatre parties. La partie introductive comprend trois communications importantes, soit celle de Dalibor Vesely, qui ouvre le débat en mettant en lumière les origines épistémologiques de la technologie dans le contexte de la naissance de l'Europe moderne sous l'éclairage de la philosophie herméneutique continentale, et celles des éthiciens Lukas Sosoe et Margaret Somerville, qui proposent deux autres visions de l'éthique et de ses applications possibles en architecture. S'inscrivant dans la tradition de la philosophie analytique, Lukas Sosoe met en garde contre une éventuelle incompatibilité entre les catégories de l'éthique et la production de l'architecture. À son avis, dans la pratique architecturale il suffit souvent de se conformer au sens commun et à un code déontologique. À l'opposé, Margaret Somerville, forte de son expérience en médecine et en biotechnologie, prône l'adoption d'une éthique appliquée. Pour éviter des conséquences désastreuses, les spécialistes doivent introduire une dimension éthique dans toutes les décisions qui ont trait à leur discipline, même dans celles qui ne soulèvent, en apparence, que des questions purement techniques ou rationnelles. L'architecture, discipline hautement technique et publique à la fois, doit faire intervenir des considérations éthiques dès le départ, car le monde des images et des formes reflète lui aussi un système de valeurs et l'architecte, qui ne contrôle pas seulement les systèmes et les formes mais parfois aussi les moyens de production et les programmes, est en partie responsable de la santé psychique et somatique de la culture humaine.

Les trois principales parties du recueil suivent en gros le déroulement du colloque. Les sujets abordés lors des séances étaient principalement liés aux intérêts des architectes praticiens, qui ont présenté leurs œuvres et donné le ton au débat.

Dans la première partie, consacrée aux rapports entre « l'éthique et la représentation architecturale à l'âge de la simulation », sont soulevées certaines questions importantes quant à la manière de représenter les édifices et les valeurs éthiques transmises par cette représentation. Dans le contexte de la théorie instrumentale, on a longtemps postulé que les plans, les sections et les élévations étaient des projections neutres, qu'ils véhiculaient l'information nécessaire à la construction d'immeubles mais ne renfermaient aucun jugement de valeur. Mais on pourrait aussi affirmer que ce postulat a été remis en question depuis que Piranèse a produit sa série d'eaux-fortes *Prisons (Carceri).* Ce questionnement s'est fait encore plus intense dans l'œuvre d'architectes comme Aldo Rossi, Daniel Libeskind et John Hejduk.

Rappelons que la construction des cathédrales gothiques n'était jamais précédée d'une planification intégrale et qu'elles étaient rarement conçues comme des « images » complètes du bâtiment. Longtemps après avoir jeté les premières fondations, architectes et mécènes débattaient souvent des concepts géométriques qui permettraient d'ériger les façades. À la Renaissance, Filarete insistait sur le parallèle entre la création architecturale et la maternité: telle la mère qui porte son enfant, l'architecte doit porter son projet durant une période de gestation de neuf mois. Après la naissance du concept, l'immeuble projeté doit être nourri jusqu'à sa maturité, en passant par une série de transmutations depuis la conception mathématique originelle jusqu'au tracé du dessin, à la maquette et, finalement, à l'érection de la structure elle-même. Ces transformations n'étaient jamais automatiques; l'essence mathématique ou l'ordre symbolique du projet devaient demeurer les mêmes sous des aspects parfois radicalement différents, mais les changements étaient attendus et bien accueillis à chaque étape. On privilégiait toujours le contact avec la réalité et l'expérience par rapport à une solution platonique idéale ou à une quelconque planification déterminée a priori.

Dans le projet relatif au Vieux Port de Montréal, présenté par l'architecte Peter Rose, la place privilégiée accordée au site et à l'importance des vestiges historiques témoigne d'un mode différent d'idéation architecturale qui composer avec une stratification des traces archéologiques. Dans un monde où domine l'espace homogène de la communication électronique, le fait de considérer les sites industriels marginaux comme une part légitime de notre héritage culturel prend valeur de paradigme; ces ruines industrielles sont peut-être les seuls lieux où peuvent se révéler les origines mystérieuses de la technologie. Il se peut même que de leurs sombres espaces, qui sont habituellement reliés à la périphérie de la ville plutôt qu'à son centre, se dégage la notion d'un but immanent pour notre culture urbaine. En révélant les traces (et les cicatrices) d'un site et en démasquant les applications cosmétiques des urbanistes, l'architecture devient un acte de résistance héroïque.

On peut lire le texte poétique de Philippe Madec comme une « représentation » de l'architecture qui reconnaît néanmoins l'existence d'une distance entre la représentation et l'œuvre construite, d'où la nécessité d'une *traduction.* Les mots qu'il emploie évoquent une architecture située au-delà du formalisme et de la technologie. L'architecture de Madec pose le principe d'une œuvre ouverte, capable d'accepter (plutôt que de réprimer) les prédispositions de la vie au poétique, et qui amorce tout doucement une polémique subtile: il est possible de conjuguer l'habitat poétique et le « style » technologique pour déboucher sur une lecture véritablement ouverte. L'architecte ayant reconnu les erreurs de la « création » romantique et égocentrique, il (elle) est désormais capable de se munir contre les excès de l'avant-garde, mais quelles sont les limites de sa responsabilité? Comment peut-il (peut-elle) exercer son imagination? Qu'est-ce qui constitue un œuvre architecturale véritablement ouverte et démocratique dans un monde où le « panopticisme » et les valeurs de la technologie sont les principaux points de référence qu'on peut trouver à la surface de toute production?

La représentation de l'architecture englobe des valeurs qui s'expriment jusque dans son exécution, bien que le rapport entre ce que nous concevons et ce que nous construisons ne soit pas toujours dépourvu d'ambiguïté. Jean-Pierre Hardenne et Dan Hoffman abordent directement cette question, chacun exprimant à cet égard une position qui lui est propre et tirant des conséquences différentes pour la pratique de l'architecture. Hardenne préconise avec enthousiasme le recours à l'ordinateur dans la conception architecturale. Ayant étudié le potentiel de la logique binaire dans ses formes les plus évoluées et les plus interactives, il souligne combien il importe de faire appel aux outils de représentation les plus sophistiqués que nous ayons à notre disposition. Hoffman, quant à lui, critique implicitement le postulat progressif qui sous-tend la vision de Jean-Pierre Hardenne. Dans son projet, Hoffman s'efforce de considérer la technologie de l'intérieur, tout en reconnaissant pleinement l'importance de la réalisation concrète. Il se sert du processus de la construction et de sa représentation photographique pour exprimer le merveilleux d'une architecture potentiellement « différente ». Il évite ainsi de tomber dans le piège d'une conception réductrice qui ne ferait que proposer des images d'édifices reconnaissables. Hoffman rappelle que l'architecture est une action plutôt qu'une chose, un verbe plutôt qu'un nom; il érige un mur qui ne peut jamais être terminé et met donc l'accent sur le processus de construction plutôt que sur le produit fini. Son projet est une réactualisation du mythe de Sisyphe dans un contexte contemporain, conférant une valeur de paradigme à la position profondément éthique du métier de l'architecte à une époque nihiliste et technologique.

La seconde partie du recueil, qui examine les liens entre l'éthique et le poétique dans le contexte de la production technologique, s'attache en particulier à la relation entre les valeurs éthiques et la production architecturale.

Quel est le rôle de l'éthique philosophique – c'est-à-dire d'une structure rationnelle de catégories – dans la pratique architecturale? Même si on pouvait faire appel à la méthode scientifique et aux enquêtes quantitatives pour déterminer (au moyen d'un quelconque consensus démocratique, par exemple) les valeurs éthiques qui conviennent à tel ou tel projet, est-il possible – ou même moralement souhaitable – d'intégrer ces réflexions à la production d'une forme architecturale dans l'espoir d'humaniser nos bâtiments et de les rendre davantage porteurs de sens?

La question de l'éthique dans la production architecturale concerne les relations entre la *pensée* et l'*acte*, conçues non pas comme des fonctions de type mécanique ou instrumental, mais plutôt comme des interrelations beaucoup plus complexes. L'éloignement ou la présence de l'architecte au processus de construction, la distance entre les conceptions originelles et l'œuvre construite, l'envergure de l'ouvrage, le savoir incarné de l'architecte quant aux matériaux et à la construction – tous ces éléments, si difficiles à identifier et impossibles à quantifier, revêtent effectivement une importance critique pour l'avènement d'une architecture capable de transcender le mutisme des procédés technologiques réducteurs. Le débat philosophique récent, notamment dans le domaine de la phénoménologie existentielle, a fourni aux architectes des mots pour décrire, tout au moins partiellement, l'expérience traditionnellement vécue par les véritables *bâtisseurs*, mais que l'époque technologique avait oubliée ou ignorée. Cette dimension paraissait subjective et inconvenante après qu'aux méthodes traditionnelles et à l'apprentissage en architecture se soient substitués les écoles d'architecture et des processus de production jugés plus efficaces. Or, ceux-ci se fondaient précisément sur une relation mécaniste entre les idées et leur exécution, qui apparentait l'architecte au cerveau informatisé du robot et le constructeur, à ses membres mécaniques. L'argumentaire moderne contre cette dichotomie cartésienne entre l'esprit et la matière a été admirablement défendu par Maurice Merleau-Ponty, notamment. Bien qu'il ne convienne pas de résumer ici les grandes lignes de cette argumentation, on peut néanmoins souligner que l'intuition d'une perception non dualiste de la réalité, où la conscience incarnée n'est plus conçue comme la seule adjonction du cerveau et des organes du corps humain mais bien comme un tout, doit être poussée plus loin pour y englober les mystérieuses « mains qui pensent » de l'architecte/constructeur – le véritable médium de l'imagination éthique de l'architecte.

Les bâtiments traditionnels fabriqués avec soin, même s'ils sont par ailleurs ordinaires ou peu originaux, incarnent une certaine éloquence, comme on peut encore le voir dans les parties plus anciennes de nos villes. Cela étant, de quels recours disposons-nous dans cet univers technologique qui est le nôtre, au-delà de la simple contemplation nostalgique? L'œuvre de Jacques Rousseau, qui accepte la présence d'un lien intrinsèque entre l'éthique et le poétique, nous invite à méditer sur l'importance de questions telles que l'échelle et le sens irréductible d'un site, ainsi que sur les qualités d'un ouvrage qui expriment la simplicité, le silence et le doute plutôt que la vérité absolue, qualités qui peuvent révéler la présence de l'*autre* tout en ouvrant la porte à un prolongement du discours. Plus qu'une esthétique purement formelle, cette œuvre établit une relation différente entre le concepteur/architecte et le constructeur. En cette fin du vingtième siècle, on peut encore reconnaître la différence entre art et métier véritables de l'architecte et ces immeubles déshumanisés, issus de processus étrangers à la construction poétique.

Les deux critiques dont on lira les réflexions dans la deuxième partie du recueil explorent diverses possibilités de tenir un discours scientifique dans le débat sur l'éthique en architecture. Alain Findeli nous invite d'abord à réévaluer l'autonomie présumée de l'architecture au profit d'une profonde unité des questions esthétiques et éthiques qui se posent à une discipline dont la technologie a radicalement modifié la pratique. Robert Prost, quant à lui, esquisse un modèle méthodologique qui pourrait permettre à l'éthique de jouer un rôle au sein de l'architecture, conçue comme processus ou « œuvre en devenir ».

La troisième partie est consacrée à la fonction de l'histoire et de la narration en architecture à l'ère technocratique. Les communications et les discussions qui les ont suivies examinent d'autres façons de définir une position éthique à l'intérieur de discours autres que philosophiques ou scientifiques. Les valeurs objectivées peuvent paraître précises et faciles à mesurer parce qu'elles suivent leur propre logique interne, mais cette logique structurelle risque toujours d'être une source d'illusions, vu le caractère complexe de la création architecturale et l'ambiguïté inhérente de l'architecture elle-même, dont l'ordre symbolique intentionnel est lu, relu et mal lu par des générations de mortels dont la nature reste toujours imparfaite. L'histoire et le récit offrent-ils à l'architecte des solutions de rechange fécondes? Proposent-ils des moyens de supprimer les aspects négatifs de la technologie tout en conservant ses réalisations les plus souhaitables? En quoi cela pourrait-il affecter la Cité? Des philosophes et des écrivains aussi éminents que Hans-Georg Gadamer et Octavio Paz ont montré que les grands thèmes de la philosophie (et de la morale) ont « émigré » vers la littérature depuis le début du dix-neuvième siècle. Et les travaux de Martin Heidegger et de Paul Ricœur ont permis de mieux comprendre la relation qui s'établit entre l'action et le langage, et entre le langage (la métaphore, notamment) et l'imagination. Leurs intuitions laissent entrevoir la possibilité d'employer les formes narratives du discours pour véhiculer efficacement les intentions éthiques dans la pratique de l'architecture.

Le rôle et l'emploi culturels de l'espace en ont traditionnellement défini la dimension temporelle. La convenance d'un bâtiment à ce « scénario », à cette « liturgie » – la *commoditas* de Vitruve – était, jusqu'à la fin du dix-huitième siècle, une donnée culturelle au même titre que l'était la conviction que les proportions et les ordres classiques pouvaient communiquer une signification transcendante. Le « caractère » demeura un aspect important jusqu'à l'époque de Claude-Nicolas Ledoux et d'Étienne-Louis Boullée, mais il fut éventuellement rejeté comme étant une préoccupation futile par Jean-Nicolas-Louis Durand, le premier véritable architecte moderne, reconnu comme le père du fonctionnalisme. De nos jours, le programme d'un projet architectural se résume habituellement à une liste de pièces et de superficies. Cela contribue évidemment à réduire considérablement le domaine de l'architecture et soulève une question fondamentale quant aux valeurs éthiques incarnées dans le bâtiment. Quel est le rôle de l'architecte dans l'élaboration et l'exécution d'un programme fondé sur une optique différente, qui ferait place à un habitat poétique dans le contexte des institutions de la fin du vingtième siècle?

Richard Henriquez aborde ces questions et d'autres questions connexes dans une œuvre qui explore la possibilité de faire appel à des « rituels » contemporains (une cérémonie de collation des grades à Trent University, dans l'exemple proposé) pour identifier des sites architecturaux potentiels. Conjuguant la fonction de la transmission du savoir à l'évocation de l'intégrité écologique de la nature, il rappelle aux jeunes scientifiques leurs propres responsabilités éthiques, au-delà des limites définies par leur spécialité. L'architecture est vue ici comme un événement doté d'un cadre physique qui lui est propre. À travers son œuvre, Henriquez explore les possibilités de « re-créer » l'histoire et de concevoir une architecture destinée à des rituels inventés. Au même titre que les exigences fonctionnelles, de tels rituels peuvent se révéler essentiels pour témoigner d'un sentiment durable de l'ordre *dans le temps*, nécessaire à la préservation de notre humanité et notre identité culturelle.

Cette partie du recueil comprend également un essai de Lili Chi, qui s'interroge sur l'emploi de la narration dans l'œuvre d'architectes comme Aldo Rossi et John Hejduk, où elle puise des idées intéressantes sur l'incarnation poétique de l'éthique et sur le concept de scénarisation en architecture. À cette réflexion s'ajoute, en fin de recueil, celle d'Irena Latek qui, mettant à profit ses travaux récents avec des étudiants de l'Unité d'architecture urbaine de l'Université de Montréal, s'interroge sur certaines incidences pédagogiques d'une architecture basée sur la narration.

LE DERNIER MOT N'EXISTE PAS, CELA EST ÉVIDENT. Les essais que renferme ce recueil montrent la complexité des enjeux qui se posent aux architectes en cette fin de millénaire. Les crises et les maux vécus par notre monde technologique ont un effet profond sur l'architecture, vue comme expression de l'imaginaire personnel et activité sociale et politique. Mais l'architecture est aussi merveilleusement bien placée pour tirer parti de toutes les possibilités qui pourraient découler d'une compassion humaine plus grande et plus universelle, d'une conscience accrue des limites qui rendent possible la vie humaine et d'une conception nouvelle de l'identité humaine, respectueuse des particularités culturelles. Plusieurs de ces essais laissent entrevoir la nécessité de redéfinir le rôle de l'architecte afin qu'il serve des intérêts culturels plus vastes, de même que la possibilité de nouvelles perspectives à notre époque post-moderne, située « à la fin de l'histoire ». L'architecte, qui doit tenir un pari de taille – c'est-à-dire réconcilier les spécificités de la culture et du site avec l'universalité de la technologie et de l'espace –, pourrait bien devenir une figure centrale dans l'avènement d'un ordre humain qui irait au-delà de la tyrannie et de l'anarchie – un ordre fondé sur une relation entièrement nouvelle entre la pensée et l'action.

Montréal, Canada, mai 1992.

ARCHITECTURE AND THE QUESTION OF TECHNOLOGY

DALIBOR VESELY

La technologie apparaît aujourd'hui comme une nécessité liée à notre destinée; elle est un mode majeur de production, qui exerce en même temps une influence profonde sur notre façon de penser et sur l'ensemble de la culture moderne. En même temps, son rôle omniprésent dans nos vies suscite une inquiétude grandissante qui vient s'ajouter à la conviction qu'il existe une différence (ou qu'il devrait y en avoir une) entre la production en tant que telle (l'invention) et la véritable créativité. Plusieurs tentatives ont déjà été faites pour comprendre la nature de cette différence, mais les résultats obtenus jusqu'à présent sont relativement peu probants.

Que la technologie soit la manifestation la plus radicale de la volonté de puissance n'explique pas entièrement la différence entre production et créativité. En revanche, cela montre bien où s'exerce cette volonté de puissance et où cette différence semble être la plus fondamentale et la plus déterminante. Cette dernière se révèle être ici entre le mode émancipateur de la production, qui est un but en soi (l'autonomie), et le mode participatif, qui n'est toujours qu'un moyen d'atteindre une créativité plus authentique (la réconciliation). C'est le premier mode qui a mené à l'émancipation de la technique (étroitement liée au phénomène de la magie) et, éventuellement, à l'avènement de la technologie moderne. Contrairement à une opinion largement répandue, cette émancipation s'est produite à une époque relativement récente (à la période hellénistique), mais on ne peut l'identifier de façon claire et explicite avant le seizième siècle. C'est pourquoi il est plus juste de parler de *tendance* vers la technique plutôt que de technique ou de technologie, au début de l'histoire européenne. À la fin du Moyen Age et au début de la Renaissance, cette tendance s'est accentuée grâce à l'influence des sciences intermédiaires (*scientiae mediae*), notamment l'astronomie, l'optique et la mécanique, qui connurent alors d'importants développements.

Les *scientiae mediae* désignent le savoir, mais parce que ce savoir est avant tout productif, elles appartiennent à la famille des arts plutôt qu'à la science pure. La traduction de la réalité en langage mathématique, processus accéléré par les *scientiae mediae*, n'était que partielle et toujours indirecte. Mais, même indirecte, cette transposition a suffi à abolir la frontière entre les savoirs productif et théorique, ce qui a permis d'envisager la possibilité de traduire la réalité directement et entièrement en langage mathématique et, au dix-septième siècle, a conduit à la création de la science mathématique.

La mathématique ou, plus exactement, la physique mathématique (la mécanique) a été le point de départ de la technologie moderne et, d'une certaine façon, elle est devenue son essence même. La principale finalité de cette longue évolution n'était pas utilitaire mais relevait plutôt du désir de comprendre le mouvement du cosmos et son origine divine, et de contempler les moyens de participer activement à ce mouvement. C'est ce qui explique peut-être l'intérêt profond soulevé par les sciences mathématiques et la croyance en leur bien-fondé général. Ce nouveau savoir, conçu comme universellement productif, n'était en réalité qu'une construction hypothétique, une invention, une façon de s'approprier la réalité et le pouvoir grâce à la représentation instrumentale. Celle-ci ne renvoie plus à l'univers tel qu'il se présente (l'univers naturel), mais uniquement à un modèle de l'univers dont le paradigme est l'expérimentation et l'ultime cadre de référence, l'atelier ou le laboratoire. L'art renvoie obligatoirement à son contexte culturel, qu'il représente de façon symbolique. La technologie, au contraire, renvoie à son propre domaine expérimental, qui lui permet de manipuler l'univers grâce à la représentation instrumentale. La tension entre la réalité du monde naturel et la réalité expérimentale de la technologie est à l'origine des interrogations sur le rôle, encore relativement énigmatique, que joue celle-ci dans le monde moderne.

Dans le domaine de l'architecture, la technologie se manifeste dans cette tension équivoque entre les modes de représentation (ou les façons de fabriquer) symbolique et instrumental. Mais il ne serait pas utile de se pencher ici en termes généraux (spécificité des individus, tendances, etc.) sur un domaine aussi diversifié et encore relativement controversé, de sorte que je m'en tiendrai à l'examen de quelques exemples choisis.

There is a strong feeling that the multitudinous traditional ways of making and creativity are slowly being absorbed into one dominant way of making and thinking. This process of homogenization is not new, but it has reached unprecedented levels today. To see the difference, it is enough to recall the nature and the depth of discussions, in the early decades of the present century, about creativity in different domains of culture, about the relation between art and science or technology, about the nature and status of the applied arts, industrial design, and so on – in such movements, for example, as the *Werkbund*, *l'Esprit nouveau*, *l'Architecture vivante*, Futurism, Constructivism, etc. While some awareness of the distinction between invention, creativity, and pure production remains, it is no longer clear how this distinction should be established; that may be one of the reasons why the current debate is mostly confusing, unsatisfactory, and frustrating.

The topic of this debate, to which most other questions are usually reduced, is the merit of technical efficiency vs. that of aesthetics. Even issues of cultural meaning or social and political relevance, or issues that directly affect the long-term well-being of our society, are often discussed in such simplistic terms. It is not too difficult to discover that this oversimplification has its roots in the dogmatically accepted belief in the universality of technical (instrumental) thinking. As a result, not only technical thinking itself but also a technical way of making have become the standards against which any kind of making is measured. This tendency is usually referred to as the technical or technological imperative. We hear often about the inevitability of technological development and progress, about the historical destiny – the "mission" – of technology. Despite the growing number of sceptical voices and despite the amount of literature devoted to the question of the technological transformation of modern culture, our understanding of the nature of technology remains surprisingly limited. One of the main obstacles to a better understanding is our inability to discuss technological problems from a non-instrumental point of view. In the current scientific parlance, this is often considered to be non-scientific – a verdict that seals the issue and encloses it hermetically in a vicious circle of understanding/non-understanding. A typical example is found in the recent attempts to study the problems of the natural environment by extending the existing technological knowledge into wider fields without changing the primary criteria, conditions, and goals of research. The illusory nature of such studies and their inevitable limits have been very clearly summarized in the following analogy: "With its seemingly unlimited growth of material power, mankind finds itself in the situation of a skipper who has this boat built of such a heavy concentration of iron and steel, that the boat's compass points constantly at herself and not north. With a boat of that kind, no destination can be reached, she will go around in a circle, exposed to the hazards of the winds and the waves."[1]

Instrumental thinking tends to impose its hegemony by creating a world that it can fully control. Control of that nature requires not only a special kind of knowledge but also a particular kind of will. And the knowledge that meets the conditions of the will to control is "knowledge as power."[2] Because this kind of knowledge must be subordinated to the will, we can speak here simply of a "will to power," which as a consequence becomes a "will to will." It is well known that knowledge as power represents the essence of modern science – its metaphysical foundation – but it is also the essence of modern technology.[3] This leads to a deeper insight into the hegemony of technical reason and into the nature of the vicious circle of our "understanding" of technology. The difficulty in breaking that hegemony – and in understanding that technology as the fulfilment of the will to power is not unconditional – is well summarized by Heidegger:

> Because the will to will absolutely denies every goal and only admits goals as means to outwit itself wilfully and to make room for its game, the will to will may not appear as the anarchy of catastrophes that it really is. However, if it wants to assert itself in beings, it must legitimate itself. The will to will invents here the talk about 'mission'. Mission is not sought with regard to anything original and its preservation, but rather as the goal assigned from the standpoint of fate, thus justifying the will to will.[4]

The need of the will to justify its role and its fulfilment reveals that the will itself is not absolute, that it is always situated and cannot completely disguise its own "situatedness."[5] References to mission and fate are a clear manifestation of a deeper intentionality and deeper historical circumstances, in which the will appears as a historical possibility, but always in contrast with other possibilities. If the will represents a movement towards the appropriation of power, culminating in modern technology, the other possibilities represent a movement towards participation that has most consistently been preserved in the domain of the arts. The existence of other possibilities – and their replacement by simple will – must be taken as a point of departure for any understanding of the apparent fatality of technological progress and of the belief that this kind of progress is our historical destiny. It is true, as we have seen, that such a belief belongs to the essence of modern technology; but it is also true that, in itself, this kind of belief is nothing technological: "Because the essence of technology is nothing technological, essential reflection upon technology and decisive coming to terms with it must happen in a realm that is, on the one hand, akin to the essence of technology, and on the other, fundamentally different from it. Such a realm is art."[6]

That art is akin to, but at the same time fundamentally different from, technology is a result of historical development, in which the two domains originally shared a common ground but became differentiated later into the arts and technology as we know them today. Art originates in *techne*, which in its Greek sense is a knowledge related to making and is always known in its final sense as *techne/poietike*. *Techne*, as a relatively new kind of knowledge, superseded spontaneous knowledge and intuitive skills, which demanded a close contact with objects and tasks but could lead to the discovery of what is common and permanent in all of them. This emancipated knowledge teaches us a general lesson about things and can be used a priori, without direct reference to the things themselves. As a project of possible knowledge, *techne* receives most of its knowledge from accumulated experience but elevates it to a priori knowledge that can be taught.

What exists a priori and can be taught was for the Greeks a *mathema* – hence mathematics as a special form of such knowledge. *Mathema*/mathematics is the true origin of the transformation of *techne* into technique and finally into modern technology. In the Greek experience, however, *techne* was still far from becoming such a project. It was a drama situated between the new possibilities of knowledge and the intimate understanding of the inner possibilities of nature (*physis*). *Techne* was not yet seen as a human possession but as a power of nature, which humans could possess only to a limited extent. This may explain why "the first man who invented art [*techne*] beyond common sense was looked upon by his fellow man as a wonder, not only because there was something useful in this discovery, but also because he was thought wise and superior to others."[7]

The fact that *techne* is only a transition to technique is reflected in its relation to making (*poiesis*). Broadly speaking, making means to bring into being something that did not previously exist. *Poiesis* takes place not only in human effort but also in nature: "All things that come into being are generated, some by nature (*physis*), others by art (*techne*)."[8] Art originally received its legitimacy and meaning from the universal divine order, which was seen as the product of an ultimate craftsmanship. "When a thing is produced by nature, the earlier stages in every case lead up to the final development in the same way as in the operation of art, and vice-versa."[9] In this rather dense formulation are already present all the future definitions of art as a reality that complements nature, as a completion and fulfilment of nature's inner possibilities, or as imitation of nature. The imitation of nature, in particular, is a creative process that contains a large residuum of mystery. The Greeks were very much aware of it and referred to it as chance (*tyche*). Aristotle made this very clear in a well-known passage: "Art dwells with the same objects as chance ... chance is beloved of art and art of chance."[10] And he wrote elsewhere: "Some hold that chance is the genuine cause of things, but one that has something divine and mysterious about it that makes it inscrutable to the human intelligence."[11]

It should be emphasized here that making is based on productive knowledge but that such knowledge is never complete. It always depends on a prior understanding that has its origin in the spontaneity of making. The inscrutable element in making to which chance refers has its main source in *mimesis*. Because *tyche* is inscrutable to our intelligence, *mimesis* is equally so. In principle, it is possible to say that *mimesis* is a creative imitation where something that exists potentially is recognized and re-enacted as something actual. For example, movement can be recognized and re-enacted as a significant gesture; sound, as song or music; visible reality, as image or picture; and ideas, as an articulated and structured experience. In its most original sense, *mimesis* is a re-enactment of elementary order: "Testifying to order, *mimesis* seems as valid now as it was in the past, insofar as every work of art, even in our own increasingly standardised world of mass production, still testifies to that deep ordering energy that makes our life what it is. The work of art provides a perfect example of that universal characteristic of human existence – the never-ending process of building a world."[12] The role of *mimesis* in the process of making reveals the mystery of order as a tension between the potential and actual existence of order, which in its ultimate form always points towards the ultimate order – the cosmos. It is in that sense that the re-enactment of cosmic order can be seen as the most primordial form of making.[13]

The mimetic mode of making, which precedes the formation of *techne*, takes place most often in the domain of ritual. This is apparent not only in such rituals as dance or music but also in the rhythm and movement of the process of making itself, thus showing that the making of order and the making of things belong together. In both cases, the result of the mimetic action becomes a vehicle for participation in the overall order of things. The participatory meaning of *mimesis* and ritual – the need to come to terms with the universal order of reality – is challenged and, to a great extent, upset by the tendency to replace participation by the appropriation and manipulation of the order-creating powers. This tendency has its origin in the efficacy of traditional ritual, often wrongly identified as magic. It is obviously a great mistake to see magic where we are dealing only with the efficacy or instrumental aspect of traditional rituals.[14] It is well known that certain gestures or objects used in rituals may have the power to produce certain desirable results, but this does not mean that they can be described as magic or as primitive techniques. The power to influence the order of things in a culture that does not yet see a difference between the natural and the supernatural always depends on the reference to reality as a whole, which cannot become a domain of manipulation: "There is an important difference between two kinds of actions, actions done by man and actions done by man in the belief that their efficacy is not human in any reducible sense, but proceeds from elsewhere. Only the second kind of action can be called any sort of a religious rite."[15] The difference between the efficacy of ritual and magic is manifested in the nature of magic itself. Magic differs from all other forms of religion in that the desire to dominate the world belongs to its essential nature.

The emancipatory, appropriative tendency of magic, in contrast to the participatory nature of ritual, could become an important phenomenon only under certain historical conditions because "the domination by will has one essential condition: before the world can be thus controlled it must be transferred inwards and man must take it into himself. He can actually dominate it only when it has in this way become an inner realm. For this reason all magic is autism, or living within oneself."[16] Historically, this became possible for the first time and in any real sense during the Hellenistic period, when the disintegration of the cultural and political institutions of the *polis* led to the disintegration of traditional corporate rituals and left man to this own resources and in relative isolation. "Magic is commonly the last resort of the personally desperate, of those whom man and God have alike failed."[17] The emancipation of magic was closely linked to the growing interest in other esoteric disciplines (such as astrology, alchemy, or theurgy), as well as to the new interest in mechanics and technicity in general.

In the introduction to his book on mechanics, Pappus of Alexandria recognized the link between mechanics and magic quite explicitly: "The ancients also describe as mechanicians (*mechanikos*) the wonder-workers, that is the magicians (*thaurnasiourgos*) of whom some work with air as Heron in his *Pneumatica*."[18] It was under such circumstances that *techne* came to exist as technique in its most elementary form. This is confirmed by, among others, J. P. Vernant, who, at the end of a long study on the possibilities and limits of technical thinking in ancient Greece, came to the same conclusion: "Only in the work of the Alexandrian engineers, especially Heron, is there any evidence of interest in the instruments and machines as such, and only here was their construction undertaken with an attitude that we can describe as truly technical."[19] What makes this attitude truly technical is not only a new type of knowledge, but rather a new interest and will. In a typical Hellenistic definition of a machine, we can see that "machine is a continuous material system ... moved by appropriate revolutions of circles which by the Greeks is called *kiklikekinesis*."[20] That circular movement is not a purely mechanical phenomenon, however; the text points to its origin in the regularity of the celestial movement, which is also imitated in ritual and dance but is represented here in a more tangible form by the body of the machine.

The incomprehensibility of the movement of nature, manifest most explicitly in the movement of the celestial bodies, has been identified by modern anthropology as the deepest motif of technicity and described as a "fascination with automatism."[21] The nature of this fascination is a continuous attempt to grasp what is most incomprehensible through something that we understand and can construct and manipulate. One can also describe these attempts as a "technization" of the original mimetic reenactment and participation. The machine is a tangible model of such a process and, as a consequence, is also a model of the inscrutable cosmic order. A model is comprehensible because we have made it. It is surprising to see how close the Hellenistic authors themselves came to such an understanding:

> All machinery is generated by nature and the revolution of the universe guides and controls it. For first indeed, unless we could observe and contemplate the continuous motion of the sun, moon and the five planets; unless these revolved by the device of nature, we should not have known their light in due season nor the ripening of

The regularity of the supralunar world; from C. Cesariano's 1521 edition of Vitruvius's *De Architectura*

the harvest. Since then our fathers have observed this to be so, they took precedence from nature, imitating them and led on by what is divine, they developed the necessities of life by their inventions.[22]

It is at this historical stage that technique becomes, at least potentially, a methodical operation that can be carried out in such a way that it can accomplish a particular predictable end. Unlike *techne*, which is always rooted in the concrete life of the *polis*, magic and technique are, to a great extent, emancipated from the political and cultural context. In the ethical sense, they represent individual or group egocentrism, based on the acquisition of power and on domination. The emancipation of magic and technique from the ethically oriented life of the *polis* creates a situation of new freedom in which there is no room for good or evil and for the sense of guilt or sin. It is in such a situation that the question of truth is replaced by the question of practical achievement. Because this is true for both magic and technique, it is very difficult to draw a clear line between them. On the other hand, it is possible to say that magic recedes into the background, leaving a certain residuum of its original power in the more rationalized forms of technique. It is for this reason that it would be more appropriate to speak of an element of magic than about magic itself in the development of modern technique. And it would also be more appropriate to speak about a technical tendency in the domain of the existing arts (*artes technai*) than about technique when we refer to the act of making or production. This would certainly simplify the confusing and very often misleading discussions about the role of magic in the formation of modern technology.[23] We have to keep in mind that the traditional understanding of art includes every kind of making – from the making of shoes or tools to arithmetics and geometry in the *quadrivium*. The difference between the arts was their involvement with matter and manual labour and with their theoretical status, which was most often expressed only through adjectives – the mechanical arts (*artes mechanicae*), usually situated at the bottom of the hierarchy because of the labour involved; the liberal arts (*artes liberales*), which include the *trivium* and the *quadrivium*; and, finally, the theoretical arts, sometimes known as *scientiae*, consisting of theology, mathematics, and physics.[24] That the arts represented not only experience and skills but also an important mode of knowledge, is reflected in the ambiguity of their relation to science.[25]

The sciences that contributed to the formation of modern technique and eventually to technology were mainly astronomy, optics, and mechanics, known as *scientiae mediae* (the "middle sciences"). The reason for that des-

Abraham Bosse's illustration of the effects of perspective on perception and the transformation of the world into a picture; from *Manière Universelle de Mr Desargues pour pratiquer la perspective*, Paris, 1648

ignation was not their "mixed" nature, as is sometimes thought, but their position halfway between metaphysics and physics.[26] The *scientiae mediae* should be seen as a branch of mathematics – physical mathematics – that prepares the way for the development of mathematical physics but is, in principle, radically different from it. It is important to bear this in mind, particularly in view of many current interpretations of Renaissance perspective and mechanics. These interpretations do not always seem to recognize the fundamental ontological difference between the indirect and direct "mathematization" of reality. In the domain of Renaissance art, mathematics plays a role of approximation, mediation, and symbolization. It still represents, on the one hand, the essential (i.e., intelligible) structure of reality (being) and, on the other hand, the visible manifestation of such structures. It is the mediating and symbolic role of mathematics – and not only its precision – that gives it such a prestigious place in early modern thinking. The process of the indirect mathematization of reality – the main characteristic and contribution of the middle sciences – can be seen particularly clearly in the role played by medieval optics in the development of perspective as well as in the mechanical inventions of the sixteenth century. The attempts to bring the physical reality of vision and movement into the sphere of mathematical reasoning were, for a very long time, faced with a paradox of apparent success and real failure. Each successful step in mathematization revealed a new area of reality that would resist completion of the process. This was expressed very often in the frustrations of sixteenth-century artisan-"engineers," who became only too aware of the gap that separates speculative mathematics from the concrete reality of the artisan. The concepts with which the mathematician works "are not subject to those impediments, which by nature are always conjoined to the matter which is worked on by the artisan."[27] It is for these reasons that Renaissance perspective and mechanics cannot be seen as true sciences in the modern sense. The middle sciences, like the arts, can be called sciences only by analogy. True sciences are concerned with universal reality and require absolute proof. Perspective and mechanics, on the other hand, are concerned with particular situations, with human works and operations, and with contingent things. If we take into account how perspective and mechanics were really practised, and not just how they are presented in textbooks or in projects, we may see them as arts, deeply influenced and informed by science. But unlike the sciences or the emancipated techniques, the arts deal with direct experience and with the probable. They belong to the primary mode of embodiment – to the visible world, which is the ultimate criterion of their meaning, relevance, and success. Indirect or partial mathematization could not change these conditions.[28] This also shows the clear limits of the mathematization and technicization of the traditional arts. As long as the arts were situated in the life of nature and society, they could not become a subject of mathematical understanding and control, and to that extent their technicization remained inevitably partial and limited. Only a total mathematization of reality could remove these limits.

It was in the second half of the sixteenth century that such a project became, for the first time, a real possibility. The initial inspiration came from the middle sciences, where the old and jealously guarded boundaries between mathematics and physics were crossed.[29] However, the most decisive changes took place in the domain of mathematics itself, particularly in the sphere of algebra, which had developed into a "universal mathematics." This was complemented by similar changes in the domain of metaphysics, where the *prima philosophia* became a "universal science." Universal mathematics became the mathematical equivalent of traditional logic. Because universal mathematics operates with the pure essences of things, which are taken for simple magnitudes, the formal essence becomes identical with pure mathematical essence/magnitude.

It is under these conditions that universal mathematics can claim to cover the same area of knowledge as traditional logic – in other words, the area of all possible knowledge.[30] The new idea of all possible knowledge is very different from traditional dialectical or demonstrative knowledge. It aims to explain things only in terms of order and measure, regardless of their material and qualitative determination. It was because of the universality of such a claim that universal mathematics earned the title, as long ago as the sixteenth century, of "queen of sciences" (*regina scientiarum*), sometimes elevated to *scientia divina* or *ars divina.*[31] These lofty definitions would obviously not be convincing without some supporting evidence that must come from the understanding of the physical world. In a similar way as in mathematics, the development of knowledge in sixteenth-century physics went through a radical change. The traditional distinction between divine and human knowledge was weakened to such an extent that it became possible to speak about physics and metaphysics in the same language and in terms of the same principles.[32] The affinity between the metaphysical interpretation of physics and universal mathematics was reflected in the new understanding and use of *scientiae mediae* and, in particular, of mechanics. It is very important to realize that, contrary to a widely held opinion, the usefulness of mechanics was secondary to its primary meaning – the understanding and representation of movement in the created world.

The continuity of movement between the celestial and terrestrial domains played a critical role, first in Aristotelianism and later in scholastic metaphysics; as we know, it was the latter which played a decisive role in the formation of modern mechanics.[33] It is only with great effort that we can understand today the complexity and importance of movement (motion) in the seventeenth-century vision of reality. The enigma of creation, the manifestation of the divine order in the terrestrial world, and the continuity of this order were all related to the phenomenon of movement. Movement was seen not only as a universal principle of reality but also as the efficient cause of everything that persists in life. The divine origin of movement was not yet in doubt, nor was the tradition in which divine reality manifested itself as an eternal truth that could eventually be grasped as mathematical truth. Descartes made this clear when he wrote: "Mathematical truths which you call eternal were established by God and depend on him entirely like all other created beings. Do not hesitate to assert and proclaim it everywhere that it is God who set up these laws in nature as the king sets up laws in his kingdom."[34] Attempts at understanding these laws were strongly influenced, if not determined, by the new idea of knowledge – knowing by doing or by construction. In other words, universal reality can be known by the art whereby it was made. In Descartes's own words, "God's will, understanding and creation are one and the same thing; none is prior to another, even conceptually."[35] The identity of understanding and creation was the last condition needed to open the door for mechanics to become the critical discipline in the formation of science and technology.

It is important to see that it was not utilitarian and purely technical interests but a metaphysical quest that gave mechanics such a privileged position. It was in the domain of mechanics that the mathematization of physical movement could be investigated or explored and finally accomplished. The tendency to treat physical reality and movement as inevitable and potentially mathematical was most certainly motivated by the growing desire to discover more tangible links between human and divine reality – which, in Galileo's time, meant more tangible links between physical and mathematical reality. In Galileo's *Dialogues*, we find the following statement: "I still say with Aristotle that in physical matters one need not always require a mathematical demonstration. Granted, where none is to be had, but when there is one at hand, why do you not wish to use it?"[36]

What can possibly motivate such a wish? Galileo himself answers this question:

As to heaven, it is in vain that you fear for that which you yourself hold to be inalterable and invariant. As for the earth, we seek rather to ennoble and perfect it when we strive to make it like the celestial bodies, and, as it were, place it in heaven, from which you philosophers have banished it. Philosophy itself cannot but benefit from our disputes, for if our conceptions prove true, new achievements will be made.[37]

It is well known that the key to Galileo's achievements is the mathematical demonstration performed in a domain that had traditionally been considered to be only contingent. This demonstration, which was radically new, can best be described as a dialogue between an a priori mathematical formula and idealized physical reality. In this dialogue, the mathematical formula, used as a hypothesis (as an argument *ex suppositione*), is followed by an approximation and anticipation of the physical results.[38] On the physical side of the experimental dialogue, phenomena are simplified through abstraction to the point that the approximate mathematical form is free of all difficult material impediments and circumstances. When Galileo speaks about the conditions of the free fall, he comes to the following conclusion: "A more considerable disturbance arises from the impediment of the medium by reason of its multiple varieties, this is impossible [to subject] to firm rules, understood and made into science. No firm science can be given of such events [as] heaviness, speed and shape which are variable in infinitely many ways. Hence, to deal with such matters scientifically, it is necessary to abstract from them."[39]

Galileo's experimental method and its potential rigour include a zone of deep ambiguity that can only be eliminated when physical impediments can be successfully abstracted. But this is not always possible – certainly not in the same degree. To that extent, Galileo's mechanics remains a promise and, even in its best moments, a rigorous hypothetical discipline rather than a rigorous science. It contains an enigmatic element that will never be completely eliminated. The enigma has much to do with the process of mathematization and, in particular, with the nature of the experimental dialogue. Paradoxically, the main source of the enigma is the nature of experimental reasoning, which substitutes an implicit demonstration for an explicit one. In the implicit demonstration, it is not necessary to take into account or to know all the circumstances, conditions, and causes of a particular phenomenon or event (irregular movement, for instance). What is not necessary to know remains enigmatic because, when it comes to understanding, this negligence also remains unknown – and therefore enigmatic.

It is for these reasons that it would be more appropriate to see the experimental dialogue as the result of intellectual craftsmanship rather than a rigorous philosophy of science. As a consequence, the *topos* of the workshop or laboratory is a more appropriate vehicle for understanding the nature (essence) of modern technology than the *topos* of study. The laboratory is a place where nature is systematically transformed into mathematically idealized models. In a world that has been transformed into a laboratory model, construction and making become the privileged form of knowing.

As ideal places for the conduct of experimental dialogue, the workshop and the laboratory represent a new, secondary mode of reality where new rules of knowledge can be developed and cultivated. Unlike traditional knowledge, which was cultivated in a dialogue with the primary conditions of reality, the new rules are articulated in the relatively closed world of the experimental dialogue. The imaginary nature of this new world is well described in Descartes's own words: "For a short time, therefore, allow your thought to leave this world in order to come to see a wholly new one, which I shall cause to be born in the presence of your thought in imaginary spaces."[40] As for the nature of knowledge or science that can be developed in the new "imaginary spaces," Descartes again tells us what is possible and what is also seriously anticipated: "By science I understand skill at resolving all questions and in inventing by one's own industry everything in that science that can be invented by human ingenuity (*ars inveniendi*). Whoever has this science does not desire much else foreign to it, and indeed is quite properly called *autarches* – self sufficient."[41] The science invented by human ingenuity is a construct. It is a productive science, motivated by an ambition to be nothing less than *creatio ex nihilo,* traditionally linked only with divine creativity. However, what is traditionally true for the divine is now considered to be also true, or at least possible, for man. In other words, we know, and can create, at least in principle, exactly as God knows or can create.

This new, unusual confidence has its origin in the drastically simplified representation of reality, which became possible because of the deep metaphysical faith in the mathematical nature of reality sanctioned by divine presence. The result, most likely unintentional, was a method for the construction of productive knowledge, based on the unlimited possibilities of experimental dialogue. "'Idea' was the term I used because it was the familiar philosophical term for the forms of which the divine mind is aware (*formas perceptionem mentis divinae*)."[42] In terms of our own interpretation, however, the idea also represents a new type of knowledge – a primary force of production and the origin of modern technology. The unlimited possibilities of invention opened through experimental dialogue have their source in the infinity of will, which for Descartes is a single analogy of the human and the divine. The full meaning of the infinity of will is "most visibly displayed in the programmatically anticipated infinity of artifices through which the new sciences are to prove their credentials."[43] In the openness to future possibilities lay the foundations of the ideal of progress and, on a deeper level, the intra-mundane eschatology of modern technology.[44] The convergence of the infinity of will and the infinity of artifices completes the ambition to understand given reality as a priori and whole, and from a clearly defined position. "Applying knowledge through construction to the whole world was as inevitable as it was dangerous. It was dangerous because it makes mankind be like God, knowing good and evil. Many seventeenth-century philosophers shunned its inevitable consequences, but only a few had the courage to deny categorically that this kind of knowledge reveals reality."[45] In a sense, that is still true today. We do not yet fully understand the real nature of the experimental knowledge on which modern technology is based because it is difficult to follow the transformation of reality and the nature of its representation in a picture (model) from which all but efficient causes have been eliminated and where the qualitative diversity of phenomena has been reduced to a mathematical interpretation of matter in motion. There is quite clearly a gap between the domain of situated knowledge and productive knowledge. This gap, which represents a radical discontinuity with the natural world, reduces the gnostic value of productive knowledge and makes it merely a technical tool. The fact that a technical tool can represent the most sophisticated achievements at the same time is demonstrated, for example, by nuclear research and its results or by the aspirations of current genetics or electronics. From the very beginning, however, the overwhelming success of productive knowledge was limited to phenomena susceptible to mathematical treatment. This has also determined the selective and uneven development of modern technology. Architecture itself can serve as a very good example here.

The area where technology had the greatest influence was in the calculation of structures; this, as a consequence, led to a more inventive use of certain materials and new types of construction. But as we know, extending the role of instrumental thinking in architecture was a very slow process. It is perhaps not surprising that it was only in certain, rather limited areas that technicization had some success. Factories, railway stations, exhibition halls, and generally structures that could be treated as an engineering problem can be seen as good examples. On the other hand, there were whole areas that proved extremely difficult to mechanize. These were mostly areas of greater complexity or areas dominated by values more deeply rooted in cultural tradition. Because of the particular development of modern European culture, the public domain became rather indifferent to private interests, could therefore be simplified, and as a consequence became rather anonymous. It was for these reasons that technology could be applied more easily in the public domain. As a result, the modern city shows more clearly the true impact of technological thinking than do private homes or residential areas.

In contrast to the earlier, rather slow and partial improvements, the transformations that took place during the nineteenth century were, for the first time, truly systematic and comprehensive. What made these transformations fundamentally different was the possibility of interpreting whole segments of reality in terms of self-referential models and systems. "System" was not a new term, but it had already received a new and very different meaning during the seventeenth century. It was at that time that a system ceased to be a representation of the essential structure of the given reality and instead became a simulated equivalent – an a priori instrumental representation with the ambition to become a universal matrix. That ambition could not be fulfilled everywhere but only in those situations where it was possible to represent the given reality through a model and its purely formal language. Because the formation of systems followed the paradigm of the laboratory experiment, the principles of non-contradiction and sufficient reason – the only criteria of experimental reality – also apply to the instrumental representation of any reality that might eventually be represented in that manner (i.e., as a system). The intrinsic conditions of a particular system cannot determine how far it can be extended and what kind of reality can be incorporated into the instrumental representation. This always remains an open question that can only be decided in light of the actual conditions of specific cases.

An example may show this more clearly. The development of railways in the last century – and, in particular, their extension into the cities – stands in sharp contrast not only to the surrounding landscape or urban fabric but also to the earlier forms of transport, such as roads or canals. Unlike roads or canals, built in an open dialogue with the situational conditions of the given world, railways were designed as comprehensive systems from the very beginning. Designing a comprehensive system requires an a priori plan in which everything is designed beforehand and in the language and logic of the system chosen. This must be done in such a way that nothing outside the system can interfere with its coherence and its working. In the case of railways, this amounted to nothing less than creating a relatively complete and autonomous reality within the given world. As a result, the movement of trains, the functioning of stations and signals, etc. must be predictable and reliable.

The relative perfection of a system is not unconditional, however. A whole set of conditions must be met if the system is to exist and work. The first is the spatial environment in which everything that we make must be situated. The second, closely linked with the first, is the cultural environment in which every system must be absorbed, incorporated, and reconciled with everything that is already there. Only under such conditions can technological production be creative in any way. However, the distance or gap that, in most cases, separates the systems produced from the given world illustrates how limited is our understanding of their conditional nature, how strong is the faith in their autonomy, and how difficult it is, therefore, to bridge the gap. The gap is very often discussed as a problem of adaptation. But what should adapt to what? Today, it is rather taken for granted that the given world should adapt to the imperative of technological possibilities. This shows the limit of our understanding of what is really taking place, what is the nature of the given reality, and what technological interference and manipulation really mean. It is this lack of understanding that is the source of the confused belief that we live in a "technological world." And yet, to understand what is the true nature of the world in which we live is probably the most difficult task. We certainly live in a world that is profoundly influenced and shaped by technology. But this is very different from living in a technological world, if we understand by "technology" what has been established in this paper. What is at stake here is not a semantic difference but the very nature of our current civilization, which we may or may not understand. If we do not understand it, we will never be able to see the ambiguity, tension, and very often deep conflict that exist between the being of technology and the being of the world. In that case, we will also be unable to recognize that ambiguity, tension, and conflict are, in fact, essential characteristics of the world in which we really live.

Only when we take into account the reality that is not directly affected by technology – the primary conditions of our embodiment, the finitude of our life, and so on – can we understand the true nature of the so-called "technological world."

NOTES

1 Werner Heisenberg, "Rationality in Science and Society," in *Can We Survive our Future?*, ed. G. R. Urban (London: Bodley Head 1971), 84.

2 This was programmatically formulated for the first time in the well-known passage of Francis Bacon's *Instauratio Magna* (Sp. 1; 132 [V, 21]): "I am labouring to lay the foundation not of any school of thought, but of human utility and power."

3 "The basic form of appearance in which the will to will arranges and calculates itself in the unhistorical element of the world of completed metaphysics can be stringently called technology"; Martin Heidegger, *End of Philosophy* (London: Souvenir Press 1975), 93.

4 Ibid., 101.

5 This is particularly apparent in the situatedness of will in the context of time, formulated for instance by Nietzsche as the attempt to overcome time through the "eternal return."

6 M. Heidegger, *Vortrage und Aufsätze* (Pfullingen: Neske 1959), 43.

7 Aristotle, *Metaphysics*, Loeb edition, trans. H. Tredennick (Cambridge, Mass.: Harvard University Press 1933 *et seq.*), 981b14.

8 Aristotle, *Physics*, Loeb edition, trans. Rev. P. Wicksteed and F. M. Cornford (Cambridge, Mass.: Harvard University Press 1937 *et seq.*), 199a7.

9 Ibid., 199a9.

10 Aristotle, *Nichomachean Ethics*, Loeb edition, trans. H. Rackham (Cambridge, Mass.: Harvard University Press 1926 *et seq.*), 1140a20.

11 Aristotle, *Physics*, 196b5.

12 Hans-Georg Gadamer, *The Relevance of the Beautiful* (Cambridge, U.K.: Cambridge University Press 1986), 104.

13 Examples that may illustrate this point can be found in ancient Near Eastern cosmogonies and in the Greek understanding of creation, as well as in the articulated cosmologies and in the role of the craftsman as *demiourgos*, which also correspond to Heidegger's understanding of metaphysics.

14 "For a science of religion which regards only instrumental action as meaningful, magic is the essence and origin of religion"; W. Burkert, *Greek Religion* (Oxford: Blackwell 1985), 55. "Science views religion and its manifestations according to its own image and regards everything which refuses to succumb to its techniques as 'magical' and primitive"; R. Granger, *The Language of the Rite* (London: Darton, Longman & Todd 1974), 90.

15 Granger, *Language*, 78.

16 G. Van der Leeuw, *Religion in Essence and Manifestation* (Gloucester, Mass.: P. Smith 1967), 548.

17 E. R. Dodds, *The Greeks and the Irrational* (Los Angeles: University of California Press 1968), 288.

18 Pappus, *Greek Mathematical Texts*, Loeb edition, v. 2, trans. Ivor Thomas (Cambridge, Mass.: Harvard University Press 1939 *et seq.*), 61.

19 J. P. Vernant, *Myth and Thought Among the Greeks* (London: Routledge & Kegan Paul 1983), 295.

20 P. Vitruvius, *Ten Books on Architecture*, book X.C.1.1, Loeb edition, v. 2, trans. F. Granger (Cambridge, Mass.: Harvard University Press 1931 *et seq.*), 275.

21 Arnold Gehlen, *Die Seele im technischen Zeitalter* (Hamburg: Reinbeg 1957), 15.

22 Vitruvius, *Ten Books*, book X.C.1.4.

23 See, for example, the discussions that followed the explicitly formulated opinion about the role of magic in the development of modern science and technology by Frances Yates, recently summarized by B. T. Copenhauer in *Reappraisals of the Scientific Revolution*, ed. D. C. Lindberg and R. S. Westman (Cambridge, U.K.: Cambridge University Press 1990), 261–303. A similar difficulty seems to arise in discussions about the relationship of art and technique before the seventeenth century, particularly during the late Renaissance. The most interesting here are the discussions about the contribution of such personalities as Leonardo da Vinci and his role as artist, engineer, or scientist.

24 For a more detailed discussion of the nature of the arts and their status in the Middle Ages and in the early modern era, see P. O. Kristeller, *Renaissance Thought and the Arts* (Princeton, N.J.: Princeton University Press 1980), 163ff; and *The Seven Liberal Arts in the Middle Ages*, ed. D. L. Wagner (Bloomington, Ind.: Indiana University Press 1986).

25 It is characteristic that the arts have been very often referred to as science (*episteme*), not only in classical but also in medieval scholarship. This is illustrated in the well-known debate, at the end of the fourteenth century, on the completion of Milan's cathedral, where the main question raised is known as *ars sine scientia nihil est.* For details, see J. S. Ackerman, "'Ars Sine Scientia Nihil Est': Gothic Theory of Architecture at the Cathedral of Milan," *The Art Bulletin* 31 (1949): 84–111.

26 In the classical ontology, mathematics (and geometry in particular) is seen as a mediating link between metaphysics (theology) and physics, just as the soul (*psyche*) is a mediating link between the intelligible and the sensible realities.

27 Buonaiuto Lorini, *Della Fortificazioni*, book V (Venice, 1575).

28 Further illustrations of the limits of mathematical mechanization and the discussion of the achievements of Leonardo da Vinci can be found in E. J. Dijksterhuis, *Mathematisation of the World Picture* (London: Oxford University Press 1961), 37–50.

29 R. E. Butts and J. C. Pitt, *New Perspectives on Galileo* (Boston, Mass.: D. Reidel 1978), 187; W. A. Wallace, *Galileo and His Sources* (Princeton, N.J.: Princeton University Press 1984), 126–49.

30 F. Vietta, "Introduction to the Analytical Art," in *Greek Mathematical Thought and the Origin of Algebra*, ed. J. Klein, trans. E. Brann (Cambridge, Mass.: MIT Press 1968), Appendix.

31 Ibid., 181.

32 Like universal mathematics, *prima philosophia* refers ultimately to the principle of non-contradiction and sufficient reason. The new algebra of a meta-mathematical kind and physics of a metaphysical kind have the same characteristics.

33 Étienne Gilson, *Études sur le rôle de la pensée médiévale dans la formation du système cartésien* (Paris: J. Vrin 1967); and Wallace, *Galileo.*

34 Descartes, letter to Mersenne, 15 April 1630, in *Philosophical Writings*, ed. and trans. E. Anscombe and P. T. Geach (Edinburgh: Nelson's University Paperbacks 1954), 259.

35 Descartes, Letter to Mersenne, 27 May 1631, in ibid., 261.

36 G. Galilei, *Dialogues Concerning the Two Chief World Systems*, trans. Stillman Drake (Los Angeles: University of California Press 1967), 14.

37 Ibid., 38.

38 "I argue *ex suppositione* about motion, so even though the consequences should not correspond to the events of naturally falling heavy bodies, it would little matter to me, just as it derogates nothing from the demonstrations of Archimedes that no moveable is found in nature that moves along spiral lines. But in this I have been, as I shall say, lucky: for the motion of heavy bodies and its events correspond punctually to the events demonstrated by me from the motion I defined"; cited in Butts and Pitt, *New Perspectives*, 234.

39 Ibid., 232.

40 Descartes, *Œuvres*, tome XI (Paris: Tannery 1975), 31.

41 Ibid., tome III, 722.

42 Descartes, *Philosophical Writings*, Objection V, 136.

43 D. R. Lachtermann, *The Ethics of Geometry* (London: Routledge & Kegan Paul 1989), 140.

44 J. Ladrière, "Technique et eschatologie terrestre," in *Civilisation, technique et humanisme* (Paris: Aubier-Montaigne 1968).

45 A. Funkenstein, *Theology and the Scientific Imagination* (Princeton, N.J.: Princeton University Press 1986), 327. Among those who had this courage are Malebranche and the *occasionnalistes* (Guelinkx, for example), and, most explicitly, Leibniz.

QUE PEUT ÊTRE OU FAIRE L'ÉTHIQUE EN ARCHITECTURE?

LUKAS K. SOSOE

The paper explores the conditions under which ethics might be applied to architecture. Regardless of the definition of ethics that is adopted, it seems difficult to find specific ethical problems in either architectural theory or practice. There are several reasons for this.

First, it is difficult to see how an ethics of architecture might emerge from architectural theories. Not only are similar theories found in other disciplines, which means that the definition of an architectural ethics would be problematic or at the very least non-specific, but it is difficult to imagine what such an ethics might encompass.

Second, even if one posits that the architect's work has an impact on the use of space by human beings and that one might thus speak of the possibility of an ethics of architecture, it is still difficult to imagine what would be specific to architecture in such an ethics.

Third, the socio-political and cultural context in which we live is such that an ultimate justification of ethical standards in architecture would be confronted by the issue of the principles on which contemporary ethics are based. Once that issue is settled, it might be possible to discuss ethical questions in specific terms in the context of architecture.

But finally, regardless of their magnitude, these questions would concern general ethical issues. Thus we may conclude that ethics in the architectural context would have to be conceived in general terms and cannot hope to be specific. That the question of ethics is raised at all has more to do with the great pressure for standardization in contemporary society than with architecture as such.

« Que peut être la religion d'un peuple éclairé? », se demandait Hegel. On peut actualiser la question et se demander: que peut être l'éthique et, surtout, que peut être l'éthique appliquée dans une société moderne à l'âge post-moderne? Plus précisément, que peut être une éthique appliquée à un domaine tel que l'architecture, quelle qu'en soit la définition. Selon Melvin Charney, l'architecture est une réalité imaginaire qui dit et quelquefois suggère le non-dit, dont les racines théoriques se retrouvent aussi bien dans les sciences sociales que dans les sciences économiques, mais qui demeure également un art dont la longue histoire est marquée par différentes sensibilités culturelles, différents goûts, différentes modes et même par des théories concurrentes.

S'il est difficile de répondre à cette question, c'est que l'actualité de l'éthique et de l'éthique appliquée se réalise dans un contexte sans aucune tradition précise, un contexte où co-existent de multiples traditions qui n'ont qu'une chose en commun, soit la rapidité des changements qui les transforment et président à leur naissance, à leur disparition ou à leur retour. Question redoutable, donc, mais question à laquelle il faut tout de même tenter de trouver une réponse dans le cadre de la philosophie contemporaine, caractérisée notamment par la diversité des positions en éthique.

Je voudrais, dans un premier temps, situer l'éthique dans le contexte de la philosophie contemporaine en insistant plus particulièrement sur les problèmes qu'elle pose et que l'on pourrait formuler sous forme de paradoxes. Dans un deuxième temps, je ferai quelques remarques sur la question de l'application de l'éthique à différents domaines, pour enfin conclure par quelques réflexions sur ce que pourrait être, à mon avis, une éthique appliquée à l'architecture.

Précisons d'abord que la notion d'éthique employée ici n'est pas un concept heuristique. Elle ne saurait s'identifier ni se substituer à une description ou explication, quasi déterministe ou non, de la succession des différentes modes de la conduite humaine dans les divers domaines de la culture en général. L'éthique, telle que je l'entends ici, ne saurait donc s'appliquer aux différents styles et époques qui se suivent dans l'histoire de l'architecture, non plus qu'elle ne saurait être perçue comme une catégorie interprétative de l'œuvre d'art ou de la culture. Elle ne saurait prétendre assumer cette fonction, qui revient plutôt à l'anthropologie ou à la sociologie, voire même à l'histoire critique de l'art ou à la philosophie de l'art. Dans la mesure où elle se définit essentiellement comme une connaissance normative, l'éthique ne s'intéresse à la conduite humaine que dans ce qu'elle a de proprement moral, c'est-à-dire de juste et d'injuste, de bien ou de mal, et ainsi de suite... et non à la détermination d'une intelligibilité réelle ou possible de l'œuvre d'art.

Quant à l'éthique appliquée, j'utilise le terme dans le sens de la distinction qui s'est imposée dans l'éthique contemporaine. Bien que l'éthique ne soit pas une doctrine ou une somme de doctrines qu'on peut employer à la solution de problèmes concrets, on peut tout de même parler d'éthique appliquée. Le concept d'application, dans ce contexte précis, désigne le souci de traiter d'un problème d'abord individué comme étant un problème éthique dans un domaine particulier de la vie humaine – dans une profession, par exemple. Dans le cas présent, c'est dans le domaine de l'architecture que nous cherchons un champ d'application de l'éthique.

La conception de l'éthique comme une forme particulière de connaissance normative pourrait s'appliquer, à quelques exceptions près, à toutes les différentes théories proprement éthiques qui proposent des préceptes visant à guider la conduite humaine suivant des critères rationnellement déterminables. Quand bien même il nous serait possible de définir l'éthique en ces termes, il faudrait toutefois nous garder de la considérer comme un corps homogène de connaissances. En d'autres termes, malgré sa longue histoire, l'éthique n'est pas faite d'un corps précis de doctrines, de pensées. Elle est, au contraire, une suite ininterrompue de positions différentes et quelquefois même opposées les unes aux autres. C'est depuis ses débuts, me semble-t-il, que l'éthique s'est présentée comme une suite d'arguments et de contre-arguments, pour ne pas dire une série de controverses théoriques, de paradigmes – si tant est que ce terme puisse avoir un sens en philosophie, en éthique plus précisément, du moins pour ce qui concerne les fondements de celle-ci.

Dans son ouvrage *After Virtue*[I], Alesdair MacIntyre estime que lorsqu'on examine sérieusement les théories éthiques et politiques des Modernes, on est bien forcé d'admettre que la situation est proprement désastreuse, dans la mesure où on est confronté à une multiplicité de positions contradictoires qui prétendent toutes à l'objectivité. Peut-il exister une guerre juste? Est-il permis d'avorter? Les disparités de revenus sont-elles compatibles avec la notion de justice? Ce sont là autant de questions qui donnent lieu à d'interminables querelles – signe que nous vivons dans une culture où il est devenu impossible de réaliser un accord sur des conclusions rationnellement justifiables. En d'autres termes, on constate que personne n'arrive plus à proposer aux citoyens des critères de justification rationnelle permettant de conférer une certaine unité à leurs diverses convictions. C'est ainsi que le débat public sur les questions touchant à la justice et à la rationalité pratique prend non pas la forme d'une recherche rationnelle commune mais d'un échange interminable d'assertions et de contre-assertions dont les prémisses sont rigoureusement incompatibles[2].

Dans un ouvrage plus récent, le même auteur revient à la charge pour de nouveau mettre en relief l'impossibilité dans laquelle se trouve notre culture moderne démocratique de trouver un accord sur les problèmes éthiques. Bien que les théories politiques libérales renferment des options particulières qui les distinguent les unes des autres, elles se caractérisent toutes, estime-t-il, par la recherche de critères, en matière de justice et de rationalité pratique en général, qui sont indépendants des diverses traditions. Ce projet requiert que soient trouvées « une caractéristique ou des caractéristiques d'une attitude morale ... applicable aux êtres humains, indépendamment de toute tradition sociale et culturelle particulière ». Pour ce faire, il a fallu recourir à une « sorte d'universalité et d'impersonnalité » qui ont pu être considérées comme un point de vue moral « indépendant de toutes traditions, de toutes déterminations particulières ». MacIntyre termine en faisant remarquer ironiquement pourquoi de telles tentatives de justifications de l'action humaine ne peuvent qu'échouer: non seulement le produit d'une telle démarche visant à justifier ou fonder l'action humaine en dehors de toute forme traditionnelle et conventionnelle de la pensée est-il « trop étroit et trop maigre » pour fournir ce dont nous avons besoin, mais encore l'évidence de l'échec d'une telle justification de type moderne est contenue dans les revues philosophiques professionnelles. « Les pages de ces revues sont les cimetières de la philosophie constructive académique; les doutes que l'on pourrait concevoir sur la question de savoir si, après tout, un consensus ne serait pas susceptible d'être atteint en philosophie morale académique peuvent être écartés pourvu qu'on les lise régulièrement[3] ».

Ces accents négatifs doivent nous inviter à reconsidérer notre volonté de faire de l'éthique et même de vouloir l'appliquer aux sérieux problèmes qui caractérisent le contexte social démocratique et technologique dans lequel nous vivons et surtout le paysage philosophique où notre discours de l'éthique et de l'éthique appliquée est appelé à se déployer. Ces accents ne sont nullement isolés et, par delà la différence de perspective, font écho aux philosophies de type heideggerien, nieztschéen et foucaldien, bref aux philosophies de la déconstruction, dans la mesure où celles-ci aussi rendent impossible toute pensée proprement éthique. Accents qui contrastent, enfin, avec l'urgence des problèmes éthiques auxquels nous tentons de trouver des solutions. D'une part, il est plus que jamais question de l'éthique et surtout de son application aux problèmes d'ordre public; mais d'autre part, cette même éthique est rendue impossible, du moins en apparence, par le genre de culture pluraliste dans laquelle nous vivons et par des courants philosophiques dont les catégories ne permettent pas de poser la question de l'évaluation et de la recherche de critères indispensables au déploiement d'un discours proprement éthique. C'est donc sous la forme d'un paradoxe (bien évidemment pas au sens logique de ce terme) que se pose la question de l'éthique contemporaine, pour peu qu'on veuille, ou bien éviter un discours éthique naïf à outrance et excessivement dogmatique, d'une part, ou bien poser le problème avec un minimum de sensibilité à la situation culturelle, de l'autre.

MacIntyre nous invite donc, pour que l'éthique soit possible, à abandonner nos conceptions modernes et à revenir à une conception plus située de nous-mêmes, de la rationalité et de la société ainsi que de nos actions, y compris l'œuvre d'art, notamment l'œuvre architecturale. Cela revient à proposer une thématique de la narrativité dans l'œuvre architecturale, telle qu'elle a été formulée par Alberto Pérez-Gómez et, pour une moindre part, développée par Lily Chi.

Mais comment peut-on alors, face à ce qu'il convient ici d'appeler le paradoxe de l'éthique, repenser la question que nous nous posions au tout début, à savoir: que peut être l'éthique appliquée – plus particulièrement, que peut être une éthique appliquée à l'architecture?

Il importe de rappeler un fait que l'on oublie trop souvent dans les débats sur l'éthique appliquée: l'éthique a toujours été appliquée, en ce sens que tout au long de sa longue histoire, les philosophes se sont penchés sur des problèmes précis. Aussi le concept d'éthique appliquée n'est-il pas nouveau; il provient de deux sources indépendantes: tout d'abord, de l'histoire récente de l'éthique anglo-américaine, où depuis Henry Sidgwick, G. E. Moore et Wittgenstein on s'est exclusivement consacré à l'analyse du discours moral et à la description du langage moral. Comme on supposait que celui-ci reflétait une neutralité axiologique, cela permettait, en toute bonne conscience, de reléguer les questions éthiques substantielles à l'univers de l'imaginaire. Mais la notion d'éthique appliquée provient, en outre, du moins pour ce qui concerne la philosophie continentale, de la critique de ce qu'on appelle les métaphysiques de la subjectivité, critique entreprise par Nietzsche et dont la continuité a été assurée dans la philosophie contemporaine par Heidegger et par les philosophies critiques de la culture et de la morale. Aussi bien la critique des métaphysiques de la subjectivité que l'éthique analytique ont, en quelque sorte, mis fin à la longue tradition de l'éthique philosophique, qui s'est toujours définie comme une philosophie de la pratique humaine dans son ensemble et donc, nécessairement, une réflexion appliquée à cette pratique. C'est pourquoi lorsqu'on parle d'éthique appliquée, il ne faut pas envisager l'émergence d'une nouvelle branche de l'éthique philosophique mais au contraire se rappeler une situation historique précise, celle qui s'étend du début de notre siècle aux années soixante, soit une période courte, très particulière, au cours de laquelle l'éthique philosophique a perdu de ce qui la définissait essentiellement, c'est-à-dire sa dimension de réflexion sur l'action humaine dans son ensemble. Il n'y a donc pas lieu, pour nous, de considérer les éthiques appliquées comme une branche nouvelle et séparée de l'éthique philosophique et défendre ainsi une « thèse séparatiste »; il faut plutôt les concevoir, au-delà de la nouveauté et de la diversité de certains problèmes, comme cette réflexion particulière qui engage la philosophie à se pencher sur la dimension proprement normative de l'action humaine[4].

Même si l'éthique a toujours été appliquée, pourrait-on objecter, le besoin de l'appliquer n'a jamais été si grand, si important. Ce besoin trouve son explication dans le fait que les progrès techniques et scientifiques atteignent des proportions qui nous inquiètent de plus en plus et qu'entre eux et nous surgit la question des limites – plus précisément des limites que nous voudrions imposer aux libertés constitutionnellement garanties, aux libertés individuelles, parfois conquises à un prix très élevé. L'objection est certes pertinente, mais elle ne vise pas l'éthique philosophique vue comme une réflexion sur la pratique humaine, c'est-à-dire sur ce que doit être cette pratique.

Mais si, comme nous le suggérons, l'éthique a été toujours appliquée, que peut être une éthique appliquée à l'architecture? Que demande-t-on à l'éthique de faire en architecture? Comment concevoir une éthique appliquée à l'architecture?

On pourrait s'attendre ici à une réponse qui prend tout d'abord en compte les diverses théories concurrentes en architecture et essayer de voir dans quelle perspective on pourrait légitimement chercher à savoir si le fonctionnalisme, le constructivisme ou toute autre conception de l'architecture se défend au plan moral. Dans la mesure où, une fois réalisée, chaque conception de l'architecture a des conséquences réelles sur l'utilisation de l'espace, ne détermine-t-elle et ne structure-t-elle pas, d'une certaine façon, les comportements des êtres humains dans leur environnement, dans l'espace urbain, par exemple? On pourrait concevoir cette façon d'aborder le problème comme la recherche des possibilités d'application d'une éthique spécifique à l'architecture.

Disons-le autrement: les incidences morales des différentes conceptions de l'architecture formeraient un ensemble d'éléments ou de problèmes constitutifs d'une éthique particulière qui serait l'éthique de l'architecture. Cette éthique se constituerait, non pas tant parce que l'œuvre architecturale se présenterait de telle ou telle façon, mais avant tout parce qu'elle influence le comportement des individus dans leur espace habitable. Ainsi le problème de l'éthique en l'architecture ne serait-il pas architectural mais il serait quelque chose d'externe à l'architecture.

On comprend mieux ce que l'idée ou le projet d'une éthique de l'architecture aurait d'étrange, d'absurde et d'incongru à la lumière des trois raisons suivantes:

1 Les tendances que l'on rencontre dans l'architecture ne lui sont pas spécifiques mais se retrouvent dans beaucoup d'autres sciences – humaines, sociales et économiques, notamment. On voit mal comment une nouvelle éthique spécifique à l'architecture se justifierait dès lors que les catégories qui la caractériseraient ne lui seraient nullement spécifiques.
2 Il ne suffit pas que, dans la structuration de l'espace, la réalisation de l'œuvre architecturale ait des conséquences réelles pour les individus pour qu'une éthique fondée sur ce rapport soit spécifique à l'architecture. En effet, on trouve non seulement d'autres disciplines scientifiques mais encore toute une série d'activités humaines qui structurent un espace habitable et imposent, tout aussi bien que l'architecture, des contraintes à l'action humaine. Et l'architecte, pas plus que le géographe ou même le paysagiste, n'a pas le monopole des questions touchant à l'organisation de l'espace.
3 En vue d'empêcher que soit discrédité son travail et surtout que le législateur, sous la pression populaire, n'impose des lois trop restrictives qui risqueraient de l'empêcher de déployer ses talents, ses capacités artistiques, l'architecte pourrait prendre les devants et provoquer un débat éthique. Tout en corrigeant l'image de l'architecte aux yeux du grand public, l'éthique répondrait ainsi au grand désir de moralisation, de normalisation qui se fait jour dans notre culture, désir qui, au-delà de ce que nous pouvons tenir pour très urgent, est une attitude critique négative, pessimiste, face aux progrès scientifiques et techniques. Ici également, on l'aura noté, l'éthique de l'architecture ne saurait être proprement architecturale, car l'architecte ne saurait être plus soucieux de défendre son

« image » que ne le sont certaines compagnies ou hôpitaux qui disposent de commissions et comités d'éthique ou ont souvent recours aux experts en matière d'éthique.

Mais après avoir perçu les limites d'une éthique spécifique à l'architecture, comment concevoir une éthique qui, tout en tenant compte du paysage culturel et philosophique contemporain, pourrait répondre à un souci de moralisation qui se substitue au manque de traditions dans le contexte d'une société moderne? En d'autres termes, si, contrairement à ce que supposent bon nombre d'éthiciens, nous n'avons pas besoin d'une éthique nouvelle qui réponde aux développements technologiques récents et que les théories éthiques dont nous disposons déjà suffisent à cet égard, comment peut-on réinterpréter celles-ci à la lumière des problèmes auxquels sont confrontés l'architecte et les utilisateurs de l'espace qu'il contribue à créer? Comment parvenir à la conception d'une éthique dont on peut espérer qu'elle soit applicable aux problèmes susceptibles de se poser dans le domaine de l'architecture?

Posée en ces termes, la question appelle plusieurs réponses possibles.

La première comprend, me semble-t-il, deux parties. Si l'architecture est une activité humaine, c'est qu'elle obéit à une ou plusieurs formes de rationalité, dont certaines sont constitutives d'une attitude éthique.

Par exemple, l'architecte qui tient compte de la rareté de l'espace, qui le respecte comme un bien public et ne le perçoit pas uniquement comme le lieu du déploiement de sa volonté de créer, cet architecte adopte déjà une attitude éminemment normative. Il en est de même de celui qui part de concepts, aussi peu précis soient-ils, tels que les besoins et désirs des consommateurs.

Quelles que soient les réalisations qui découleront de ces attitudes, on voit mal comment on pourrait refuser de les qualifier d'éthiques. La rationalité qui les sous-tend coïncide donc avec la normativité éthique et accorde ainsi à l'œuvre architecturale son efficacité. Or, qui dit ici efficacité dit réalité, et qui dit réalité dit rationalité, celle-là même qui a présidé à la facture de l'œuvre architecturale et lui donne son sens. Il ne restera donc qu'à conclure avec Hegel: ce qui est rationnel est réel et ce qui est réel est rationnel. Le problème de l'éthique à l'époque contemporaine se résorbe donc dans la rationalité qui préside à la réalisation de l'œuvre architecturale. Vouloir poser un problème éthique en dehors de cette réalité de l'œuvre architecturale, vouloir le concevoir en marge des traditions qui ont nourri l'architecture, c'est reposer une question à laquelle nous avons toujours déjà donné une réponse;

celle-ci se trouve dans le souci qui guide le travail de l'architecte et lui confère forme et intelligibilité.

Une autre variante de la même attitude consisterait à voir dans les restrictions juridiques formelles imposées à l'architecte et au déploiement de son art, dans l'existence des pouvoirs publics et commissions nationales de protection du patrimoine, des limites ayant une valeur éthique, comme le souligne Robert Prost. Même si l'on dispense ainsi l'architecte d'adopter volontairement une attitude éthique, celle-ci se trouve prise en charge par le droit. Prise en charge parfois redoutable, mais non totalement irrationnelle, du problème éthique, et qui permet de dire que l'éthique est toujours déjà présente dans l'architecture. Ce que nous cherchons, à savoir des principes éthiques, faut-il le répéter, se trouve toujours déjà présent dans les cadres constitutifs de ce que nous faisons. Autrement dit, il n'y a d'éthique de l'architecture que le savoir de l'architecte même et le cadre socio-juridique où ce savoir est produit et réalisé sous la forme de l'œuvre architecturale[5].

Cette position pourra déconcerter celui qui réclame ou revendique une éthique pour l'architecture à l'âge de la technologie et de la démocratie, mais elle est néanmoins défendable, si l'on considère les difficultés que soulève la définition d'une éthique spécifique à l'architecture. À cela s'ajoute le fait que l'architecte, comme toute personne socialement compétente, n'a pas besoin d'un discours éthique pour savoir ce qui est bon ou mauvais dans ses relations avec ses clients ou dans sa façon de traiter l'environnement. Même l'introduction de nouvelles techniques hautement perfectionnées dans le domaine de l'architecture ne me semble pas avoir rendu caduque cette compétence socialement normative, contrairement à ce que pensent d'autres éthiciens[6].

Il reste que si, face à la question de l'éthique en architecture, on se contente de défendre les usages et traditions établis dans la profession, cela ne donnera qu'une réponse partielle, peu satisfaisante, car si la question de l'éthique appliquée se pose, c'est justement parce que ces usages et traditions ne semblent plus répondre aux dilemmes auxquels sont confrontés les praticiens[7].

Pour résoudre ces dilemmes, il faut, me semble-t-il, que soit tout de même envisagée la possibilité d'appliquer l'éthique à l'architecture. Mais cela ne nous renvoie-t-il pas à la difficulté de départ, celle-là même que souligne MacIntyre lorsqu'il déclare que la société libérale démocratique est incapable de trouver un fondement aux normes éthiques? Il y a toutefois de bonnes raisons de croire que la situation n'est pas aussi alarmante que le pensent MacIntyre et d'autres philosophes à sa suite.

Tout d'abord, parler de l'impossibilité de la société libérale démocratique à se mettre d'accord sur des principes éthiques m'apparaît comme un diagnostic plutôt superficiel et très hâtif; en effet, ce ne sont pas les grands principes qui sont mis en cause par le caractère interminable des débats consacrés à l'éthique, mais plutôt l'accord sur certains faits, ainsi que le statut et l'importance qu'il faut leur accorder.

Deuxièmement, le prétendu caractère interminable des débats éthiques n'est lié qu'indirectement à l'éthique en tant que telle. Il est lié de façon plus directe à la nature même de la philosophie et des attentes que nous pouvons avoir de l'éthique en tant que discipline philosophique face aux nouveaux développements technologiques.

Enfin, si l'on suppose que les intervenants sont guidés par le souci de l'échange et de la formation d'un consensus autour des grands problèmes sociaux, le caractère interminable des débats dans les démocraties libérales peut être perçu comme le refus de tout dogmatisme, l'expression du respect de l'opinion d'autrui, la recherche commune d'un accord sur les limites que nous voulons poser dans nos sociétés. Loin d'être un élément négatif, la longueur des débats n'est que l'expression même de notre finitude et la vocation proprement humaine de se dépasser[8].

Trois raisons, donc, qui suggèrent qu'un accord sur des principes éthiques devrait être possible. Aussi faut-il s'interroger sur le rôle de l'éthique et de l'éthicien dans la formation de cet accord, aussi minimal soit-il, sur des principes éthiques. En un mot, si l'éthique est possible et si nous pouvons aborder certains problèmes dans le domaine de l'architecture dans la perspective qui est celle de l'éthique – et, plus précisément, de l'éthique appliquée –, à quoi pouvons-nous nous attendre?

Disons-le clairement: l'éthique ne peut rien faire d'autre que de clarifier les problèmes, d'aider à les formuler d'une façon conceptuellement acceptable[9]. Cela ne signifie pas encore que l'on souscrive ici à la thèse d'une neutralité axiologique. À mon avis, il existe un nombre restreint de principes éthiques dont on peut prouver la justesse par des procédures argumentatives précises. Parmi eux se trouvent les principes qui forment le cadre de la discussion nécessaire à l'« individuation » d'un problème et au choix des critères qui en font un problème véritablement éthique dans le cadre de l'architecture. Qu'est-ce qu'un problème éthique en architecture? Quels rapports existe-t-il entre l'esthétique architecturale et l'éthique? Quelle signification proprement philosophique pratique la poursuite de la grande aventure esthétique de l'humanité pourrait-elle revêtir pour nous?

Autant de questions qui renvoient à des choix, tantôt publics, tantôt individuels, et qui justifient ma conclusion. À la question, « que peut être ou faire l'éthique en architecture? », je réponds, « rien ou alors presque tout ». Réponse somme toute modeste, peu constructive pour une conclusion, mais réponse quand même soucieuse d'éviter le dogmatisme aussi bien que la naïveté ou même l'enthousiasme éthique. Car il faut, après Montesquieu, le répéter: « Même la vertu a des limites ». Vouloir s'acharner à la réaliser à tout prix pourrait bien aboutir à sa défiguration, au « jacobinisme » éthique.

NOTES

1 A. MacIntyre, *After Virtue*, South Bend (Indiana), Notre Dame University Press, 1981.

2 Ibid., p. 7.

3 A. MacIntyre, *Whose Justice? Which Rationality?*, Londres, Duckworth, 1988, p. 334.

4 Voir à ce sujet Joseph P. DeMarco et Richard M. Fox, *New Directions in Ethics: The Challenge of Applied Ethics*, New York, Routledge & Kegan Paul, 1986; et *The Applied Turn in Contemporary Philosophy*, Bowling Green Studies in Applied Philosophy, Bowling Green, Ohio, Bowling Green State University, 1983.

5 Nous retrouvons ici également la distinction, faite par certains analystes, entre problème interne et problème externe. Suivant cette distinction, l'éthique en architecture serait faite du savoir de l'architecte et du cadre socio-juridique, constitutif de la production et de l'application de ce savoir revient à dire que poser un problème d'éthique de l'architecture, c'est au plan herméneutique, se poser une question dont la réponse est évidente.

6 L. Bourgault, *L'Éthique et le Droit à l'âge technologique*, Montréal, Presses Universitaires de Montréal, 1991.

7 Pour une critique exhaustive de cette position, voir Alan Cewirth, « Professional Ethics: The Separatist Thesis », *Ethics*, vol. 96 (1986), p. 282–300.

8 Sur cette critique de MacIntyre, lire Lukas K. Sosoe, « La pensée communautarienne – Généalogie ou subversion des Modernes? », *Cahiers de Philosophie Politique et Juridique de l'Université de Caen*, vol. 20 (1992), p. 215–231.

9 Voir Nicholas Rescher, « Problems in Applying Philosophy », in *The Applied Turn in Contemporary Philosophy*, p. 1–15.

ETHICS AND ARCHITECTS: SPACES, VOIDS, AND TRAVELLING-IN-HOPE

MARGARET A. SOMERVILLE

L'introduction présente quatre concepts relativement nouveaux en matière d'éthique, auxquels les architectes devraient être sensibilisés. Le premier fait référence à deux notions de l'éthique, soit l'éthique « mince », qui exige seulement de satisfaire à un minimum de conditions légales, et l'éthique « épaisse », qui recouvre un champ plus étendu. Le deuxième concept englobe l'éthique dite « appliquée », celle qu'on pratique dans la vie. En troisième lieu vient l'éthique « individuelle et sociétale », concept double puisque les principes qui guident les comportements individuels ne sont pas forcément identiques à ceux qui régissent l'éthique sociétale à un niveau plus général. Enfin, la notion d'« identification personnelle et sensibilité éthique » fait référence à la nécessité d'élargir à une gamme plus étendue de manquements à l'éthique la sensibilité de l'individu aux fautes d'éthique qui le touchent personnellement. Les principes qui sous-tendent ces notions sont évoqués à plusieurs reprises au cours de l'exposé, dans lequel sont examinées cinq questions relatives aux liens entre l'architecture et l'éthique.

Tout d'abord, d'où vient le souci actuel envers l'éthique? La réponse pourrait être que nous éprouvons un sentiment de vide à la fois dans nos vies individuelles et dans notre vie sociale; la poursuite de l'éthique représente une démarche en vue de combler ce vide. Les professions et leurs praticiens, y compris l'architecture et les architectes, sont des acteurs importants dans cette recherche.

Mais cette démarche a-t-elle donné des résultats? La quête de l'éthique représente en elle-même un changement important mais elle a aussi suscité d'autres modifications. Lorsqu'on examine une question touchant l'activité humaine, par exemple, c'est d'abord à ses aspects éthiques qu'on songe aujourd'hui, les autres aspects passant au second rang. Cela confère à l'éthique une place dominante dans plusieurs de nos activités personnelles et sociétales, ce qui n'était pas le cas auparavant.

En troisième lieu, en quoi la recherche de l'éthique et les motifs qui l'animent concernent-ils l'architecture? Les professions, y compris celle de l'architecte, jouent un rôle important dans l'élaboration d'une symbolique dans les sociétés modernes et séculières. De concert avec d'autres professions, l'architecture confère ainsi à la société une sensibilité profonde où se conjuguent l'éthique et le juridique.

Quatrièmement, pourquoi le développement de la technologie moderne a-t-il donné lieu à un débat public sur l'éthique? Les raisons en sont multiples: grâce à la technologie, nous pouvons aujourd'hui susciter des événements qui ne se produiraient jamais sans elle dans la nature, nous détenons sur la nature et sur le destin humain des pouvoirs sans précédent, nos valeurs subissent des transformations sensibles. En outre, la technologie accroît aussi bien nos certitudes (et avec elles, la crainte du connu) que nos incertitudes (et avec elles, la crainte de l'inconnu), non seulement à cause des risques qu'elle présente mais aussi en raison de l'angoisse que suscite la multiplicité des choix que la technologie rend possibles; ces craintes ont même mené au développement d'une véritable « technophobie ».

Enfin, quel rôle l'éthique peut-elle jouer dans nos rapports avec la technologie? Elle contribue à réduire le sentiment d'angoisse associé à la technologie parce qu'elle nous transmet un mécanisme permettant de porter des jugements de valeur.

En conclusion, nous avons besoin de versions séculaires et modernes des vertus traditionnelles de la foi, l'espérance et la charité. Nous avons toujours besoin d'avoir la foi ou de faire confiance en quelque chose ou quelqu'un. Nous avons toujours besoin d'avoir des rêves ou des aspirations qui vont au-delà de nous-mêmes et que nous visons à réaliser. Nous avons aussi toujours besoin de faire reposer notre existence individuelle et sociétale sur l'affection, la sollicitude envers l'autre et l'optimisme – qualités étroitement liées à la notion de charité.

INTRODUCTION

"Thin" and "Thick" Concepts of Ethics[1]

It is often interesting to free-associate about the titles chosen for conferences. My free association about the title of this conference – "Architecture, Ethics and Technology: An Interdisciplinary Symposium" – led me to wonder whether ethics was perhaps the ham in the sandwich, squeezed perfectly flat by architecture, on the one side, and technology, on the other. This image, in turn, brought to mind what has been called a "thin" or minimalist concept of ethics. According to this concept, persons act ethically as long as they do not breach the law, and the law should enshrine only those requirements on which there is general societal consensus.[2]

The second free association provoked by this title is that ethics is a bridge between architecture and technology. This is a much more expansive concept of ethics – one suggesting that architecture, in utilizing technology in order to realize its creations, will reach across the bridge of ethics. Conversely, the impacts – on persons and on our world – caused by the application of this technology rebound upon architecture in a kind of feedback loop, again across the bridge of ethics. This metaphor could symbolize architecture's concern to meet fully all ethical requirements with respect to the impact of the technology it uses. This would represent a "thick" or "rich" concept of ethics, which goes beyond the minimal requirement not to break the law and takes into account the harms and risks surrounding both the professional and technological activities deployed by architecture (for example), with respect not only to present generations but to future ones as well.

"Thin" and "thick" concepts of ethics can also be seen as end-points on a continuum. Persons, including architects, may have different views of what is required in terms of the ethical position that should be taken on this continuum in any given circumstance. All are likely to regard their positions as being ethical, which raises the question whether some ethical positions are more ethical than others.

Applied Ethics

Ethics has long existed as a subdiscipline of philosophy. "Applied ethics" has evolved recently and can be seen as a subdiscipline of ethics, particularly relevant to professional ethics.

The distinction between ethics and applied ethics can also be expressed as the difference between "talking about ethics" – which, by and large, has been the preoccupation of philosophy in relation to ethics – and "doing ethics."[3] The latter involves taking ethics out into the world in order to apply it to concrete situations – for example, to issues raised in the context of a hospital, an architect's office, a space research centre, a building site, a university campus, or even a battlefield, especially when we are confronted by a globally televised war. (During the Gulf War, discussions took place concerning military ethics and the ethics of war – terms which, for some, represent an inherent contradiction.)

Fundamental to applied ethics and to the ethical analysis that this entails is that it must be undertaken as a transdisciplinary activity – that is, it must use structured methodologies and analyses that will facilitate and organize the input of the various disciplines relevant to a given issue and result in an integrated "knowledge product."[4]

Few ethical issues have "right" answers – that is, single, clearly correct answers. This is reflected in an essential feature of the role of ethics consultants – namely, that they are largely non-directive in helping others to analyse the ethics of situations that they encounter. "Good" ethics depends on "good" facts; and good facts depend on adequate input from a sufficiently broad range of persons. This "right" process usually does not occur spontaneously; it needs thought, research, and organization. We need to be aware of this, even though it is not always possible to engage in an ideal process for ethical analysis in all situations.

Individual and Societal Ethics

We also need to recognize that ethics is relevant at both the individual (micro) level and the societal (macro) level. But the principles and processes that apply at one level may not necessarily apply at the other.[5] For example, maximization of efficiency and effectiveness may be a valid consideration in decision making at the governmental or institutional level, but giving priority to these principles may not be acceptable at the individual level. One of the clearest examples of this proposition pertains to health care. It may be ethical for a government to decide not to allocate resources to certain treatments or to fund one kind of care but not another in order to maximize efficiency and effectiveness, but it would be unethical for a physician to make the same decisions, for the same reasons, relative to an individual patient with whom he or she is in a treatment relationship. The physician has a primary obligation of personal care to the patient, which requires that the patient's best interests be given priority. Similarly, individuals may have valid claims to rights to privacy, but claims to the same rights by multinational companies, for example, many constitute an abuse when assessed from the perspective of ethics.[6]

Personal Identification and Ethical Sensitivity

My final introductory comment concerns the relationship between one's personal identification with ethical issues – in particular with respect to breaches of ethics – and ethical sensitivity in general.[7] This can be illustrated by an example cited during this conference – that of the recent destruction in the city of Dubrovnik. We all identified with the horror of what was happening; the architects, in particular, mourned the tragic loss of many beautiful historic buildings. As architects, not only did you think about what was happening to that city, you "felt for" that city. Similarly, there is a difference between thinking about ethics and feeling about ethics. We need to feel concern, sadness, anger, and sometimes horror in relation to breaches of ethics with which we identify personally, and then we need to generalize those feelings to other breaches of ethics with which we may not identify so readily or so personally.

At this conference, our concern for the buildings in Dubrovnik was immediately amplified to include concern for the persons who were harmed by the same destructive activity; indeed, many of us felt a strong sense of guilt that we had not thought of the persons first. (As an aside, extensive television coverage of wars often warps our priorities and blunts our sensitivities. Yet ironically, at the same time, it provides opportunities for greater sensitivity through personal identification with the human suffering that is inflicted.) Interestingly, you might not have felt quite so deeply or so personally about the Dubrovnik tragedy, had your special sensitivity to what was happening to the buildings not been aroused. This is a valuable lesson; we can promote ethical conduct by encouraging people to enlarge their personal identification with certain ethical concerns into a wider sensitivity to other ethical concerns with which they might not, spontaneously, personally identify.

QUESTIONS

In the remainder of this paper, I wish to address, briefly, five questions. First, why, suddenly, is there this widespread concern with ethics? Second, has our concern for ethics caused any change? Third, what relevance could this search for ethics, and the causes of this search, have for architecture? Fourth, why has the development of modern technology given rise to a societal discussion of ethics? And, fifth, what role can ethics play in dealing with technology?

Why Are We Concerned With Ethics?

Why, suddenly – at least, it appears to be sudden – is there this widespread concern with ethics? For example, why is this conference on "architecture, ethics, and technology" being held? I suggest that the search for ethics is associated with a new feeling that we have of a space – a void, even – in our individual and collective lives, and with a need to fill this appropriately. The need to fill a space or void will no doubt be a familiar feeling to architects. This newly recognized void results from the concurrence of a variety of factors.

Those factors include a decline in adherence to organized religion. Even if religious adherence was more of a façade of spirituality than a reality for many persons in the past, religion fulfilled important functions both for society and for individuals. One such function was that participation in religious ritual provided some commonality of culture among the members of society. In this regard, sociologists might claim that museums have replaced churches as the symbols of some common culture and provide an opportunity to experience that commonality.

The decline in adherence to organized religion has also led to the recognition that there is no consensus on values in our society. Again, there may not have been such a consensus before, but at least there was the appearance of a consensus when we purported to have a dominant religious tradition. This recognition also contributes to the perception that a space or "values vacuum" has opened up.

We may also feel that it is now more difficult to know what is and is not ethical conduct, in part because of a powerful realization that good economics is not necessarily good ethics. We used to assume, and even preach, that good economics was good ethics. For example, the puritan or Calvinist ethic was based on a belief that God offered economic rewards to the good person and, conversely, punished the bad person by withholding such rewards. Consequently, there was an intimate link between doing well economically, being seen to be a good person, and having one's conduct characterized as morally acceptable. We have come to realize that many situations may be examples of exactly the opposite phenomenon – in other words, that good economics not only is not necessarily good ethics but indeed has often proven to be very bad ethics. We need only look at serious environmental damage – a major concern of those working in applied environmental ethics – to find such examples.

When listening to many of the papers given at this conference, it seemed to me that there was a heavy emphasis on one important, but relatively narrow, stream of ethics – namely, the relationship between ethics and aesthetics. Moreover, it seemed to be suggested that, like economics and ethics in the past, aesthetics and ethics are necessarily consistent aims – in other words, that in architecture good aesthetics mean good ethics. I suggest that this is not necessarily so and that one must consider whether, in some instances, giving predominance to the value of aesthetics could indeed turn out to be bad ethics.

Other factors giving rise to our sense of a space or void include the fact that we may have developed a greater sensitivity to harming others and even ourselves. There is a renewed recognition, I believe, that there are "rights" and "wrongs" in many situations. This certainly does not mean that we need to believe that there are absolute, universally applicable rules about what is right and wrong in all, or even most, situations. If one examines various schools of applied ethics, some of which are still evolving, at one end of the continuum there are what could be called the deontological schools, which are based on substantive rules or principles as to what is right and what is wrong. But other schools of applied ethics – for example, hermeneutical or situational schools, which can be regarded as being at the other end of the continuum – propose that what is right and wrong is a matter of interpretation or analysis in each situation. In some senses, these latter schools rely more on the use of a "right" process than on substantive rules about right and wrong. The process needs to be applied in each situation in order to determine a range of acceptable behaviours and to determine which behaviours would cross the borderline, indicating they are unacceptable and wrong. The important point that needs to be made is that no matter which school of ethics is adopted, all recognize that there is right and wrong, although they differ from each other with respect to how one determines this and may differ as to what constitutes right and wrong in given circumstances. It may seem an unusual statement, but the very notion that there is right and wrong may need to be reaffirmed; at the same time, the fundamentalism, fanaticism, and extremism often associated with such affirmations must be carefully avoided. There has been a tendency in recent times to abandon the notion of right and wrong, together with religion. The search for ethics may be a corrective mechanism in this respect.

Finally, part of the space or void that we are experiencing and seeking to fill stems from a growing distrust of many traditional societal institutions – especially, I think, of politics, politicians, and political institutions.

In summary, then, there is a feeling of a space or a void in both our individual and collective lives, which we are seeking to fill. I suggest that even in a secular world, some of the essential components required to fill that space are the equivalents of the old virtues of faith, hope, and charity. We still need to have faith or trust; we still need to have dreams, visions, and aspirations beyond just ourselves, to believe in and pursue so that we can travel through life in hope; and we still need to base our individual and societal lives on caring, empathy, and optimism – qualities that are closely related to the notion of charity. We may have come to a secular realization that we need these virtues in their broadest sense, and it may be the absence of the façade of religion that has caused us to become more acutely aware that we have a space for a spirit of some kind. This does not necessarily mean that we need to engage directly in religious activity. But it does mean that we have needs beyond our immediately temporal and materialistic ones.

We are using the arts and sciences, and, secondarily, the professions as one means to fill this space. It should come as no surprise that architecture is included among the space-filling professions. In creating or filling physical spaces, architects are also creating positive and beneficial spaces for our spirits, including our societal spirit.

But these same spaces for spirit could also be filled by wrong-doing and harm – it has even been suggested, by a new Hitler. The current neo-Nazi movement, particularly among young people, is frighteningly present. At a more everyday level, as we saw in slides presented at this conference, there are many disillusioned young people, just in our own city of Montreal, who are searching for something that will give meaning to their lives. What they ultimately find could be seriously harmful to themselves and to society. That possibility is the downside of the presence of the space to which I refer. In short, this space is open for enormous good, but it could equally be filled by evil, and the scope of both this opportunity and this danger may have been augmented relatively recently. Thus one function of the search for ethics is to establish a protective mechanism against this space being wrongfully filled.

We are also searching for ethics because of the immensely complex situations and possibilities to which our new technologies give rise. These range from the exploration of vast outer space to that of vast inner space through genetic and biotechnological interventions that include mapping the human genome – the genetic blueprint that makes up each of us – and undertaking biopsy on a test-tube human embryo in order to determine, before we place the embryo *in utero*, what the resulting person is likely to die of in adulthood. Such information is mind-altering; it places us in a situation where we can choose an embryo's life on the basis of its future death. Similarly, our explorations of vast outer space provide us with knowledge to which our individual and collective psyches will need to adjust. We no longer think that our world is the centre of the universe, but when we realize just how very minute we are in comparison with our universe, a mind shift is required in order to accept the implications. I have suggested elsewhere that these two new poles of the continuum of our exploration of knowledge may cause a paradigm shift not only in science but in society itself.[8]

We may also be searching for ethics to counterbalance the trends towards uniformity and depersonalization of, and disidentification from, our world and the people in it. These trends are related, in part, to the development of new technologies because the latter challenge our concepts of who and what we are. The trends towards individualism, self-determination, and autonomy – which are epitomized in theories of individual rights – are often regarded as counter-trends relative to those listed above, but they are not necessarily corrective ones. Paradoxically, they may even, at times, result in similar depersonalization and disidentification by causing us to treat individual persons as isolated from the communities of which they are members. Ethics may be part of a search to establish a structure within which we can strike an acceptable balance or a new type of accommodation between respect for individual rights and protection of the community.

It may be that, in the future, we will evolve new structures that will accommodate matters currently viewed as "impossible combinations." At present, we often perceive ourselves as having to choose between respecting individual rights or protecting society, but new insights or approaches may enable us to do both at the same time. For example, there has been loud and acrimonious debate about whether we should respect the individual rights of persons with AIDS or act to protect the health of the community in general. It is only recently that we have recognized that both of these aims are consistent, because respect for individual rights is most often the approach that best protects the community by reducing the transmission of HIV.[9] Similarly, we often consider that there is necessarily conflict between reducing the cost of health care and providing the best health care to people, but that view may not be correct. One study has shown that one group of low-risk pregnant women who were cared for at lower cost by family medicine practitioners had fewer complications than a matching group of women cared for by more expensive, specialist obstetrician gynaecologists.[10] This outcome probably resulted from less routine use of high-tech interventions by the family practitioners and from the avoidance of the risks associated with these interventions, especially where these risks outweighed potential benefits. In short, there was no conflict between reducing the cost of health care and providing the best health care in those circumstances.

Examples such as the above should make us aware that we need to ensure that we do not have what can be called standard or Pavlovian reactions to situations, such that we always and automatically equate certain matters and disassociate others. New insights, new theories, new perceptions need to be gained in order to create the paradigms that will guide our world in the future, and these may well cause us to see that world differently. Ethics will be an important informing principle in evolving these new paradigms.

Has Our Search for Ethics Caused Change?

Our findings from our search for ethics have caused many important changes in many aspects of our individual and societal lives. These will not be discussed here. Rather, I wish to point out that the search for ethics is, in itself, an important change that has caused further changes. Among the most important and fundamental of these changes is that in the order of our analysis. To take law as an example, increasingly we are changing from an analytical process that can be described as going from law to ethics (or indeed, sometimes only considering the law and excluding ethics) to one that goes from ethics to law. In other words, we tend increasingly to start by asking the question, What is required in terms of ethics? We then examine whether our law is in accord with this and, if it is not, try to do something about the law. This contrasts with the former approach, which consisted of asking, What is the law? and, often, just assuming that what was legal was necessarily ethical.

If this mode of analysis were applied to architecture, one would ask, first, What are the ethics of situations that involve architecture and architects? and then, How can the ethical requirements be translated into good architecture? The architectural drawings for the concentration camp at Auschwitz were discussed earlier at this conference. If the primary question had been about the ethics of what was being planned rather than about the camp's architecture, those "beautiful" drawings would never have been made.

We need to realize, in the context of this discussion, that starting points for analysis – our initial presumptions – are not neutral. They colour the whole discourse of an analysis. They establish the base rules from which we work – what the rules are and what, if any, are the exceptions. This means that initial presumptions establish burdens of proof because persons who rely on an exception normally have the burden of proving that the exception is justified. We need to be aware that when we take the starting points of our analyses as givens, we are not necessarily correct in doing so. We have already made a choice – often a very important value choice – in determining these starting points. The extent to which ethics will predominate in our decision making relevant to architecture will depend, in part, upon whether ethics constitutes a basic premise of the analysis of what is being done in and by architecture.

What Relevance Could the Search for Ethics, and the Causes of This Search, Have for Architecture?

Ethics is not just a fashion. The search for ethics represents our generation's revolution in consciousness,[11] and a major part of this search is being carried out in the professions, which include architecture. In this secular, end-of-the-twentieth-century, western world, professions are value-forming, value-carrying, value-affirming, and value-destroying institutions. Our actions have these effects on values, not just for the individual professionals or other persons directly involved or affected by these actions or for the professions concerned, but for society as a whole. This can be described as the horizontal analysis – the horizontal impact of our actions. We also need to recognize that what we do as professionals and professions not only has a conscious or practical aspect – in particular, in relation to values – but also that it can arise from unconscious sources and establish symbolism. In other words, vertical analysis makes us aware that there are effects at multiple levels. The tangible products of the exercise of the profession of architecture, in particular, have highly important symbolic effects that often span many generations.

Applying the insights that Freud gave us, could we postulate that not only do individuals have a psyche but so, also, does society? If so, how does architecture contribute to, affirm, or detract from, society's psyche? Symbolism is a very important factor in setting what can be called the ethical and legal "tone" of a society. What role does, and should, architecture play in establishing symbolism for society and, thereby, its ethical and legal tone? Would architects think somewhat differently about what that they do if they were more conscious that they are not dealing just with architecture but with establishing, maintaining, or changing the structure of society itself?

Architecture builds walls. If we accept that architecture also sets symbolism for society, we need to ask not only, What is architecture walling in and walling out in a material or tangible sense? but also, What is it walling in and walling out in a symbolic sense in any given case? The poet Robert Frost wrote: "Something there is that doesn't love a wall, that wants it down."[12] Some unethical walls have been brought down by architecture – an example that comes most immediately to mind is that of the impediments that obstructed access to buildings by handicapped persons. Bringing down such walls, both in fact and symbolically, is a major contribution of architecture to the promotion of human rights. As this example shows, both real and symbolic walls are relevant to architecture and to ethics.

No discipline or profession can exist in a vacuum, although we often act as though they do. We constantly influence others, and are influenced by them, with respect to concepts, theories, and paradigms. Take, for instance, the concept of a space. I have suggested that our search for ethics is linked to a certain perception in this respect. The concept of a space is also relevant with regard to our own perceptions of ourselves. Freud caused us to recognize that each one of us can be divided into two spaces – an inside and an outside – of which we were not necessarily conscious before the development of psychoanalysis. It may also be that "invasive" procedures in medicine – especially major surgery, which, as a routine intervention, is a twentieth-century phenomenon – contributed to this perception that we have an inside and an outside. We can ask whether this radical change in our vision of ourselves could be associated, for example, with the development of notions of public and private space, and of the right to privacy. Often, we take for granted concepts such as privacy and the right to have it legally protected, without realizing how very modern these concepts are. Moreover, we may not be fully aware of the range of important protections that are associated with a concept such as privacy and that have evolved only because of its development – for example, protection from sexual harassment or from coercive, invasive advertising techniques.

We could then ask, What role do concepts of public and private space play in affirming, or detracting from, concepts, claims, and rights of the individual, compared with those of the group or society? And how is the balance to be struck between individuals and society when they are in conflict? Much of the modern law of privacy has been developed by focusing on a differentiation between public and private space and providing legally protected rights of privacy within the latter and denying them in the former. The most recent arguments in this respect involve the proposition that homeless persons are being wrongfully deprived of a right to privacy because they have no private space within which they can live and from which they can keep others from entering through the exercise of a privacy claim, and that this is a breach of their fundamental rights.[13] In short, there may be no right to privacy outside the private space that one is entitled to inhabit.

In summary, architects need to ask, What does, and should, architecture contribute to society's psyche and, through this, to its ethical and legal tone? To answer will necessarily force architects to reflect on ethics.

Why Has the Development of Modern Technology Given Rise to a Societal Discussion of Ethics?

Modern medical technology, and especially the techniques that have made heart transplants possible, have had a major impact on individual and societal psyches. Transplantation was a dramatic, non-natural event: it did not simply seek to repair or supplement nature but gave rise to an outcome that could never happen in nature – one person living with the support of another's (a dead person's) heart. We saw ourselves as achieving the impossible, doing the unthinkable.[14] It is difficult to believe, unless one sees archival films on heart surgery, that until the end of the 1940s it was thought that the heart set the limit of modern surgery: the heart would always be the one organ in the body on which a surgeon could not operate. Not surprisingly, the first openheart surgery operation was regarded as a miracle, but by the late 1960s surgeons had transplanted a heart from one person to another.

Technology also gives control. It changes events that were formerly matters of chance and makes them matters of choice. With choice comes a greater feeling of responsibility. Technology means that there is now more that we can do than we will do or can afford to do – and that among the things that can be done with technology, there are some that we ought not to do. The advent of these possibilities has resulted in many difficult choices for both individuals and society.

Technology is not neutral in its impact on values. One need only consider the debate surrounding birth and death technologies, among others. In addition, the impact of our approaches to technology in a given area is not limited to the values relevant to that area. Moreover, discussions of values may sometimes be undertaken through discussions of the use of technology, even though the issue to which the values are relevant may not involve technology. A case in point is the current debate on euthanasia, which is often strongly attached to a discussion of technology, in part because people sometimes regard modern medical technology as a means of prolonging dying rather than prolonging living. But the debate could and should be carried out independently of this issue. The right of competent adults to refuse treatment, which is the relevant concern with regard to the overuse of medical technology, is now almost universally recognized. Euthanasia – an intervention the primary aim of which is to kill the person subject to it – can and should be debated quite separately.

Technology increases the sense of uncertainty and the anxiety that accompanies uncertainty. This occurs because technology enables us to increase our knowledge base: the more we know, the more we know that we do not know – as the radius of the light expands, so does the circumference of darkness – and that leads to uncertainty and anxiety. In the field of embryology, for example, it was thought at the beginning of the 1970s that we knew practically everything there was to know about human conception, but by the end of the 1980s we knew much more – enough to tell us that we knew hardly anything.

Technology provides information about harms and risks. As just mentioned, this can raise levels of uncertainty and anxiety, but the certainty of the information that technology provides can also make us more anxious. Our individual or collective psyches may find it difficult to cope comfortably with information that technology gives us regarding the harms that are likely to befall us (as in the case of information provided by genetic testing, for example).

We may also, inappropriately, seek to reassure ourselves that we can avoid harms and eliminate risks through the use of technology. One example is the widespread concern raised in relation to HIV-positive health-care professionals and the perceived risk that they might represent to patients. Here is a situation in which there is no substantial risk but on which many people have chosen to focus their anxieties about HIV infection and AIDS in general. They may have done so in order to reassure themselves that they can eliminate any risk of HIV infection to themselves through the use of technology – in this case, HIV-antibody testing of all health-care professionals.

Technology can be used to control us. In particular, our privacy can be invaded in a controlling manner. "Smart cards" – cards with computer chips containing a person's entire medical record, for example – are one current example of a technology that raises this type of concern.

Technology presents risks: our world could be destroyed by technology "gone wild." Some of our major environmental problems are current examples of this; and there is increasing concern, for instance, about the release of genetically engineered micro-organisms – a realistic concern, given the progress and potential of modern genetic technology. Even technologies that have overwhelming benefits carry harms and risks; the risks of modern medical technology are a good example in this respect.

Technology also gives us the anxiety of choice – for example, the anxiety that would be associated with choosing the sex of one's children, again a possibility with modern technology. One study has reported that if we were able to choose the sex of our children, many people would choose a birth order that would match that of the children in their favourite television "sitcom." We need to be aware of such possibilities so as to ensure that the technology that gives rise to them is not misused.

We may even have developed technophobia.[15] We may have selected some technologies, and not even the most harmful, as the focus of this phobia. In a paper presented at a recent international meeting, for example, it was reported that in some parts of the United States there was a strong demand from parents that asbestos already in place in school buildings be removed.[16] And yet, the risk is virtually non-existent as long as the asbestos is not disturbed; at the same time, there is some risk in removing the asbestos as this can cause the dispersion of asbestos fibres into the air. Moreover, it was found that there was not the same level of concern regarding the use of pesticides in the yards of the same schools, although the risks from these substances far exceeded those presented by the asbestos. There are opportunity costs in acting on such phobias; for example, the money spent on removing the asbestos could have been spent in much more productive ways to reduce risks for the children concerned.

One of the insights that the recognition of technophobia provides is that even though our technologies depend for their development on rational, logical science, we humans are not exclusively rational beings and cannot treat our relationship with technology as though we were. In particular, we need to be aware that our reactions to technology cannot be judged just from a rational and logical perspective, if they are to be properly understood, assessed, regulated, and utilized.

Finally, our spirits can be annihilated by faith in "technology as God." This can result in a de-humanized and depersonalized world. At the individual level, an example of this type of phenomenon is provided by children who are unable to form friendships because their major relationships have been only with the characters in video games.

What Role Can Ethics Play in Dealing with Technology?

Ethics can help us to deal with technology and, in particular, with the uncertainties, anxieties, and fears that are associated with technology. It may allow for hope and realistic optimism in situations where these attributes are otherwise difficult to maintain. Without hope and optimism, we certainly cannot live life in a full sense; hope and optimism are the wellsprings of our creativity, imagination, enthusiasm, and industriousness. Ethics also helps to reduce anxiety by providing us with a value-assessment mechanism, the very presence of which is reassuring.

Consideration of some concepts of medical ethics – which is, at present, the most highly developed field of applied ethics – could give some idea of how applied ethics could work in architecture. In particular, the way in which medical ethics has dealt with concerns raised by medical technology could provide insights for architecture. For example, the doctrine of informed consent would merit exploration.[17] Issues that would be relevant here include whether a population can or should be asked to consent to the harms or risks attached to the use of technology – whether, for example, we may impose risks or harms on persons who are not aware of such risks and whose consent it is not possible to obtain in advance because they cannot be identified prior to the harm being caused.[18] If consent is relevant, what should we do about the fact that future generations are unable to provide consent when the risk or harmful activity is undertaken?[19]

We could also consider the change from "blind trust" (a paternalistic model) to "earned trust" (an egalitarian model) in medicine.[20] These concepts are related to the doctrine of informed consent, in that the requirement for informed consent promotes egalitarian relationships at the expense of paternalistic ones. This is true because the sharing of information and the subsequent consensus, which are required by the doctrine of informed consent, are antithetical to a paternalistic, blind-trust relationship. It would be fruitful to investigate which type of trust – earned trust or blind trust – underlies the relationship between architects and their clients, and between architects and society.

Other concepts from medical ethics that it would be relevant to consider in the context of architecture include the duty to warn and the principle of proportionality. Professionals may have a duty to warn, even well after a professional relationship has been terminated, when harms and risks related to professional activity later become apparent. These duties may also require ongoing monitoring of the results of professional activities so as to be aware of circumstances that would warrant a warning to others. The principle of proportionality means that benefits and potential benefits must outweigh harms and risks, recalling that some harms are never acceptable, no matter what the benefit promised. Value judgments are involved here as to what counts as benefit and as harm, and as to what weight should be given to each. This requires answers to such questions as, Who should make these value judgments? In case of conflict, whose values should predominate? Which procedures should be followed? To what purpose?

Other areas of medical ethics that could provide insights for architecture include those of privacy, access to information, duties of follow-up care, experimentation involving human subjects, and allocation of resources. For instance, in the context of allocation, efficiency is not always an overriding goal in medicine – certainly not at the level of individual patient/physician relationships; rather, humane, individualized care is more important. One may ask what are, should be, and should not be, overriding goals in architecture. Within this context, one may consider whether priority should be given to values of aesthetics, efficiency, equity, need, or relief of suffering, for example. One could also consider concepts of human rights in medicine – for example, to basic medical care – and compare this with a basic right to shelter and a home. Within this context, the plight of the homeless must be given urgent consideration. Moreover, as is true in medicine, there will also be a need, in some situations, to strike a balance between claims of individuals and those of the community.

CONCLUSION

The search for ethics is an exciting voyage of discovery. The great danger is that we might trivialize or routinize it. The survival of our individual and collective spirits depends upon this not occurring. Architects and architecture will play an important role in designing and building the tangible and intangible spaces – the "spaceships" – in which we can travel in body, mind, and spirit into the future. These spaces should, indeed must, be designed to enable us to undertake these voyages with hope, wisdom, courage, humanity, and a true sense of humour.

NOTES

1 M. Walzer, "Moral Minimalism," in *From the Twilight of Probability: Ethics and Politics*, ed. W. R. Shea and A. Spadafora (Canton, Mass.: Science History Publications 1992), 3–14.

2 H. T. Engelhardt, "Recapturing the Distinction Between Ordinary and Extraordinary Care: Towards Defining a Concept of Basic Health Care," Astra lectures in ethics, McGill Centre for Medicine, Ethics and Law, McGill University, Montreal, April 1991.

3 For a discussion of the difference between the use of language to "talk about" an activity (e.g., law or psychiatry) and "to do" that same activity, see M. A. Somerville, "Labels versus Contents: Variance Between Psychiatry, Philosophy and Law in Concepts Governing Decision-Making," *McGill Law Journal* (forthcoming).

4 M. A. Somerville, "La transdisciplinarité, vague de l'avenir: Comment préparer nos rivages à l'accueillir" ("Transdisciplinarity, the Wave of the Future: Building the Foreshore"), in *Entre Savoirs*, proceedings of the UNESCO international symposium on interdisciplinarity, Paris, France, April 1991 (Toulouse: Erès 1992), 117–36.

5 C. Fried, *Medical Experimentation, Personal Integrity and Social Policy* (Amsterdam: North Holland 1974).

6 D. Thompson "The Politics of Professional Ethics," Astra lectures in ethics, McGill University, May 1992.

7 M. A. Somerville, "Human Rights and Medicine: The Relief of Suffering," in *International Human Rights Law: Theory and Practice*, ed. I. Cotler and F. P. Eliadis (Montreal: Canadian Human Rights Foundation 1992), 505–22.

8 M. A. Somerville, "New Perceptions, Old Values from Inner and Outer Spaces" ("Spacing-out and Spacing-in: Searching for the Purple-Pink Middle"), *Canadian Speeches: Issues of the Day* 6 (1992): 65–8.

9 J. Osborn, "AIDS: Politics and Science," *New England Journal of Medicine* 318 (1988): 444–7.

10 E. E. Rosenberg and M. Klein, "Is Maternity Care Different in Family Practice?: A Pilot Matched Pair Study," *Journal of Family Practice* 25, no. 3 (September 1987): 273–42.

11 M. A. Somerville, "A Generation's Revolution in Consciousness: The Search for Ethics," Falconbridge lecture, Laurentian University, Sudbury, Ontario, October 1990.

12 "Mending Wall," *The Poems of Robert Frost*, The Modern Library (New York: Random House 1946), 35–6.

13 M. D. Granston, "From Private Places to Private Activities: Toward a New Fourth Amendment House for the Shelterless," *Yale Law Journal* 101 (1992): 1305–30.

14 M. A. Somerville, "Biotechnology: Doing the Unthinkable," keynote address, 57th annual Couchiching conference, Geneva Park, Ontario, August 1988.

15 I am indebted for this idea to Dr Joan Kahn of the McGill Centre for Medicine, Ethics and Law.

16 C. Uram, "Environmental Code 2000," paper read at the fifteenth conference on the Law of the World, World Jurist Association, Barcelona, Spain, October 1991.

17 M. A. Somerville, "Structuring the Issues in Informed Consent," *McGill Law Journal* 26 (1981): 740–808.

18 See, for example, Federal Centre for AIDS Working Group on Anonymous Unlinked HIV Seroprevalence, "Guidelines on Ethical and Legal Considerations in Anonymous Unlinked HIV Seroprevalence Research," *Canadian Medical Association Journal* 143 (1990): 625–7.

19 M. A. Somerville, "Justice Across the Generations," *Social Science & Medicine* 29 (1989):385–94.

20 J. Katz, *The Silent World of Doctor and Patient* (New York: Free Press 1984), 100–3.

PART ONE / PREMIÈRE PARTIE

ETHICS AND ARCHITECTURAL REPRESENTATION IN THE AGE OF SIMULATION

ÉTHIQUE ET REPRÉSENTATION ARCHITECTURALE À L'ÈRE DE LA SIMULATION

MASTER PLAN FOR THE OLD PORT OF MONTREAL

PETER ROSE

Le plan directeur pour le Vieux Port de Montréal vise une parcelle de terrain entre le fleuve Saint-Laurent et le quartier historique de la ville. Le projet, présenté dans le contexte d'un concours annoncé en 1989 par la Société du Vieux Port de Montréal, comporte un dévoilement archéologique des ruines industrielles accumulées sur le site. Le concept élaboré ici révèle également les ambitions d'une époque où les nouvelles technologies étaient perçues avec optimisme, une époque enthousiasmée par les possibilités qu'offraient l'expansion et le développement de l'industrie.

Le projet soulève une question d'éthique dans son attitude face au programme. Toute intervention sur un tel site industriel à caractère historique demande une attitude qui cherche d'abord à révéler cette couche historique avant d'appliquer un programme approprié, au lieu d'imposer sur le site une solution programmatique qui ne ferait qu'accommoder les caractéristiques des lieux.

In this paper I will present the thoughts, the ideas, of a practising architect rather than those of a theoretician or an architectural historian, concerning a project on which I have worked for several years – the design of a master plan for the Vieux Port de Montréal.[1]

The project addresses a number of issues relevant to a consideration of architecture, ethics, and technology. It uncovers an archaeology of technology and reflects an era when new technologies were viewed with optimism and

Montreal Harbour from the Custom House, c. 1885 (Notman Collection, McCord Museum of Canadian History)

enthusiasm, and when a new sense of scale and industry presented heroic possibilities. It also raises an ethical question concerning this submerged technology by suggesting that any intervention in a historic industrial site such as this should first seek to reveal its history and then develop an appropriate program, rather than impose a programmatic solution that would simply make accommodations for the site.

The project was submitted for a competition announced in 1989 by the Société du Vieux Port de Montréal – the authority that has jurisdiction over the site. The Vieux Port sits on reclaimed land between the St Lawrence River and the historic *quartier* of the city – le Vieux Montréal. The old city and the harbour have coexisted in a symbiotic relationship for more than 150 years. Because of limitations of infrastructure and changes in the nature of the shipping trade in North America since the 1950s, the port of Montréal has gradually shifted its operations downstream. Now the historic port no longer functions as a harbour facility. By the late 1970s, most of its structures had been demolished, leaving a vast, vacant space between the city and the river.

There have been many attempts during the last decade to do something with the site. Most of these attempts have been centred on programs intended to accommodate specific recreational or cultural uses. And because politics and fashions change, they have shown a tendency to evaporate over a ten-year period.

The project designers took the view that, given the nature of the site, historical research rather than a predetermined program should be the starting point for design. Thus our design process was aimed at developing an architecture that would reveal and reinterpret the site, that would support diverse and changing programs but would also be capable of existing independently of any program.

The site of the Vieux Port marks the location of the founding of the city in 1642, downstream from the Lachine Rapids. The rapids (and later the Lachine Canal) made Montreal the transfer point between the interior of Canada and the sea lanes to Europe. In great measure, the development of the port drove the economy of the nineteenth-century city. In the twentieth century, grain arriving by ship and by train from western Canada was stored in the great elevators of Montreal and transferred by a monumental network of conveyors to the holds of ocean-going vessels headed for Europe. The decline of the Vieux Port since mid-century has been caused largely by the opening of the St Lawrence Seaway and by the growth of container shipping, which requires huge parcels of land for its operations, as well as by the development of air transport.

In 1981, ownership of the site was transferred to the Société du Vieux Port de Montréal. Its jurisdiction covers a territory a mile long and 350 feet wide between the waterfront and the city, as well as four piers 300 feet in width by 1,200 feet in length. It also extends westward to include the first two locks of the Lachine Canal. The canal, rendered obsolete by the St Lawrence Seaway, was filled in along this portion in the 1970s, but it is now gradually being restored along its full length as a linear park.

As a waterfront, the Vieux Port enjoys a physical relationship with its urban environment that is rare for a North American city. Unlike Boston, Toronto, or New York, whose ports are severed from the city by railway

Aerial photograph from the Port of Montreal (Archives du Port de Montréal, 1975)

yards and highways, the Vieux Port provides a link between the city and the river. The port also relates to one of the oldest and most authentic historic districts anywhere in North America. Largely bypassed and ignored in the twentieth-century development of the city, Old Montreal still has a texture, pattern, and structure that were established in the eighteenth century, and a physical character that has remained largely unaltered since the late nineteenth century.

As I studied the site, it seemed to me to have a special intensity, partly because of its location: it is the last point of calm water at the end of an up-river journey – the last navigable point, whether by canoe or ocean liner. It is the place where ice dams used to form when the ice on the river broke up in the spring, causing water build-ups of twenty feet or more that would flood the city. The port is pinned in this one spot and has been since its be-

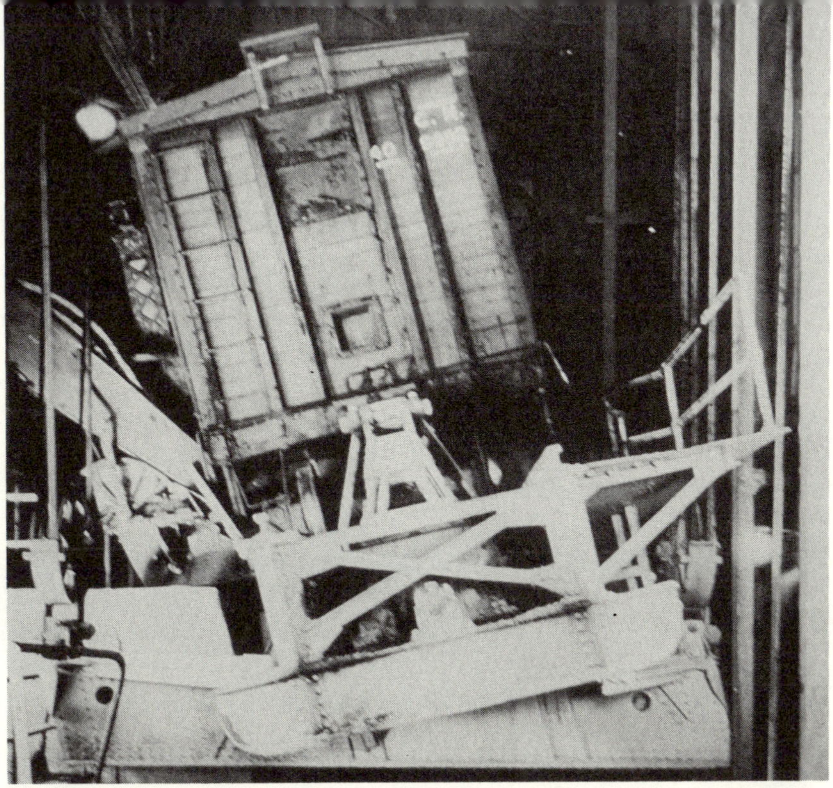

Grain elevator machinery to unload wagons, July 1923 (APM M187)

First extension to grain elevator no. 1, under construction, 14 September 1915 (APM XI-A-2-d [24])

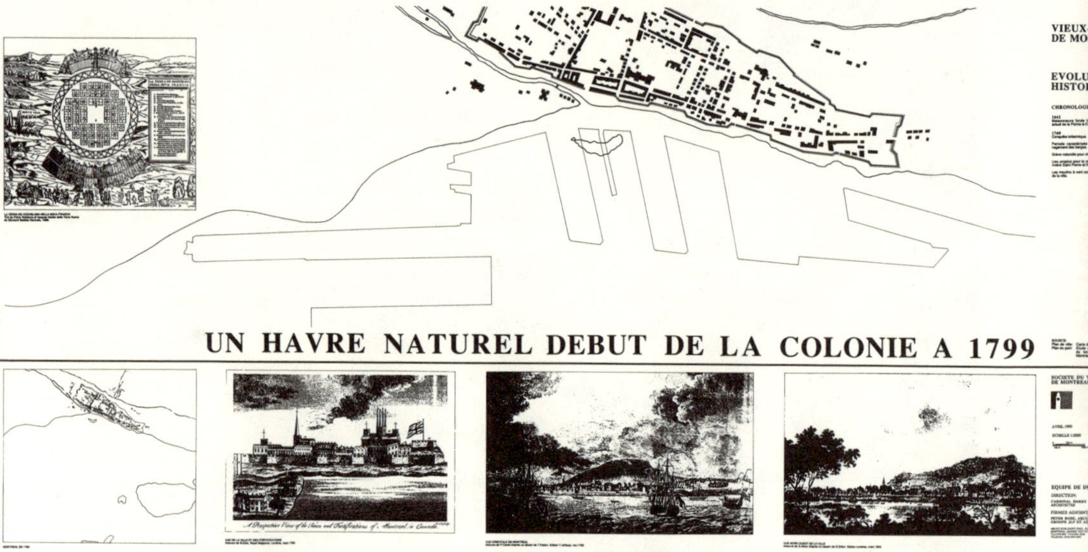

Historical research panel no. 1 with 1760 figure-ground of Montreal showing the fortification walls, April 1990 (office of Peter Rose Architect)

ginning, but in another sense the port seemed to me to have a value and a presence, even though its original function had disappeared. In its early days, it overlapped with the old city and shared its infrastructure. Its twentieth-century history reflects the birth and flourishing of Canadian industrial power. Machines there once loaded grain into twenty ships at a time. The grain elevators that once stood there, built of concrete and steel by hundreds, even thousands, of workers, were structures of heroic scale – at a height of 250 feet, the largest and most powerful objects on the Montreal skyline at the time. In the 1920s they inspired Le Corbusier and Walter Gropius, and became icons of modern architecture. Today, their foundations are the archaeological remains of a vanished technology that was once central to the economy of Montreal and of Canada. It could almost be said that the old port relates to Montreal today as the Roman Forum relates to modern Rome.

The architecture and infrastructure of the Vieux Port from 1806 to 1981 were extensively documented in materials that are now held by the Archives du Port de Montréal. The design process for the master plan started with this remarkable collection of photographs, architectural and engineering

Historical research panel no. 5 with 1912 figure-ground of Montreal showing Harbour at its peak, April 1990 (office of Peter Rose Architect)

drawings, and written records of construction, modification, and demolition.

Over a one-year period, we made a series of figure-ground studies of the site of the Vieux Port, starting in 1760, the date of the earliest map of Montréal that contained sufficient physical detail for our purposes. These studies show the port in the context of the city at significant stages of its development, from the walled city of the eighteenth century, medieval in pattern, to the early twentieth century, when the port had grown to a scale that dwarfed that of the city and had become almost entirely independent of it physically.

A final study showed all of these plans superimposed, using broken lines to represent the shifting natural and artificial shoreline and solid lines to represent buildings or other structures that had existed on the site at various point in time. This plan shows the port as an artificial creation, built on filled land and conceived at an immense scale. (It also shows the evolution of the technology of surveying and cartography: islands and piers whose position did not change over the years appear to dance across the superimposed plan.)

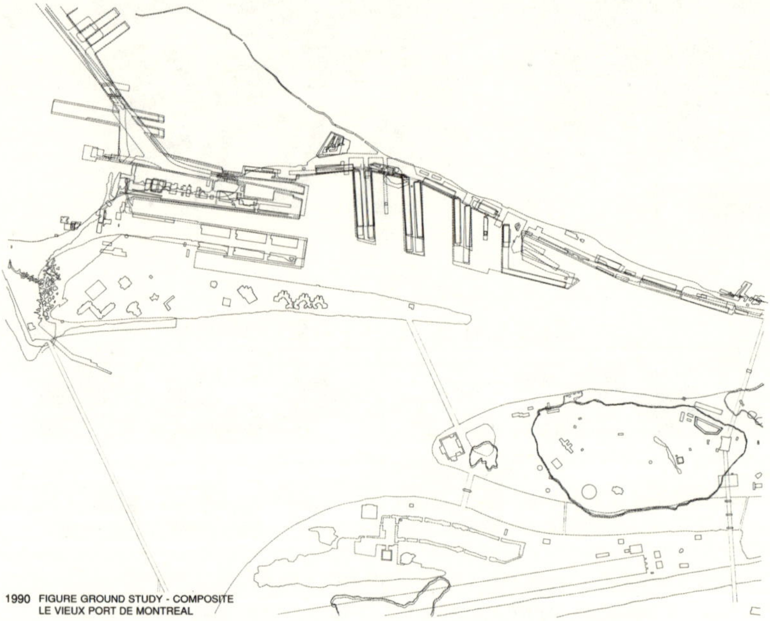

Composite drawing recording evolution of shore line from 1760 to 1989, August 1989 (office of Peter Rose Architect)

Our research revealed that demolition contracts for vanished structures had generally excluded the foundations. To trace the latter, we also prepared a plan showing the successive layers of development of the port that could reasonably be expected to exist below grade. In its sheer density, this plan confirmed that the history of the port could be used to restore a comprehensible scale to the site. An accompanying plan drawn from municipal, federal, and utility-company sources as well as from the port archives showed all of the utility lines (water, sewer, electricity, and gas) located below grade. This plan suggested an order in the planning of the port that was not apparent even in street plans.

The most striking fact revealed by our research was the continuity of the port's development. This continuity is nowhere more apparent than in the twentieth-century study, which shows the port as a huge machine dedicated to the efficient transfer of grain from one mode of transport to another. Its complex network of large and small piers, conveyors, sheds, elevators, and railway lines was conceived and arranged in accordance with a simple and legible order, even though at a scale that overwhelmed that of the old city immediately adjacent. Equally legible were connections between the perception of the waterfront area today and constructions that date back to the city's colonial era. The gentle undulation of the wall of buildings along la rue de la Commune follows the outline of Montreal's fortifications, which were demolished in the early nineteenth century. A later retaining wall running the length of the waterfront, with a network of ramps down to the piers, and then a stone flood wall built in this century, marked the seam between the city and port, giving a physical edge to the city.

The flood wall and the port structures served to define a scale for the buildings of la rue de la Commune. They also gave the impression that the river was nearby, even if it was not actually accessible. When the port was closed and most of its structures were demolished, this edge was erased, creating an ambiguous and vast esplanade elevated well above the level of the water beyond. The buildings along la rue de la Commune are now overwhelmed by this huge foreground, the water's edge actually seems farther away than it did before the port closed.

In developing our scheme, we realized that a legible and singular gesture was required to restore coherence to the space. We began with the wall that had existed in many iterations over time – as seawall, concrete flood wall, and now as a line beside the street and just under the ground. We decided to dig down to expose six feet of the historic wall and then to build on top of it six feet of new wall, making a raised promenade on the nineteenth-century river's edge, a continuous linear belvedere along the length of the waterfront. At the same time, we proposed to lower the embankment by excavating it on a slope leading from the wall to the water.

The raised promenade and the lowered embankment would work together to manipulate perspective and view. The promenade would emphasize and reinforce the gentle concave curve of la rue de la Commune. It would mark the original water's edge, provide a foreground for the buildings, and restore an appropriate scale to the street. (In addition, it would define a corridor for traffic, including the one railway line that still runs through the port.) From its vantage point, the slope of the embankment down to the pier would create the illusion that the river is closer than it is.

This gradual excavation would also reveal the archaeology of the site – stonework of enormous scale that formed the foundations of sheds and grain elevators built at the turn of the century. These include the structure known as Grain Elevator no. 2, designed in 1910 by the Chicago engineering firm

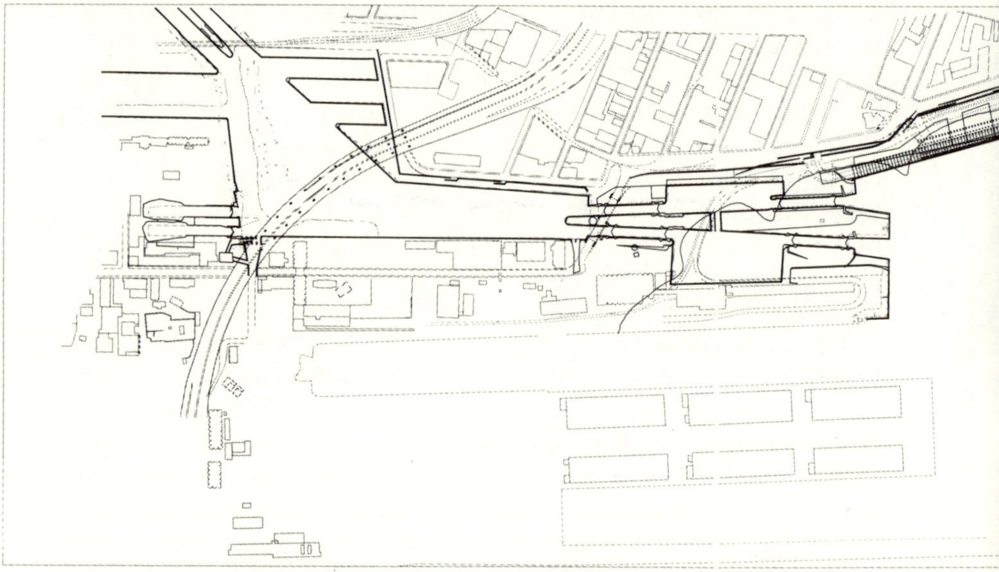

Archaeological plan of Port of Montreal showing foundations of grain elevators and sheds and structure of piers, June 1990 (office of Peter Rose Architect)

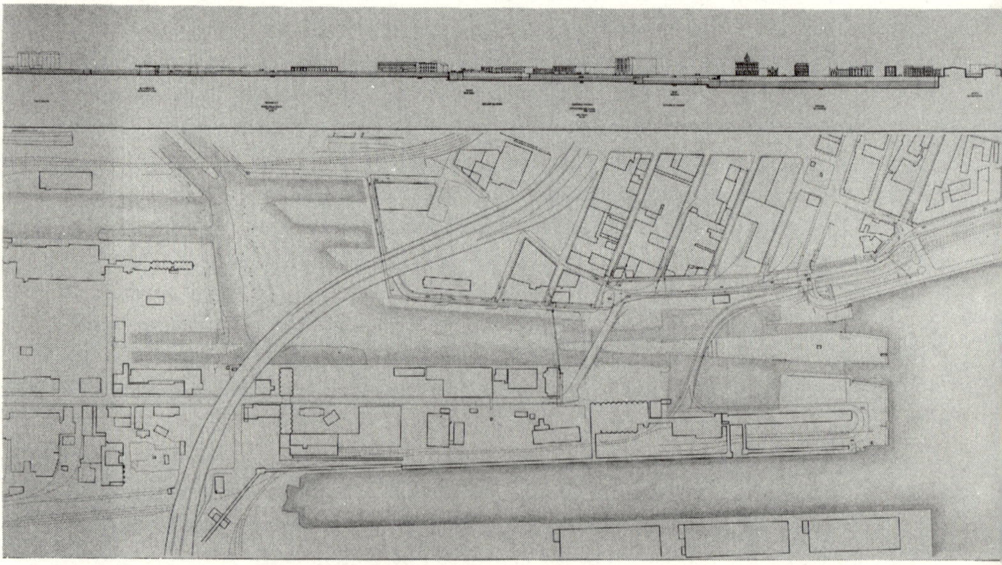

Service plan of Port of Montreal, July 1990 (office of Peter Rose Architect, with Cardinal Hardy Architectes)

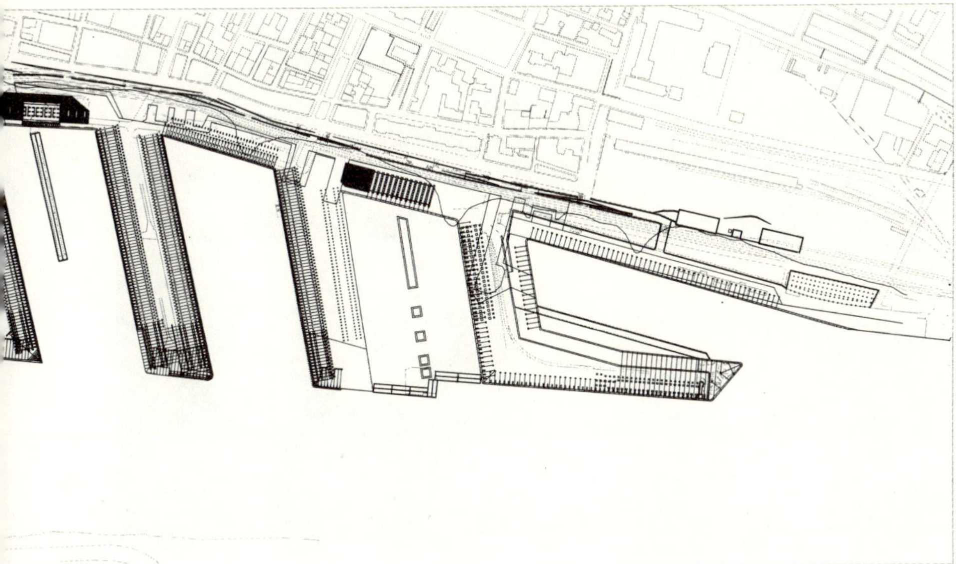

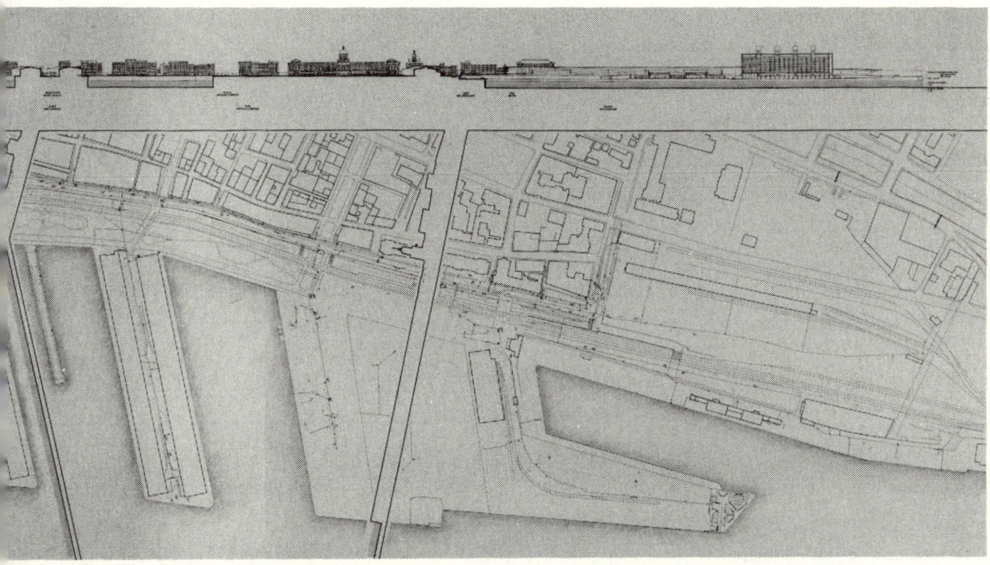

Model of proposal, summer 1990 (office of Peter Rose Architect, with Cardinal Hardy Architectes; photo by Dennis Farley)

of J. S. Metcalf and completed in 1912 at a height of 225 feet. Le Corbusier used a photograph of it to illustrate *Towards a New Architecture*, cropped to exclude Bonsecours Market and mistakenly identified as an American grain elevator. (A surviving elevator, still in operation, provided a poignant counterpoint to our exploration of the ruins.) Along with archival plans and photographs, we used computer drawings and models to develop a hypothesis about the foundations that remain in place, showing that, once revealed, these ruins would help to establish a density and a scale for the space. The results of our research were confirmed by a test excavation carried out in 1990.

On the basis of similar archival research, we also proposed to excavate the filled-in portion of the Lachine Canal that lies within the jurisdiction of the port. This part of the canal forms the last missing link in the linear park along the canal that would connect the Vieux Port with Lac St-Louis in Lachine, six miles to the west. Our excavation would reveal the canal as it existed at mid-century and expose some of the remaining fragments of its earlier versions.

Detail from model showing foundations of grain elevator no. 2, summer 1990 (office of Peter Rose Architect, with Cardinal Hardy Architectes; photo by Dennis Farley)

Grain elevator no. 2 with Bonsecours Market in the background, 4 August 1912 (APM XI-A-3-a [25])

Detail from model showing Lachine Canal proposal, summer 1990 (office of Peter Rose Architect, with Cardinal Hardy Architectes; photo by Dennis Farley)

The principal elements of the design – the promenade and the excavated slope of the embankment – were then developed in a prototypical section that could be modified to accommodate various conditions in plan. The extension of this section along the length of the waterfront follows the curve of the rue de la Commune, recalling historical images of the port. The continuity of the promenade and the sloping embankment throughout the section was reinforced by detaching the piers from the embankment and connecting them back to the street with bridges, making them islands to accommodate programs for cultural and recreational use.

Although our proposal was informed by historical research, it did not seek to take the port back to any one point in its past. Nor was it determined by unreliable and hypothetical programs for the future. Instead, it attempted to establish a physical and perceptual framework for development. We tried simply to edit what was there and to let the program follow. Above all, we tried to simplify – as someone said of the writings of the American author Raymond Chandler – to create a hush, not to hear less, but to hear better.

I wish to thank Cynthia Ware, Nicholas Olsberg, and Eve Blau for their assistance in preparing this material for the symposium and for publication.

NOTE

1 The work described here was carried out by Peter Rose, Architect (Peter Rose, principal; Mark Poddubiuk, project architect; Louis Brillant, Helmut Klassen, Joanne Leung, Filiz Onguc, and Greg Wernicke, project team). Architects involved at later stages included Cardinal Hardy et Associés, Montreal; Chan & Krieger Associates, Cambridge, Massachusetts; and Jodoin Lamarre Pratte, Montreal. Others who participated included George Sexton Associates, lighting consultant; Peter Walker and Partners; Landscape Architecture Inc.; Pauline Desjardins, archaeologist, Société du Vieux Port de Montréal; Raymonde Gauthier, historian; and Ernest Labelle, archivist, Archives du Port de Montréal. The project was carried out in somewhat different form than described here for the 350th anniversary of the founding of Montreal in 1992.

POUR QUE LA VIE AIT LIEU (FRAGMENTS)

PHILIPPE MADEC

Architecture is neither an art nor a science; nor is it the sum of art and science.

Architecture arises from man's need to create a dwelling, to inhabit his world. It stems from man's need to adapt his dwelling to a place, to space and time. Architecture exists so that life may take place, so that we may exist.

Architecture evolves in the immensity of the already-there. Within the homogeneous and fluid infinity of the desert, the plains, or the city, a wall or a roof slows down the movement of the world, allowing man to remain in the stillness of the place. At rest, at peace, man may inhabit the world. Then ensue his presence, the wholeness of being and place, the encompassing extension and duration.

Never again combine aesthetics and moral precepts! No more hatred of styles, for they are manifestations of life. Architecture requires an ethics rather than moral prescriptions. No more rules and systems of values, but rather attitudes and ways of being, the poetic and philosophical basis for action: the action of opening itself to life and endowing it with immanence.

I —

L'architecture n'est plus un art.

L'architecture n'est pas une science.
L'architecture n'est pas davantage une somme de l'art et de la science.

2 —

Nos grands-pères ont recherché la définition de l'architecture entre l'art et la science.
Ils ont échoué dans la tentative insensée d'une équation de l'art et de la science.
Ils ont échoué face à l'impossibilité même d'une définition.
Nos grands-pères n'ont mis au point que des règles et des principes prescriptifs.

Nos pères, eux, étaient persuadés de trouver les fondements de l'architecture dans l'activité de l'architecte, dans le projet et ses effets : les textes, les dessins et les bâtiments.

Ils se sont épuisés et s'exténuent encore au fond de l'insondable dilemme qui sépare la théorie de la pratique.

Les uns puis les autres ont été confrontés à l'immensité de la connaissance impliquée dans l'œuvre architecturale et accumulée au cours du modernisme.

Leur volonté de définir les principes du projet architectural les a conduits à des discriminations et à des choix partisans injustifiables autrement que par des raisonnements dont les fondations ont été impressionnistes, stylistiques, politiques, morales, esthétiques, c'est-à-dire idéologiques.

Nous ne savons toujours pas *ce qui fait que l'architecture est ce qu'elle est.*

3 —

Ce n'est pas parce qu'il y a un métier, ce n'est pas parce qu'il y a un savoir et une activité de l'architecte que l'architecture existe.
Ce n'est pas parce qu'il y a des architectes que l'architecture existe.

Aujourd'hui, l'architecture est écartée du projet urbain.
Elle est réduite à la production de bâtiments plus ou moins esthétiques.

Ce n'est pas seulement parce que, depuis la période classique, le champ de l'architecture est inexorablement réduit à l'apparition des nouveaux acteurs modernes: l'ingénieur, le plasticien, le designer, le décorateur, le sociologue, le paysagiste, etc.

Ce n'est pas seulement parce que la production architecturale ne représente qu'une faible part de la production du cadre bâti, et qu'elle est trop peu souvent à la hauteur de la demande qui lui est faite.

Pendant deux siècles, les architectes ont alimenté un malentendu historique grave. Passant du service du roi à celui du citoyen, ils persistent à croire que la demande d'architecture reste toujours globale. Ils ne se défont pas de la Forme et s'attachent même à étendre l'hégémonie de leur « art » au domaine public et au domaine privé.

Ils cherchent à tenir la condition humaine dans la Forme.

Cette vanité-là provoque le rejet de l'architecture.

4 —

Maintenant que nous sommes dans la civilisation urbaine –
à l'exact moment du devoir architectural –,
il convient de s'interroger sur la légitimité de l'architecture.

Autonome, indépendante d'une centralité de la personne, l'architecture est aussi largement répandue que la civilisation.

Cette condition a son revers. Trop de phénomènes, qu'ils soient d'architecture ou qu'ils viennent d'autres champs du monde, masquent la cause de l'architecture.

Nous ne pleurons pas notre origine, nous n'y reviendrons jamais sauf au risque de nous perdre.

La cause de l'architecture nous manque.

Cherchons dans le présent.

5 —

L'homme vit, l'homme est là, passant, habitant, peuple du lieu.
Flottant, il appelle l'architecture pour qu'elle l'établisse.

L'architecture provient de la nécessité pour l'homme de fonder son habitation, d'installer son monde. Elle vient de la nécessité pour l'homme d'accorder son habitation en un lieu, à l'espace et au temps.

L'architecture installe la vie.

6 —

Installer: établir l'homme dans un lieu, avec dignité.

Depuis les temps immémoriaux, l'architecture fonde dans le lieu l'accord humain de vivre ensemble.

L'architecture assoie la Cité, elle fonde noblement la citoyenneté.

7 —

La vie appelle l'architecture, et en retour, l'architecture appelle la vie.

Pour l'architecture, il n'y a pas de visée hors de ce qui l'a appelée, hors de ce qui l'appelle et qui l'appellera.

D'où une destination:
là, poser des lieux de liberté,
dans lesquels l'imprévisible peut surgir,
dans lesquels le dessein de la vie tire sa ligne,
dans lesquels le hasard et la nécessité trouvent leur champ,
dans lesquels la vie trouve son siège, sa stalle.

8 —

Là, dans un lieu, désigné, l'architecture installe l'homme.
Tout à la fois l'univers et notre jardin, ce lieu est l'endroit où la vie est née et s'est organisée.

Tout y est déjà donné qui a pris les allures d'un monde, ou plutôt qui a pris les allures du monde tel qu'il demande à continuer d'apparaître dans ce lieu.

Dans le lieu, le monde y est en son entier mouvement, mais aussi le temps et l'espace.

Tout est là, déjà là.

Là coule la totalité.
Il est nécessaire de reconnaître au lieu sa capacité à recevoir le monde en son entier,
il est nécessaire de reconnaître sa capacité à recevoir la totalité.

Il faut devenir capable de la découvrir, de la mettre à vif, de la mettre à vie.
Le monde est dans un lieu comme un trésor à inventer.

L'architecte n'inventerait-il que lorsqu'il découvre le monde qui est dans le lieu où il œuvre? Gageons-le. Cela deviendrait vivant et infini. Le monde change d'un lieu à l'autre, le monde fluctue d'un temps à l'autre.

L'architecte n'imagine pas; il trouve (s'il le peut).

Pas d'espoir; le monde tel qu'en lui-même.
Pas d'utopie; le monde installé en son lieu.

9 —

L'architecture procède de l'ensemble.

L'architecture prend place dans un conflit de forces et elle naît de leur accord.
Elle vient d'une composition des puissances, elle vient d'un compromis spatial au sein de monde, d'un compromis spatial entre le monde et le lieu, aujourd'hui: la Terre.

L'architecture est là pour que la vie ait lieu.
La vie a lieu, la vie prend place.

L'architecture permet à l'homme une demeure, dans un lieu désigné, au sein de l'ordre défini et mouvant du monde.

L'architecture ne peut pas être une utopie.

10 —

L'union du temps et de l'espace qui commande à l'architecture s'appelle l'*Arkhé.*

Origine et fin de l'architecture, elle est la stalle de la vie, elle en regroupe les conditions essentielles: étendue et durée.

Le monde n'est pas dans l'espace et dans le temps; c'est l'espace et le temps qui sont au monde.

L'*Arkhé* est le toujours-déjà-là du monde.

Au présent de l'espace, du temps, du monde et de celui qui y est, elle est la genèse appelante toujours déjà là, le commencement toujours au présent, la phénoménalité de l'origine.

Empruntée à l'espace-temps, l'*Arkhé* est hors du récit et évacue le dilemme théorie/pratique en engageant d'emblée la chair et l'esprit, le Corps en son entier.

Plutôt que la cabane comme origine perdue dans la nuit des temps d'une architecture considérée comme art de bâtir, l'*Arkhé* comme fondement toujours présent et toujours agissant.

L'*Arkhé* ne bâtit pas un objet en soi, mais livre une union à la vie, une union avec moi.

11 —

Quand tout est là, déjà là, tout advient.
La totalité donne au présent ce qui est advenu.

L'ouverture est la qualité centrale de la totalité.
Elle en est la qualité naturelle.
La maintenir ouverte est une attention, c'est un souci, c'est une préoccupation, c'est l'occupation d'avant l'architecture.

L'ouverture n'est pas le vide, elle est plus ouverte que le vide, elle est déjà apprêtée à recevoir la vie.

12 —

Le monde ad-venu et le monde à-venir possèdent l'architecture.

L'architecture est tournée hors d'elle dans une impatience et une attente que les hommes et les choses viennent creuser.

La vie ne remplit pas l'architecture, la vie creuse l'architecture.

S'attendre à tout, n'attendre rien. Attendre l'inattendu et le désir.
C'est l'envie, la vraie. C'est l'en-vie.

S'assurer que l'architecture peut installer la totalité.
S'assurer que l'architecte peut maintenir l'ouverture.
S'interroger sur leurs capacités à ouvrir à l'ouverture le vide rempli du monde.

Le vide n'est pas le vide; il est déjà plein de nous, innombrable.
Le vide n'est pas le vide, le monde y ondule.

Creuser le pli du monde pour y libérer la place de l'ouverture,
pour maintenir le vide ouvert, attentif.

Rien ne justifie, bien sûr, que l'architecte referme l'ouverture là où il œuvre.

Aujourd'hui, la demande d'où provient l'architecture est inaccessible à l'architecte. Elle est plurielle, toujours diverse et spécifique, pourtant toujours légitime.

Les programmes et leurs spécialistes ne réussissent pas à rendre justice à sa dimension historique, culturelle, ils ne réussissent pas à rendre justice à sa dimension ontologique.

Confronté à l'impossibilité d'en restituer non seulement les aspects fondamentaux mais aussi ceux d'un usage quotidien, il ne reste plus à l'architecte qu'à en autoriser le plus possible.

Il n'y a plus de programmes.
Il y a le programme de l'architecture: ouvrir à l'homme des lieux où être.

13 —

La fluidité est cette qualité du vide qui autorise la liberté du flot de la vie: elle est le mouvement lui-même.

La transparence est une qualité commune au vide et à la matière. Il n'y a pas de transparence sans opacité. Entre l'opaque et le non-opaque, c'est le compromis qu'emporte la vie.

L'abstraction, quant à elle, est l'ouverture elle-même.
L'abstraction: le seul accès possible au tout.

14 —

Face à l'abstraction chacun voit à son gré.
On y pénètre individuellement comme il nous convient.

Toutes les lectures et toutes les interprétations en deviennent respectables.

Il est si émouvant qu'un homme voie un paysage, qu'un homme aime à reconnaître une colline, alors que son voisin aperçoit la vie et qu'il en pleure.

Grandeur de l'abstraction innommable, ouverte à tous.

Nommer tue l'ouverture. Nommer la remplit d'un mot, d'un concept, d'une idée.
Nommer installe un horizon et le bouche dans le même temps.

L'ouverture est sans horizon, autre que l'ouverture elle-même.

L'abstraction donne l'accès direct au présent, au toujours-déjà-là du monde.
L'abstraction est l'accès direct à l'*Arkhé*.

En architecture, l'abstraction est le silence ou le brouhaha.
C'est ce qui n'apporte pas de sens ou ce qui le brouille.
C'est ce qui laisse venir le temps, toujours.

L'ouverture par l'abstraction est un pas vers l'étendue et la durée.
Un pas vers la sérénité, un pas vers la paix avec les êtres et les choses.

15 —

La vie, bien sûr!

Pourquoi en donner, architecturalement, une représentation?
Pourquoi figer le flux dans la forme, le mouvement dans le béton, la tension dans le métal, le temps dans la répétition?

Servir la vie, telle est la vocation de l'architecture.
L'accueillir. Lui donner son siège, sa stalle. Au cœur du déjà-là, l'établir dans un digne établissement.

L'architecture comme monument et mémoire de la vie, l'architecture comme édification banale et monumentale de la vérité de la vie.

16 —

L'architecture installe la vie, qui porte son propre sens.

L'architecture installe la vie en son sens – cette tâche est bien assez difficile –, elle n'a pas à l'afficher, à le figurer, à le représenter.

Qui connaît le sens de la vie?

La pensée, philosophie et science, échoue sur ce point. L'architecture ne connaît pas davantage le sens de la vie. Mais elle en détient – complice – le secret de l'installation.

Installer ce n'est pas connaître, c'est plus que connaître: c'est édifier.

Elle installe la vie dans l'ouverture, elle libère son sens, elle le déploie.

L'architecture ne peut pas témoigner de la vie, elle appartient à la matière même de la vie. Elle ne la décrit pas, elle en est.

L'architecture ne transmet pas de sens.

Elle s'ouvre au sens du monde en un lieu et en un moment.

Abstraite, absentée, elle laisse à chacun le soin de se constituer comme sens.

La signification en architecture ne peut pas être verbe mais silence ou brouhaha inaudible. La Cité elle-même ne se fonde pas dans le verbe, mais dans le lieu flottant et dans le corps transpercé, d'où émerge le verbe.

17 —

Longtemps l'architecture a été engagée au service de la représentation. Elle n'était pas alors engagée au service du citoyen.

La Cité a tout autant besoin de l'art que de la science. Elle demande à l'architecte de s'en servir.

Malentendu.

L'architecte se croit porteur d'une vision artistique ou scientifique. L'architecte cherche à représenter sa vision du monde.

L'architecte n'est pas un artiste. Il ne lui est pas demandé d'avoir le pressentiment du monde et de nous en donner une représentation.
L'architecte n'est pas davantage un scientifique. Il ne lui est pas demandé de justifier une représentation du monde.

Représenter, telle est, depuis deux siècles, la vanité de l'architecte.

Qu'est-ce qui légitimerait que l'architecte maintienne la vie de l'Autre dans sa vision personnelle du monde, dans une vision nécessairement restreinte, étriquée?

L'architecture n'est pas la représentation de la vision du monde de l'architecte.

L'architecture présente le monde, elle rend présent l'accord initial de la Cité, recommence à chaque instant. Voilà pourquoi toute architecture est en devenir d'urbanité? Elle en provient.

La représentation n'est pas l'objet de l'architecture.
La représentation appartient à un moment limité du travail de l'architecte.
L'architecte s'en sert au même titre que l'« ingénieur-béton » et l'artisan pour leurs plans d'exécution.

L'architecture n'est pas une idée mais se sert des idées.
L'architecture n'est pas une forme mais se sert des formes.
L'architecture n'est pas le bâtiment mais se sert des bâtiments.
L'architecture n'est pas un dessin mais désigne une impatience vitale.

L'architecture n'est pas une figure;
elle configure le lieu, le monde, l'espace et le temps.

18 —

Il ne suffit pas que la pensée et la représentation cherchent à nous engager dans la vie, il ne suffit pas qu'elles essayent de nous montrer l'extraordinaire beauté de ce monde. Il faut nous enrôler tout entier.

En tant que fondation ouverte à la vie, l'architecture offre à l'homme les conditions de compréhension du monde en sa présence, non seulement dans le présent mais aussi dans la conscience de l'histoire dont elle atteste.

L'architecture permet à l'homme de saisir pleinement le déploiement étendu et durable de son être, un déploiement au delà de lui-même.

19 —

L'œuvre de l'architecte ne peut pas donner une nouvelle réalité à l'architecture. Elle relève exclusivement d'un événement qui concerne avant tout l'architecture elle-même.

L'architecture fait de l'architecte.

L'architecte pense à des lumières, des atmosphères, des matières et des volumes: l'objet, c'est l'habitation du monde.

Chez l'architecte, point besoin de visions.
Chez l'architecte, un amour infini de l'humain et des choses, un amour de la vie en sa vérité.

Un accueil de l'autrui et des choses comme soi.

20 —

L'architecture ne vient pas de l'art ou de la science. L'architecture vient d'au-delà du droit, de cette vérité de la présence humaine née de l'adhésion de la vie à son lieu.

Le talent et le désir de l'architecte ne suffisent pas à légitimer l'architecture.

L'œuvre d'art existe de son plein droit; il n'en va pas de même pour l'œuvre d'architecture.

Si la demande d'installation autorise l'œuvre d'architecture, c'est la loi qui légitime l'action de l'architecte et qui en marque les termes et les enjeux.

L'architecture répond à une demande.

L'architecture est nécessaire, mais elle n'est pas partout justifiée.

L'architecture a un endroit légitime. Son en-droit est le lieu où la communauté s'établit, ce *là* où l'art et la technique sont convoqués à l'installation urbaine de l'Autre.

L'ailleurs, ce *là-bas* où l'Autre viendra.

21 —

Plus jamais une morale et une esthétique ensemble.
Plus de haines à l'égard des styles, ils sont les symptômes de la vie.
Pas une morale, pas de prescriptions, mais une éthique.

L'architecture s'avère bien au-delà du goût, du jugement et des valeurs.
Avant d'être une loi, une idée ou un savoir-faire, elle est une adhésion à la vie.

Plus de règles et de systèmes de valeurs,
mais des attitudes, des manières d'être,
les bases philosophiques d'une action,
de l'action de s'ouvrir à la vie et de lui dresser le plan d'immanence.

Un plan d'immanence tel un plan de composition et d'ouverture,
non pas un plan d'organisation ou de développement,
pour que s'épanouisse la piété de la vie et se libère le *pathos*, tant joie que tristesse.
Mais, seule, la joie peut rendre compte du digne établissement.

ARCHITECTURE VIRTUELLE ET INFOGRAPHIE – QUELQUES QUESTIONS POSÉES À L'ARCHITECTURE

JEAN-PIERRE HARDENNE

The introduction of new tools for architectural design and representation – including data-processing systems and, in particular, computer graphics – is revolutionizing the whole process that leads to the creation of public and private spaces in our post-industrial societies. For several years now, these new tools have been altering all of the various phases involved in an architectural project; today, they are having a major impact on the planning and scheduling of the project itself.

This transformation began almost a quarter of a century ago; with growing importance being given to project execution and management, various technical aspects of the building activity – such as working drawings, specifications, and cost control – have been taken over by data processing. Today, this process is leading to a completely new approach to spatial design, thanks to the combined techniques of computer-assisted design (CAD) and imaging systems. These technological and cultural changes give rise to a number of fundamental questions that architecture must address.

The new situation facing us bears some resemblance to the emergence of new forms of spatial representation in the past – in particular, that of perspective during the Renaissance and that of orthogonal design during the period following the First World War – that were epistemological watersheds. One major consequence of these revolutions, which fundamentally altered the perception of space and the ability to conceive it, was the emergence of the classical and modern notions of space, respectively.

Today, it is within the various phases of architectural planning, and in particular at the sketch and preliminary-draft stages, that the new changes are most likely to lead to a radical transformation in the specific features and processes of architecture. It is now possible, from the earliest sketches, to produce a three-dimensional view, to experiment with site location, to test materials, and so on. This suggests the possibility of a new type of experimentation that is totally without parallel in the use of traditional tools of formal design, such as drawings, perspective, orthogonal graphics, and so on. And we now have access to the virtual-space and kinetic-imaging techniques that will fundamentally alter both the space of architectural representation and the representation of architectural space.

A number of *secondary* issues also confront architects, as the emergence of these new tools will initiate and encourage a diversification and transformation of those areas of activity where visual representation is a major tool of design and communication – areas in which film-makers and computer-graphics specialists are already at work.

Architects must focus without delay on the epistemological problems raised by the new methods of representation and address a few theoretical and practical questions that may shed light on these issues. The challenge posed by the new tool – a vehicle not only for contemporary technology but also for contemporary culture – will undoubtedly lead to a new kind of modernity, the shape of which cannot yet be discerned.

Moreover, because of the strong attraction exerted by synthetic imaging and the virtual space that imaging produces, an entirely new relationship between the architect and his work is being created. Given the current state of development of the discipline, creative experiments in this area remain very uncommon, however.

Thus we must test the true potential of computers in the areas of design and simulation, in particular because of the new possibilities that they offer for the visualization of a project's overall concept and of its future realization – a visualization made possible by wandering through virtual space. Such experiments should lead to a new and fruitful collaboration between the various actors involved in a project.

Les nouveaux outils de conception et de représentation que sont l'informatique et, plus particulièrement, une branche de celle-ci, l'infographie, sont en train de bouleverser l'ensemble du processus de création des espaces, tant privés que publics, dans nos sociétés post-industrielles. Ces nouveaux moyens exercent leur influence sur l'ensemble des phases techniques du projet d'architecture depuis quelques années déjà, mais cette influence s'approfondit encore davantage aujourd'hui, alors que ces outils transforment désormais le moment de la projetation.

HISTORIQUE

Cette évolution a débuté, voilà bientôt vingt ans, par la prise en charge des différents actes techniques du chantier, tels le tracé des plans d'exécution, l'élaboration des métrés et le contrôle des coûts. Elle débouche actuellement, grâce aux techniques conjuguées de la conception assistée par ordinateur (CAO) et des images de synthèse, sur une véritable recomposition des pratiques de spatialisation.

Au début des années 60, les recherches théoriques et appliquées dans le domaine de l'informatique devaient composer avec une technologie encore peu développée; l'exploitation de la mémoire et des capacités de traitement des gros ordinateurs centraux dépendait de programmes lourds édités sur cartes perforées. Les quelques logiciels de représentation cartographique et d'allocation spatiale issus de ces recherches, qu'un petit nombre d'architectes tentèrent alors de mettre au service de leur discipline, se révélèrent totalement incapables de répondre aux besoins de la projetation architecturale, activité qu'on ne peut traduire sous la forme d'une expression logique et dont on ne peut résoudre les problèmes sans faire appel à des simulations complexes. C'est pourquoi ces praticiens abandonnèrent alors leur intérêt pour l'informatique, une situation qui durerait presque deux décennies.

En effet, ce n'est que vers la fin des années 70, avec l'apparition des mini-ordinateurs et des logiciels conçus spécialement pour l'architecture, que les architectes en vinrent à s'intéresser de nouveau à l'informatique; cette fois, ils étaient en grand nombre. Ce progrès n'a cependant été possible que grâce à un groupe de passionnés, issus du système d'enseignement au début des années 70, qui avaient investi temps et énergie dans les centres de calcul universitaires et privés. Ce sont eux qui formèrent le noyau des équipes qui conçurent les premiers logiciels et développèrent les premières applications informatiques axées sur l'architecture. Celles-ci, dès le début, furent cependant orientées vers les phases techniques du projet, leur objectif prioritaire étant d'automatiser les dessins d'exécution et d'assurer la gestion en temps réel des coûts de réalisation et de production.

Quelques années plus tard, l'apparition et la diffusion extrêmement rapide de la micro-informatique et de logiciels graphiques performants facilitèrent l'accès aux technologies nouvelles pour un nombre croissant de petits cabinets d'architectes. Cette période vit l'apparition du duo IBM/compatibles et Apple/Macintosh ainsi que de nombreux logiciels spécialisés (Autocad, Versacad, Star, Keops, Computervision, MacArchitrion et autres). Les prix se stabilisaient alors aux alentours de 15 000 dollars, tandis que les mémoires prenaient une expansion vertigineuse, les mémoires vives (RAM) passant de 128 kilo-octets à cinq méga-octets et les mémoires mortes de 300 k-o à 80 m-o.

Par la suite, la mini-informatique connut de nouveaux développements importants dès le début des années 80, notamment avec l'avènement des stations graphiques spécialisées (Appolo, Sun, Intergraph, etc.) et de la CAO, qui permirent à certaines firmes privilégiées d'entrevoir les nombreuses possibilités qu'offraient les nouvelles machines et les nouveaux logiciels.

Aujourd'hui, on trouve sur le marché des superstations de travail (telles celles fabriquées par Silicon Graphics ou Hewlett Packard) qui se caractérisent principalement par une unité centrale puissante et des possibilités graphiques très étendues. Ces stations supportent de nouveaux programmes (tels Alias, Vertigo, Explore, SoftImage et autres), dont les plus performants comportent des modules d'animation en temps réel. Ceux-ci offrent dès à présent la possibilité d'animer des scènes tridimensionnelles. Sans doute, ce matériel et ces logiciels sont-ils encore trop coûteux pour la plupart de ceux qui œuvrent dans le domaine de l'architecture, mais cela ne devrait pas durer. Déjà, on peut observer une réduction importante des coûts en matériels et logiciels, en même temps qu'augmentent les puissances de traitement, qui passent de quatre millions d'instructions par seconde (MIPS) à près de 300 MIPS, tandis que les capacités de stockage s'accroissent de 70 m-o à plus de 1 000 m-o.

Ce rapide historique ne semble cerner rien d'essentiel, si ce n'est des considérations d'argent et de matériel. On pourrait penser qu'il ne s'agit actuellement que de tirer parti de l'avènement d'un nouvel outil de formalisation. Et pourtant, sous le couvert de cette transformation technique, on assiste en fait à une mutation technologique et culturelle qui pose plusieurs questions de fond à l'architecture.

Cette situation nouvelle qui nous interpelle n'est pas sans rappeler les « ruptures épistémologiques » que signalèrent l'apparition de la perspective à la Renaissance et celle de l'axonométrie au lendemain de la Première Guerre mondiale. Ces deux nouveaux modes de représentation ont, chacun à son époque, bouleversé fondamentalement les perceptions de l'espace et les capacités à le concevoir, donnant naissance, respectivement, à l'espace classique et à l'espace moderne. Pour éclairer cette intuition, nous ébaucherons ci-après quelques constats qui permettront de préciser sur trois plans – celui de la dimension graphique de notre action, celui de son support matériel et celui de nos schèmes intellectuels – les situations et les enjeux auxquels nous allons être confrontés.

LES OUTILS GRAPHIQUES

Le dessin, outil traditionnel de l'architecte, a toujours rempli plusieurs fonctions qui s'enchaînaient naturellement dans le développement d'un projet: à chacune des phases de celui-ci correspondait un type de dessin. L'apparition de nouvelles images nous force à examiner de plus près l'opérationnalité et les fonctions de chacun pour essayer de déceler les changements majeurs que l'introduction des nouveaux outils devrait entraîner dans l'avenir. Aujourd'hui, dans la pratique de l'architecte, on peut distinguer cinq grands types de dessin, auxquels correspondent différentes fonctions utilitaires:

1 *outil d'apprentissage* – le dessin à main levée et le dessin d'après nature permettent de lire le réel en construisant pour chacun d'entre nous une culture visuelle propre;
2 *outil de conception* – le croquis de naissance et les esquisses préliminaires permettent d'exprimer le parti, les références majeures du projet et les citations métaphoriques, en même temps que d'expérimenter les solutions différenciées qui vont déterminer sa singularité;
3 *outil de représentation* – les dessins préparatoires permettent de préciser les dimensionnements globaux et partiels en travaillant plus finement les éléments constitutifs du projet;
4 *outil de transmission* – les dessins techniques et les plans d'exécution permettent de mettre en place et de définir l'ensemble des contraintes techniques entourant la construction;
5 *Outil de simulation* – Le dessin de perspective ou d'axonométrie permet d'évoquer et de vérifier les apparences futures d'une architecture et de ses différentes composantes.

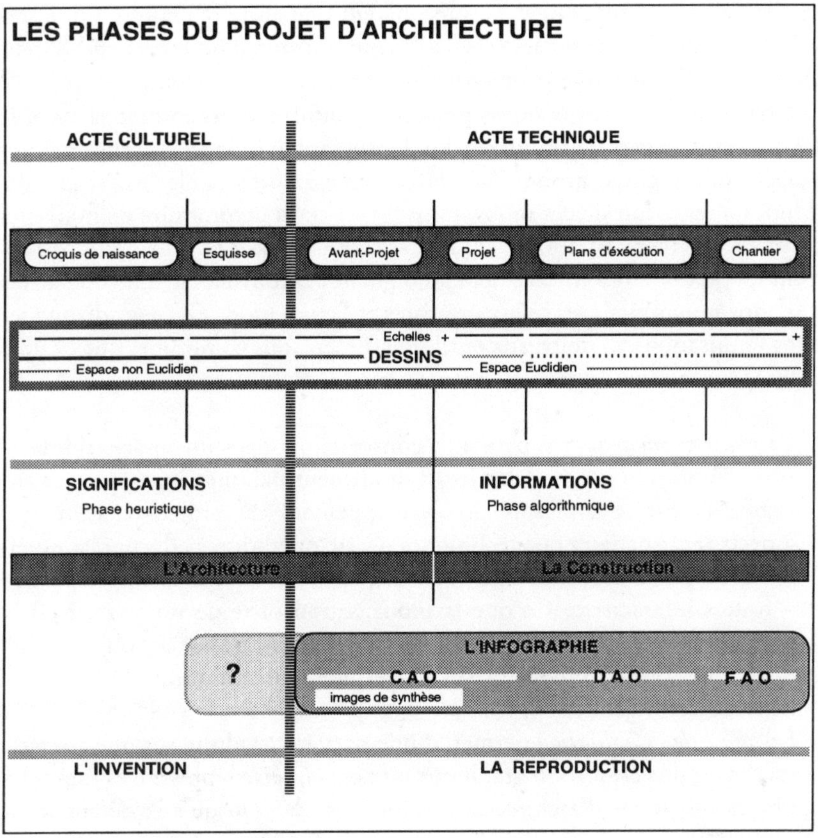

Les phases du projet d'architecture

Pour schématique que soit cette catégorisation, elle permet néanmoins de se rendre compte que la lente maturation de l'outil « dessin » a permis de construire dans le temps des modes opératoires parfaitement adaptés aux objectifs de chacune des phases du projet d'architecture, depuis les premières esquisses jusqu'aux documents techniques du chantier. Cette différenciation n'est cependant pas qu'utilitaire. Elle correspond aussi à une division du travail et à une hiérarchisation sociale et culturelle qui tend à valoriser certaines productions: les architectes en charge du projet s'approprient généralement les esquisses et les croquis de naissance, tandis que les plans techniques d'exécution sont laissés à leurs subalternes. L'enchaînement et la synergie qui existent entre ces types de dessin recouvrent une autre opération qui, à travers le jeu qui s'opère sur les échelles, mène le projet d'un espace indifférencié non euclidien à une géométrisation intégrale.

Chacun des nouveaux outils graphiques informatisés, comme la CAO, le DAO (dessin assisté par ordinateur) et l'infographie (images de synthèse), répond à une logique propre dans la pratique architecturale. Mais, aujourd'hui, l'intégration de ces innovations est en train de produire de nouvelles capacités de représentation, empruntant une partie des possibilités qu'offrent les types traditionnels et leur adjoignant les nouvelles formes opératives que sont notamment les images réalistes et l'animation. On peut distinguer dans l'utilisation des nouveaux outils des dimensions communes que chaque type opérationnalise spécifiquement:

1 La *phase de modélisation* permet la construction, dans un espace tridimensionnel infini, d'un modèle virtuel totalement paramétrisé. Celui-ci s'enrichit, au fur et à mesure du développement du projet, de différents aspects tant spatiaux que techniques. Cette opération s'effectue habituellement à partir de quatre vues – en plan, de face, latérale et en perspective – toutes interactives. Ce que la modélisation offre de nouveau, c'est la possibilité de produire, dès les premières lignes tracées, un objet virtuel directement positionné dans l'espace et une représentation bidimensionnelle de cet objet.

2 La *phase de visualisation* permet, d'une part, de produire sur une traceuse un espace de vérification graphique où peuvent être représentées toutes les phases du projet d'architecture, depuis les plans jusqu'aux détails techniques, à des cadences et à des séquences répétitives sans commune mesure avec les outils traditionnels. D'autre part, sur un écran couleur haute définition, elle permet de tester, par la production d'images de synthèse réalistes, colorées, texturées et ombrées, les différentes options architecturales tant au niveau volumétrique qu'architectonique.

3 La *phase de simulation* permet, dès les premières étapes du travail de l'architecte, de tester le parti architectural et, par la suite, son développement. Elle permet de figurer la déambulation visuelle sur le site et dans les espaces conçus mais non construits, ainsi que de représenter les incidences du soleil sur le bâtiment et son environnement, de même que le jeu de l'ombre et de la lumière dans les espaces qui définissent son architecture.

La possibilité de faire appel, dès les premières esquisses, à des visualisations tridimensionnelles, à des essais d'insertion dans le site et à des tests de matériaux offre des possibilités de vérification sans commune mesure avec ce que permettaient les outils de formalisation traditionnels (dessins, perspective, axonométrie, et ainsi de suite), sans parler des transformations profondes qu'introduisent les espaces virtuels et les visions cinétiques à l'égard des moyens de représentation de l'espace des architectes.

De plus, au fur et à mesure du développement et de la construction du projet, les possibilités de traçage répétitif et de correction quasi instantanée vont aussi transformer la pratique de l'architecture; elles vont en modifier les aspects budgétaires et techniques en plus de redistribuer les rôles des acteurs à l'œuvre dans le processus du projet d'architecture.

LES OUTILS MATÉRIELS

L'ensemble de ces opérations graphiques a traditionnellement eu comme support un objet (la planche à dessin), un lieu d'inscription (le papier calque) et un outil (le porte-mine et le rapidographe) qui, eux aussi, sont le fruit d'une longue évolution à travers le temps. Le premier résultat en a été, d'une part, un système spatial qui, compte tenu des contraintes optiques et des rapports d'échelle, fait coïncider parfaitement l'espace graphique et l'espace visuel de l'architecte au travail, favorisant un va-et-vient entre la partie et le tout, entre le plan global et le détail constitutif. Un second résultat est lié à la production d'un support graphique en strates qui permet d'intégrer totalement le processus itératif de la conception, cet aller-retour entre versions différentes qui construit finement l'ensemble du projet. Enfin, cette évolution a abouti à une hiérarchisation de la démarche créative grâce aux possibilités de modulation et de surcharge du tracé du crayon et du trait d'encre, qui restent, tout au long du processus, les traces opératoires du chemin de la conception.

L'adéquation totale des différentes composantes, tant matérielles qu'intellectuelles, de l'espace de travail de l'architecte est aujourd'hui remise en cause par l'introduction des nouveaux outils informatiques. À l'intérieur de cet espace concret, de même qu'au sein du processus d'élaboration du projet d'architecture, notamment à l'étape de l'esquisse et de l'avant-projet, les innovations qu'on peut déjà deviner à l'heure actuelle risquent de transformer radicalement ce qui fait la spécificité et la singularité de l'acte d'architecture. Pour essayer de cerner l'incidence éventuelle de ces changements, il faut comprendre comment le regard, la manualité et la trace sont modifiés par les nouveaux outils informatiques et quelles sont les conséquences prévisibles de ces transformations sur la production contemporaine d'architecture.

1 Dans le domaine de la *vision*, l'écran de l'ordinateur représente une lucarne vide, un index spatial qui délimite un regard. Son champ visuel restreint encadre une surface infinie qui supporte des dessins dont les échelles d'affichage sont, elles aussi, aléatoires et infinies. Ceci transforme notre relation visuelle en réduisant notre cône perceptuel à l'espace de l'écran, qui n'est plus qu'un élément de notre environnement visuel parmi d'autres, et en détruisant les possibilités de lecture globale d'un objet architectural, dont, du reste, la représentation ne correspond plus à aucune référence usuelle d'échelle.
2 En ce qui concerne le *geste*, le clavier et la souris matérialisent une nouvelle démarche gestuelle qui assume une double dissociation. La première sépare la vision et le geste, mais ce dernier est lui-même dédoublé entre, d'une part, la frappe du clavier, une pratique précise et linéaire associée à la production textuelle, et, d'autre part, une nouvelle manualité, le parcours incertain et pourtant rigoureusement précis de la main contrôlant la souris.
3 Dans le champ de l'*inscription*, le pixel coloré, trace indifférenciée et reproductible à l'infini du travail graphique à l'œuvre dans le développement du projet, conjugue son immatérialité et sa normalité à ses possibilités de coloration infinie, qui permettent de lire, par un travail perceptif et intellectuel, une hiérarchisation des strates graphiques.

Comme ces profondes transformations matérielles affectent l'ustensilité de l'espace de travail traditionnel de l'architecte, on comprend les réticences que peuvent éprouver même les plus créatifs d'entre eux à changer ainsi radicalement leurs manières de faire et leur *habitus*. Ils craignent que leurs habiletés ne soient ainsi affaiblies, ce qui, estiment-ils, aurait pour conséquence de perturber leur travail et leur capacité d'invention.

LES OUTILS INTELLECTUELS

Pour mieux cerner la spécificité de la création en architecture, il importe d'interroger les mécanismes intellectuels mobilisés par cette pratique tout au long du processus du projet. Ceci permet en effet de définir aussi bien les méthodes d'intégration des savoirs déjà constitués à l'activité de l'architecte que les méthodes de production des nouveaux savoirs appropriés. Cette opération s'effectue dans la relation dialectique qui existe entre l'espace de représentation et les représentations de l'espace de l'architecte. Là, s'intégreront ses propres schèmes structurels et intellectuels à partir d'une série de thèmes variés: langage et information, signes et symboles, culture et mythes, matière et matérialité. Et c'est aussi à ce moment-là que se joue l'interaction entre la pensée et la trace graphique et que s'élaborent intellectuellement les images opératives du projet, des images qui manipulent des données objectives tout en offrant la possibilité de concevoir du nouveau, de créer.

C'est donc le lieu et le temps où « savoir dessiner » ne peut être que le symptôme de « savoir concevoir » et s'incarne dans une forme qui est pure spéculation intellectuelle et graphique, strict objet de connaissance. C'est à partir de cette forme dessinée, de cette image mentale exprimée graphiquement, que le projeteur traduit son projet dans le possible d'une réalité visible.

Rappelons que le projet d'architecture n'est pas encore de l'architecture mais plutôt le moment et le lieu de sa représentation. Il fournit souvent à l'architecte/projeteur dans cet espace/temps de la projetation l'occasion de faire valoir un point de vue différent, d'insinuer un « écart pertinent » entre le discours de son projet, les déterminismes d'un programme et les possibilités qu'offre un site.

Et c'est parce qu'il est à la fois solidaire de l'état d'avancement des connaissances dans son domaine et singulier dans sa démarche créative que l'architecte peut établir un pont heuristique entre la théorie et la pratique architecturales au sein du dialogue qu'il entretient avec lui-même dans l'espace de la conception.

Le projet d'architecture comme processus d'invention et de création comporte donc une activité de médiation, le dessin, qui dévoile son opérativité grâce à la coordination dans l'espace graphique non seulement des connaissances et des savoir-faire que cette activité interpelle mais encore des habiletés cognitives et des aptitudes créatrices de celui qui en assure le déroulement. Et ce processus, à tout le moins dans les premières phases de l'élaboration conceptuelle du projet, n'est jamais linéaire non plus qu'il ne procède selon une logique déductive.

En ce sens, le dessin s'impose comme nécessité en tant que moyen de connaissance. Il manifeste ses potentialités créatives à travers son sens analytique, l'image génératrice, et sa puissance d'évocation, l'image référentielle. C'est par là que l'architecte exprime une intention projectuelle et qu'il manipule en son absence l'objet virtuel auquel il va donner forme.

En tant que support des activités conceptuelles, l'infographie est en train de révolutionner le rapport entre l'image et l'imagination. Aussi, force-t-elle l'architecte non seulement à réévaluer la fonction cognitive du dessin dans son rôle de médiation entre l'idée et l'architecture à concevoir, mais encore à repenser la succession et le déroulement des activités du projet dans le temps.

Premièrement, les supports qui servent à matérialiser les images de synthèse (tels l'écran vidéo, la table traçante ou l'imprimante) confèrent à celles-ci des qualités qui leur sont propres et qui diffèrent qualitativement des images produites jusqu'à présent par le dessin. En effet, c'est parce que les images de synthèse offrent au projeteur la possibilité de construire, dès les premières étapes de son projet, des images dont le réalisme visuel et la précision graphique lui permettent déjà de vérifier la validité spatiale et constructive de certaines de ses hypothèses conceptuelles, et même d'entrevoir leur réalisation effective, que l'infographie peut modifier l'organisation et la structure du processus projectuel. L'infographie transforme la chronologie du projet en modifiant l'agencement de ses phases d'élaboration; en outre, elle introduit, pour la première fois, le facteur temps dans la manipulation des moyens de représentation.

De plus, par la formidable puissance d'attraction des images de synthèse et de l'espace-projet virtuel qu'elles supposent, c'est une nouvelle relation qui se crée entre l'architecte et son œuvre. C'est aussi une transformation radicale de son univers référentiel, qui est en prise directe sur les transformations que subit notre culture visuelle à travers la production de nouvelles images, de vidéo-clips et ainsi de suite. Enfin, les dimensions d'hétérogénéité, de polycentrisme et de polymorphisme ne seront pas sans conséquence sur nos capacités à penser et à concevoir les espaces contemporains. Cependant, étant donné l'état actuel du développement de la discipline, les expériences de création qui s'appuient sur ces nouveaux moyens sont extrêmement rares.

Il faudra donc expérimenter les possibilités réelles de ce nouvel outil du point de vue de la conception et de la simulation – et notamment tirer parti des nouvelles capacités qu'il offre de visualiser le parti d'un projet et sa réalisation future grâce à la déambulation dans des espaces virtuels – mais surtout du point de vue des espaces et des objets architecturaux différents qu'il permet d'imaginer. Ceci devrait déboucher sur de nouvelles collaborations fécondes entre les divers intervenants du projet et ceux qui travaillent dans le monde de la production visuelle.

CONCLUSION

Comme nous l'avons vu, les nouveaux outils informatiques ébranlent les assises sur lesquelles reposait jusqu'à maintenant la conception architecturale.

Cette mutation nous concerne, car elle met en cause notre être dans le travail, dans nos émotions et dans nos connaissances. Le défi consiste donc, aujourd'hui, à savoir comment préserver notre spécificité sans rejeter les nouveaux outils technologiques et, en tant qu'individus, comment maîtriser le changement culturel inévitable de façon à favoriser le développement de nos propres capacités créatrices. C'est en ce sens que le défi posé par ce nouvel outil contemporain, vecteur technologique et culturel à la fois, débouchera sans doute sur une nouvelle modernité dont les formes sont encore à inventer.

La notion même de projet ne répond plus aux exigences théoriques d'une définition conceptuelle. Cette situation risque, à moyen terme, de réduire la dimension intellectuelle de la projetation et d'occulter les problématiques de la création en architecture. Une réflexion s'impose donc d'autant plus qu'elle permettra de mettre en perspective les problèmes que pose le recours aux nouvelles technologies informatiques, particulièrement l'infographie, dans la conception de l'œuvre d'architecture.

Il importe en outre de se pencher avec beaucoup d'attention sur les problèmes épistémologiques que soulèvent les nouveaux moyens de représentation, et de poser quelques questions, théoriques aussi bien que pratiques, qui pourraient éclairer les développements futurs.

Ainsi, l'utilisation de l'infographie en architecture met en évidence la nécessité de mettre au point une nouvelle théorie du projet et d'élaborer de nouveaux modèles logiques et analogiques d'exécution. Pour l'architecte, la question ne se résume donc pas à savoir s'il faut ou non adopter la micro-informatique.

Enfin, le problème général actuel est la prédominance des images électroniques, des espaces virtuels et des immatériaux sur le monde des espaces concrets, des objets tangibles et de la matière. À travers cette prédominance est en train de se forger une grande partie de la culture contemporaine. Cela implique un ensemble de valeurs éphémères et discontinues qui sont en contradiction avec les valeurs qui sont à l'origine de la discipline de l'architecture, car elles privilégient l'immédiateté et la virtualité sur la durée et le lieu. Avec nos modestes moyens, nous proposons une hypothèse d'action en deux temps:

1 Démontrer, en situation de projet, que les nouveaux outils informatiques peuvent servir non seulement à stimuler l'invention de formes nouvelles, mais aussi à transformer radicalement le vocabulaire formel contemporain, du fait qu'ils permettent d'établir des confrontations immédiates entre d'innombrables solutions de rechange et d'expérimenter et de

manipuler de nouveaux espaces avant même que ceux-ci soient réalisés.

2 Diffuser les résultats pratiques de ces travaux de recherche/création à l'ensemble des intervenants du domaine du cadre bâti en faisant valoir, à l'aide de médias accessibles, les possibilités de ces nouveaux outils dans leurs pratiques respectives.

REPRESENTATION IN THE AGE OF SIMULATION

DAN HOFFMAN

Une réflexion sur l'éthique et la représentation architecturale à l'ère de la simulation doit considérer ce qui est admis et ce qui est exclus dans les hypothèses de la pensée simulationniste. Ce que cette thèse n'exprime pas directement en dit souvent davantage sur la signification et la portée qu'il convient de lui donner.

Il convient de replacer l'ère de la simulation dans le contexte des concepts fondamentaux qui ont modelé l'architecture occidentale dans le passé: l'espace, le temps, le matériau (ou la matière) et la géométrie. Chaque ère se caractérise par une configuration de ces concepts qui lui est propre et qui, à son tour, influence directement l'architecture et ses représentations.

L'importance qu'une époque accorde à un concept donné peut alors s'interpréter à la lumière des relations qu'il entretient avec d'autres concepts. Ainsi, Michel Serres souligne que la découverte du principe des triangles semblables par Thalès de Milet a rendu possible la représentation d'échelles spatiales variables au moyen d'une système unique de relations proportionnelles. Serres soutient toutefois que cette découverte a rendu caduque la dimension temporelle inhérente à la géométrie égyptienne qui avait précédé Thalès. Le soleil, dont le mouvement jouait un rôle si important dans la pensée égyptienne, dut interrompre sa course dans le temps et devenir un point de référence spatial absolu, ainsi que le réclamait la géométrie grecque. Dans ce nouveau système, la lumière de la raison géométrique pénètre elle aussi la matière, qui perd son opacité et sa vitalité. L'éthique d'une représentation architecturale qui repose sur la géométrie euclidienne doit tenir compte de l'absence du temps et de la sublimation de la matière qui sont au cœur d'un tel système.

La simulation est importante car elle suppose la réintégration du temps et de la géométrie dans ses représentations (je songe ici aux phénomènes optiques conçus par ordinateur). L'observateur pose l'hypothèse que ce que fait voir la simulation se produit dans le temps « réel ». Le temps est le moteur qui anime la représentation simulationniste, il est le facteur qui unit la représentation à la soi-disant « réalité » sous la forme d'une construction géométrique animée dans l'espace. Dans la représentation, le matériau et le processus matériel sont également sublimés. Dans la simulation, les aléas et les obscurités inhérents au matériau sont inévitablement perçus comme une forme de résistance à la nécessaire vélocité des transformations de l'information.

La représentation culminera avec la tentative de reconfigurer la nature spatio-temporelle de la simulation en une position qui tienne compte des concepts sublimés du matériau et du processus matériel. C'est dans ce contexte que je propose de substituer la méthode de la « documentation » à celle de la simulation.

Le processus de documentation se situe à mi-chemin entre la construction et sa représentation de l'architecture. Dans ce processus, les ordonnances de la représentation architecturale sont considérées comme des surfaces qui unissent les conditions matérielles et temporelles. Le document construit une représentation plutôt qu'il ne représente une construction. Il est un moyen de conférer à la création architecturale le bénéfice du poids des circonstances et, par là, de lutter contre le caractère immatériel de la simulation. La dimension éthique de ce procédé tient au fait qu'il témoigne de l'aspect matériel de l'architecture dans l'immeuble. Il ne faut pas oublier que les difficultés propres à l'acte de construire, la nécessité de faire face à des impératifs matériels, temporels et géométriques sont de nature à inspirer l'effort architectural. Une représentation architecturale devrait pouvoir refléter ce conflit et, par là, transmettre la nature de sa position éthique.

As a first approach to the question at hand, I would like to consider the significance of the terms "representation" and "simulation." In the title of this segment of the symposium, we find that simulation is referred to in the historical sense of an "age" and that representation is placed within the context of this age. In this paper, I take the position that representation, understood in the somewhat narrow sense of a referential means of thinking towards an end, can itself be considered as informing a historical age – one that establishes the ground for the age of simulation.

My reading of the title therefore asks us to consider how it is that one "informing idea" of history and culture comes to supplant another and to enquire about what is lost and what is gained in the transformation. It is also an invitation to speculate on the territory that these informing ideas offer for reflection. But perhaps the most difficult question of all to consider is what impact these ways of thinking have on our daily lives. This question poses a particular dilemma at a time when the proliferation of technologies is accelerating at a pace such that considerations of this nature are left behind.

The issue of ethics is bound up with the question of what constitutes thought itself and how it is applied in the world. The difficulty arises because of our present inability to find a point of reference that would occupy a judicious space between thought and action. The motion of technology makes such a reflection subject to imperatives outside such a discourse: its simultaneity does not allow a space for reflection. By assessing this development within the historical horizon of an age, we may be able to navigate a course through these questions and arrive at a point of view from which a position on "representation in the age of simulation" may be attained.

My reading of the title implies that the age of representation historically precedes that of simulation. It follows that representation should be the first term to be considered. To begin with a definition: representation is a means of referencing in the form of a deliberate construction. A representation fixes a relationship between a phenomenon and the world as such, and it occurs in the space where differences between the two reside. Put another way, a representation is that condition which separates a phenomenon from the continuum of being. Regardless of how it is read, a representation indicates a deliberative state of mind, for what is at issue in a representation is the idea of construction itself, the *means* through which a phenomenon or presence is "re-presenced" or represented. Thus a representation is an agent that, through its construction, aids in isolating and binding an object of thought in relation to its context.

However, this agent cannot be viewed as a neutral signifier. A representation *represents* itself. In other words, it is understood as an appeal on behalf of the difference between consciousness and its being in the world. It is evidence of our intentional manner of being. For example, one can consider the representation of speech itself as an appeal on behalf of the distance between the silence of another's consciousness and the sounding environment of the world. Speech inserts itself into that distance as an appeal to occupy, an appeal to fill – an attempt to reconcile the noisy silence of all that is and the voice that articulates, in a singular way, the manner of its being.

In a similar way, the representation of geometry can be understood as an eidetic appeal to the distance that exists between the opacity of things and the possibility of their illumination as a constructed form of thought. By understanding geometry as an appeal, we are able to isolate the construction that is geometry from the intentionality that causes it to come into being.

This consideration of geometry points to another aspect of representation – its function as a "fixing." The fixing isolates the phenomenon from the flux of the world so that it may be re-inserted in the form of a representation. In this way, a representation also carries with it a teleological aspect. Its construction is a means towards an end. And since it begins with a fixing and extends itself as a construction upon and through that fixing, a representation offers the possibility of a thinking-back through to its origins, the intentionality that is served through the appeal of the representation.

Michel Serres's re-reading of the commentaries on Thales's theorem of congruent triangles is an incisive accounting of the nature of the representation that is geometry.[1] As Serres points out, the two commentaries situate Thales in Egypt in the context of an attempt to measure the height of a pyramid. They note how Thales solved this problem by establishing a relation between the height of a pyramid to the length of its shadow on the sands. The geometrical demonstration of this relationship was offered by the idea of *representing* the pyramid by a stick placed vertically in the sand and measuring its shadow at the moment when its length equals the height of the stick. But to achieve the conceptual clarity of a geometric principle – to make it a spatial relationship – the sun itself had to be considered as motionless during the instant in which the measurement of the pyramid's shadow is made. Accordingly, Serres reasons that time must stop so that a precise spatial or geometric measurement may occur. The fixing of time is at the core of his reflection on the origin of geometry. It is this very abstract fixing which enables the entire construction of geometry to unfold upon and through itself – a point from which conscious representation may be constructed.

The eidetic appeal on behalf of geometry is to illuminate the opacity of things in time so that the construction of geometry as an illumination of the mind can occur. The importance of geometry for us is that it becomes the space of construction – a fact confirmed by the paradigmatic form that geometry has assumed in the history of our architectural endeavours and their representation. The teleological character of geometry – its beginnings and endings in the form of an infinitely fixed and reduced body – makes all of the constructions that unfold from it transparent to themselves. This does not mean that time itself disappears in the considerations of geometry, for there is always the matter of the execution of the proofs and their forms as demonstrations of geometric constructions. The language of Euclid's *Elements* anticipates that the proofs are drawn out over time, that they can be inserted into the temporal conditions of language as discursive speech. The fusing-together of the fixed certainties of geometric principles and of the spoken form in which they can be delivered contributes to the transmissibility of geometry and sets in motion the proliferation of constructional knowledge and activity that has virtually built the world that we inhabit today.

Serres's relocation of the origin of geometry in Egypt underscores the historical significance of the release of this knowledge as a form of representation. The shadow of the pyramid's inaccessible interior conceals the temporal dimension of the dark knowledge of its construction without the referential certainty of geometric principles – the slow piling of stones, one atop the other, fitted according to a knowledge of the body that is confirmed by the brutal and automatic repetition of acts. Through this repetition, a certain refinement by approximation arises and evolves into a truth.

This is the manner of being that is given to the inhabitants of Plato's cave, whose only access to understanding is through the observation of the shadows that are cast on the wall of the cave in front of them. But it would be wrong to say that these unfortunate inhabitants are denied the constituting aspects of the world, for what they do have, in their ability to observe the play of shadows, is an intense and singular awareness of time. The being that leaves the cave for the light above is granted the certainty that is represented in the construction of geometric knowledge. The illuminated inhabitant now has the responsibility for conveying this knowledge to the others in the cave. This he can do, for geometry is equipped with the means to represent itself in language. His first task, then, is to represent to the inhabitants of the cave the section of the cave itself as a *construction* and, through this, the functioning of its various parts. In effect, he must teach geometry to the inhabitants of the cave so that they may be able to conceive of the structure within which they exist. Through its construction, the representation of geometry explains the phenomenon of the shadow to the inhabitants. The transparent light of the representation illuminates and erases the shadows of the cave, just as the reconstruction of the pyramid by Thales erases its programmatic, temporal function as a gnomon. The erasure of the shadow displaces the cyclical, temporal horizon for the spatial horizon of a geometric construct.

For the Egyptians, eternal recurrence was the very representation of time. This faith was not transmissible because it was not transportable outside the cultural framework of Egypt. But time has its attractions in the immortality that it promises. We are continually drawn to the "dark knowledge"; and with the eclipse of the age of representation, we find it again at our door.

The supplanting of the fixed horizon of geometry, considered as the construction of an ideal body by the infinite extension of a relational construct, is a later, historical development that, I would argue, arises with the reintroduction of temporal concerns into the problems of geometry.

Today, the age of representation is about to be integrated and transformed into the age of simulation. I would agree with Husserl's view that the process begins largely with the thinking that understands nature – the sum of all that is not touched by human artifice – to be in essence mathematical.[2] This reduction of phenomena to mathematics enables the manipulation of qualities through their symbolic formulation in numbers.

Such thinking makes it possible to incorporate time as a variable in an equation and thus give form to phenomena that are considered to be outside the limit of geometry. Concepts such as acceleration have no limit; their forms are those of perpetual transformation, threatening to unhinge the secure, representational mooring of Euclidean geometry. Rather than being understood as an enclosing form with us as its centre, the horizon is now viewed as an ever-receding (accelerating) limit. Motion – and hence time – becomes an essential constituent in the consideration of all phenomena.

Accompanying this development is the necessary intervention of instruments designed to measure and accommodate motion. Constructions that engage the temporal aspect of phenomena inevitably demand engagements of instrumental complexes, for the construction of motion is, by nature, technological. Husserl notes how the infinite horizon of mathematics now combines with the perfecting qualities of instrumentation to form what he calls an "idealizing praxis" – a technology that continually perfects and transforms the very limit-horizons from which it is constituted. The temporal aspects within this new historical age are actually more complex and contradictory than the project that Husserl envisions, for with the invention of the internal combustion engine and with the accompanying theoretical frame of thermodynamics, another idea of time is introduced. This is the third law of thermodynamics, or entropy, defined as the amount of energy available in a thermodynamic system. The acceleration of technology is now opposed to the deceleration in the thermodynamics of energy systems. The acceleration towards the speed of light is now opposed to the cold, still death of maximum entropy. And projections of an ideal ordering are opposed to the chaos that dominates in the dissolution of ordered energy systems. The totalizing embrace of this system is felt in the anxiety of an age of ever-diminishing ends that must be matched by proportionally increasing means. The combined effect of these opposing vectors is the curious, neutralizing sense of moving forward to remain in the same place. The imperatives of this state of affairs make it difficult to assume a critical distance from which to observe the conflicting energies that seem to affect our understanding of every phenomenon. It may be this problematic state of affairs which causes technology to be viewed as an autonomous construct that frustrates our ability to understand our relation to it.

Poets such as Beckett succeed in obtaining a brief distance by breaking the system down in some way. Beckett's art is that of diminishing means that bring upon us the question of ends. He would argue that we are *at* the end, or just near it. His preoccupation with ends is a reflection on the impossible beauty offered by the teleology of geometry – end as finitude, end as certainty, the end as a beginning. But all is grey in Beckett's world – or "light black," as one of his characters puts it.[3] We can no longer sustain the fixity of representation that the dualism of light and darkness would allow. In the play *Endgame,* these "ends" exist as wistful memories or vague and curious desires that the characters cling to, for there is nothing else happening except the game of reason played against the inevitable depletion of the body in time. The space that the characters occupy is ritually surveyed time and time again but does not provide the hoped-for certainty. The activity of measuring exhausts itself as the constructions of reason are reduced to the shell of paradox that would appear to constitute meaning itself. Representation is now an empty vehicle through which drip the temporal remains. As Hamm, one of the play's characters, states, "there is something dripping in my head." For Hamm, things "are taking their course." He waits for the end in that nether presence of time passing with little or no motion. The play ends in a silence that keeps us listening, waiting for the end or, possibly, the beginning.

This grey silence anticipates the grey noise that now is heard in the rapid expansion of the electromagnetic field, the field of simulation. The "striated" constructions of geometric space have now been integrated into the "smooth" reconfigurations of the simultaneities of space, energy, and time. Grey is no longer the mixture of light and darkness but the result of interference patterns or unfocused channels. The opposing temporal vectors no longer neutralize each other in silence, but they are now energized by their confluence in the electronic media. The entropy of thermodynamic systems is now the entropy of information; and the projections of idealizing praxis have been confounded by the uncertainties about the nature of light itself. The old constructions of representation are now extended through cybernetic connections into simulated environments. No longer is it possible for us to inhabit the geometry of the plan, for now the geometry of the plan inhabits us. Understood in the configural matrix of electromagnetic fields, the representation of the plan is simply one of many possible views. Constructions based on the plan are no longer felt with a representational weight. The plan is simply *made available.* It appears, and we receive it.

The simultaneity of the projection/reception loop is, I believe, the decisive transformation of representation in the age of simulation. The representation of geometry is inseparable from its construction, the time of geometry being either the sequential time of the construction or the absolute stillness of space itself. The certainty of geometry was always proof of its ability to be understood, for certainty was viewed as the key to transmissibility. In the projection/reception loop of simulation, understanding is preprogrammed or engineered into the system. Since the computer is an engineered device, a projection simply involves the release of prestructured arrays into the reception vehicle of the monitor. Time is ever present as an ambient potential of motion in the device. Its presence saturates the projected image since the electromagnetic screen is always in motion within itself. Delineated representations of movement are received uncritically because of their durational quality. The gates or openings for reception are carefully anticipated in the programs of the system, therefore insuring a smooth looping along the preprogrammed channel of receptivity.

What is significant here is that phenomena such as the sound of a voice or the appearance of a view are displaced along pre-established or highly structured channels of communication. These channels are products of a time-based technology that maintains certain aspects of the phenomena as it is transferred from one situation to another. But since the original situation cannot be repeated, an aspect of its presence is lost – a loss that is hidden by the mesmeric and instinctual attraction that time-based media exercises.

And so we find ourselves back again confronting a temporal condition. Michel Serres's timely re-reading of the foundation myths of geometry are a compelling evocation of the historical dimension of the transformations briefly referred to in the preceding discussion. The telling of the myth can also be understood as a cautionary tale, for the problems that simulation poses for us can certainly be traced back to the historical developments that established the possibility of universal *mathesis* and its transformation of the world through technology.

As an extended footnote to this discussion, I would like to offer a recently completed work for consideration. This involves a concrete block wall, eight feet high and sixteen feet long. The stacking of the wall was documented by cameras positioned so that they could photograph each block being laid into place from both sides of the wall. The photographs were then developed on the surface of the blocks, and the space in front of or behind the wall blanked out in white or black.

The result is a wall that records its own making. The traces of the building act and the surface of the block are reflected in the presence of the wall itself. It is evident that the wall resists being understood as either a representation or simulation in the manner described previously. The wall is not a representation of a wall because of its literal presence before us (recorded in an indexical manner with the camera). The photographs are of the wall itself being constructed. Their display across the surface of the wall produces a multiplicity of readings – vertically, diagonally, horizontally, none of which can claim to be the singular representation of the wall itself. It is because of this multiplicity of readings and of the circumstantial grain of the work – the presence of the wall as a pile of blocks and the grain of the block within the photographic image – that the piece also resists being understood as a simulation. The traces of the multiple productions in the work are simply too emphatic for the "gloss" of simulation to take hold.

Here, finally, the question of ethics may be addressed. It is evident that both representation and simulation evolve from a way of thinking that sees the "real" in the reconstruction (mathematical or technological) of the world. Neither recognizes that we actually *find* ourselves *in* the world – a world whose horizon always recedes from us as we approach it. The real, if it could be said to exist, is in the circumstance and situation of our existence. Our thinking and works, if they are to be true, must refer reflexively back

White wall

White wall detail

Black wall

Black wall detail

to the situation – to raise its grain, as it were. In this sense, it is unethical to consider a representation or a simulation as conditions of reality rather than as forms of culture that exist to order our individual actions.

The reflexivity of the work is the opening through which its situation may be established. This offering to interpretation becomes a dwelling with the work. Aspects of this manner of thinking can certainly be found within the endeavour of architecture. Our task is now to re-read its territory for its circumstantial (and resistant) possibilities.

NOTES

1 Michel Serres, "What Thales Saw," in *Hermes – Literature, Science, Philosophy* (Baltimore: Johns Hopkins University Press 1982), 84.

2 Edmund Husserl, *The Crises of European Science and Transcendental Phenomenology* (Evanston, Ill.: Northwestern University Press 1970), 26.

3 Samuel Beckett, *Endgame* (New York: Grove Press 1958), 32.

PART TWO / DEUXIÈME PARTIE

ÉTHIQUE ET POÉTIQUE DANS LE CONTEXTE DE LA PRODUCTION TECHNOLOGIQUE

ETHICS AND POETICS IN THE CONTEXT OF TECHNOLOGICAL PRODUCTION

ÉTHIQUE ET TACT

JACQUES ROUSSEAU

The site. The pregnant nature of the site awakens in us, users or designers, a feeling and an emotion that we have not experienced before.
Doubt. The desire to express doubt carries us to the edge of risk. We might well forget everything and not take that risk, as this might free us from the responsibility of having acknowledged its presence. Is this not what often happens?
Choice. Reflection on the intentions of the work.
Roland Barthes, writing about Dutch paintings: "There are objects on every surface, on tables, the walls, the ground: pots, overturned jugs, baskets in disarray, vegetables, game, bowls, oyster shells, glasses, cradles. All this is the space of man. He takes his own measure within it and establishes his humanity through the memory of his gestures: his time is filled with the use of objects. There is no authority in his life beyond that with which he endows inert things by creating and manipulating them."
Simplicity. As continuity and immediacy, depth and surface meet in endless confrontation today, simplicity is an authentic value through which one may recognize contemporary complexity and adhere to the essential.
Silence. Everywhere, signs and messages proliferate and become superposed; speech has become unintelligible and meaning is diffused; thus silence is the threshold where the emergence of memory is appreciated.

The project. The project seeks to understand the contemporary complexity affecting the site, but it rejects overemphasis. That is why it seeks to welcome a memory built of silence, while maintaining that silence is not muteness. It suggests that the spirit of the place becomes rooted when the space that welcomes people also remains alone. Ultimately, it knows that it will remain a place. When no one is there, it is.

Epilogue. At the end of this work, a question is posed: Is there any authority other than the one with which inert things are endowed when they are created and manipulated? Hope can only be born within us. Man is the measure of history!

Je sens ce qui se passe, mais je ne comprends pas encore parfaitement. Peut-être suis-je un peu craintif, mis mal à l'aise par la force de pénétration des immatériaux. Et pourtant (depuis près de dix ans), je me sens bien en leur compagnie. Je précise ce que j'entends par là.

Les immatériaux que je connais sont en moi, et c'est au travers d'une longue relation avec eux que je suis arrivé à les faire sortir. Ce sont ces êtres virtuels qui nous habitent et qui possèdent toutes les conditions essentielles à leur réalisation, sauf qu'ils ne connaissent pas encore la matière, et qui n'ont donc ni chair ni sens. Ce sont les idées. Au bout du compte, les miens sont identiques aux immatériaux latents dans les logiciels. J'admets ici une légère dérive, car je comprends bien qu'en fait, ils ne sont pas vraiment dans les logiciels. Mais enfin, on les trouve bien là où l'on veut.

Non, ce qui me trouble un peu, c'est qu'ils pourraient tous sortir. Ce qui me trouble un peu, c'est le danger du jeu, car l'usage du clavier n'est peut-être pas encore au point. C'est d'abord ce problème dont je voudrais vous entretenir. « Entretenir », c'est un fin mot: « tenir entre nous ». C'est là une qualité du langage que celle de contenir des sous-entendus, mais enfin!

Je tiens à vous avertir, le fil conducteur de mon exposé n'est pas toujours clair. Je ne m'en excuse pas; au contraire, j'estime que chacun de nous y trouvera peut-être un peu de soi.

Il y a trois ans, j'avais besoin d'un sac à bandoulière, j'ai acheté un ordinateur portatif. Un peu naïf, je n'ai pas acquis l'imprimante. Dans le sac, j'ai amené mes amis d'atelier – ça ne vous gêne pas, j'espère? – ceux-là dont je vous ai parlé tantôt, mes immatériaux! Je vous en présente encore quelques-uns, ainsi nous serons plus au chaud, presque toute la famille y sera.

Mes amis d’atelier ...

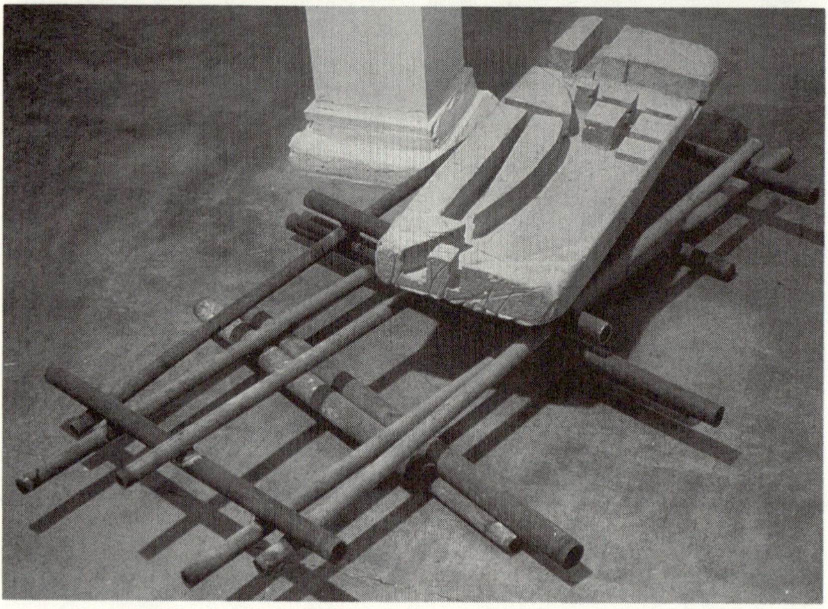

... ces êtres virtuels ...

... ces idées qui nous habitent

Ah oui, le clavier!

Mais vous vous doutiez bien que je n'allais pas vous parler du clavier en tant que tel. En fait, c'est plutôt le temps qui en précède l'usage qui m'intéresse, et ce que l'on retiendra de ce temps – comment le projet le porte avec lui.

L'approche de Peter Rose dans sa communication est juste. Il dit: « Il faut tout porter ». Soit!

Mon travail a consisté à porter ma part à vos côtés. C'est dans l'étude de la ville (Montréal) que j'ai découvert ma position, une attitude, une éthique, peut-être!

LE PROJET DE L'ARCHITECTURE – MONTRÉAL, 350 ANNÉES PLUS TARD

« Cette ville hérite de son plus profond caractère en ce qu'elle est dans sa condition d'origine: une ville du Nouveau Monde. De la réflexion sur cette nature jaillit la notion fascinante de la destinée.

En ces thèmes de l'origine et de la destinée, si intensément liés l'un à l'autre en elle, se déroule tout le combat de son enracinement au nouveau sol. À peine apparaît le sédiment des générations qu'apparaît son érosion, si bien que la mémoire y est une œuvre immédiate et fugitive.

À la fois inspirée d'un caractère auto-référentiel et douée d'une mémoire directe et mobile, cette ville a formé son corps en filigrane, marquant le vide comme le plein d'un sens en formation, laissant à y lire ce qu'elle n'y a pas explicitement inscrit. »

Des projets ont-ils présenté l'architecture comme un art conscient? L'art en tant que savoir, conscient en ce qu'il est lucide, connaît et juge son propre sens et le monde extérieur? Un art qui reconnait l'évolution et présage des changements?

De tel projets ont existé. Dans toute l'histoire, l'architecture prétend (pose devant) son état de conscience et sans arrêt cherche à qualifier son sens en tant que projet de la conscience. Très largement et positivement, nous pouvons dire que tout architecte poursuit cet engagement, que celui-ci est sa source et son but, que les idées et les formes visent à l'adéquation dans la pertinence et la vision, qu'ainsi son Art vit et sa conscience s'incarne.

Voilà en quelques mots le parti de l'architecture et l'espoir de la ville. L'intention est manifeste.

Reste l'opération.

Or, comment se fait-il que nous ayons les cités que nous avons?

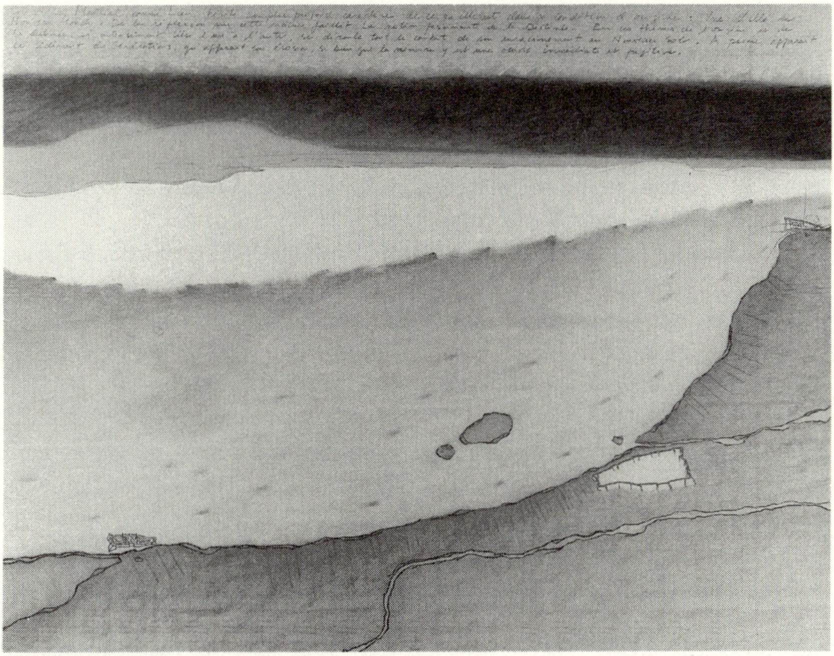

Dessin initiatique: le pays avant le pays

Alors que nous fondions nos premières cités, il y a plus de 350 ans, se rédigeait le premier chapitre de notre essai sur l'architecture. Ingénue et en toute simplicité, elle cherchait activement. Soucieuse des besoins et des moyens, elle réalisait littéralement la greffe de la culture à la nature. Elle entreprenait le long travail de l'essai, celui de sonder et développer son objet sans qu'elle ne puisse l'épuiser, réalisant ainsi un travail d'information.

Succinctement, les mutations réalisées à partir des premiers types de la Nouvelle-France, les variations sur les propositions victoriennes, les exceptions apportées par la pensée et l'espace modernes et les récentes réminiscences de symboles historiques – tout cela résume des périodes significatives de notre essai sur l'architecture qui a donné forme à nos cités, à ce jour. Voilà une première lecture.

Mais encore, au-delà de ce regard et de cet inventaire des formes de l'habitat, nous découvrons des cités très jeunes, animées d'une histoire intense et tout à la fois très courte, si bien qu'elles sont plus qu'ailleurs des structures encore malléables, pouvant absorber des valeurs changeantes et les adopter spontanément.

En effet, si la cité, de par sa nature historique, est un lent processus de reconstruction et de refondation appelant l'itératif et le différé, nos cités interrogent cette condition dans leur processus d'adaptation « en direct ».

Implicitement, il faut beaucoup de temps et de patience pour que les idées et les formes atteignent la mesure juste. Il est tout aussi vrai que l'on ne peut pas ne pas admettre l'immédiateté comme autre condition de production des idées et des formes de la cité.

Je crois que c'est à cette jonction du reconduire et de la fuite vers l'avant que se comprend dorénavant la nature de la cité et, plus loin, que s'anticiperait la nature du travail à y faire. (Ici, la notion de fuite vers l'avant servirait d'image au thème de « mémoire immédiate et fugitive » propre au Nouveau Monde.)

Reste l'opération.

Voilà un moment critique dans l'histoire de notre essai sur l'architecture, où s'opposent pour la première fois, globalement, continuité et immédiateté.

Trois cent cinquante années plus tard, nous savons que nous avons complété un cycle, depuis la fondation jusqu'à la réminiscence; la théorie et la pratique se trouvent aux antipodes l'une de l'autre, la pratique plaidant pour un processus itératif et la théorie œuvrant dans le futur immédiat. Comment, alors, l'architecture, art conscient, peut-elle entièrement s'engager, soucieuse des besoins et des moyens (le temps réel), et poursuivre le développement du rapport de la culture à la nature – cette dernière étant prise dans son sens structurel, définie comme modèle de l'espace, alors que la culture est le site idéal de projection du modèle (la cause réelle)?

C'est cette question, qui joint le temps réel et la cause réelle, qu'on propose à notre réflexion, à nous, héritiers de l'humanisme et témoins-acteurs de l'immédiat.

Pour soutenir notre propre histoire et ne pas sombrer dans la mimique d'un quelconque mouvement du passé ou dans quelque indulgente architecture passe-partout, il faut trois conditions: la pertinence, la vision et la transformation, cette dernière étant désormais l'impératif le plus percutant auprès duquel il nous est donné d'apprendre et d'agir... La poursuite de l'adéquation de ces conditions m'apparaît l'approche la plus actuelle pour incarner à la fois le processus direct de nos cités et la quête continue d'une conscience culturelle.

J'envisagerais alors que l'essai sur la transformation est potentiellement l'affaire de tous et, pour l'architecture, un outil valide et puissant qui lui permet de « poser devant », son projet de conscience.

Le pouvoir de la transformation

Reste l'opération.

Cette image veut évoquer la transformation. Le choix d'un objet et d'une situation, apparemment sans lien, veut exemplifier l'idée même de transformation. Ce lieu veut être vu comme manifeste, exacerbant le concept même que dans l'un réside l'ensemble, que la finalité de l'ensemble est inscrite dans chaque facteur isolé, et que l'ensemble est un et multiple à la fois, que la transformation, lorsque reconnue en tant que condition active, en tant qu'idée, engage la responsabilité d'en faire état et de la poursuivre.

Et voilà notre homme! Notre porteur du temps. Il pense: « Quelle hypothèse de fond, en quête de justesse et d'équilibre dans la formalisation des idées qui transforment les lieux et donc les humains, serait pertinente et visionnaire, en cette époque? Élaborer le lien entre le passé et le futur, en agissant sur le présent? »

Voilà une dynamique pointue, description et impératif intrinsèque du défi contemporain, devenue en quelque sorte une évidence. En effet, élaborer le lien entre le passé et le futur tout en agissant sur le présent, c'est

Notre homme

évident! Mais en ces temps chargés d'évidences, alors que tout semble facile à atteindre, tout ayant été reconnu, c'est une des tâches les plus ardues que d'en assumer une.

Exemple: « Faites-moi une école! Est-ce que tout n'est pas à refaire? »

Ainsi, le travail sur l'architecture, sur le temps, au présent, s'énonce d'incertitudes et nous engage en des observations labyrinthiques et des spéculations fragiles, dans des modes aussi distants que la mesure (la pertinence) et l'exploration (la vision), au point d'instaurer un sentiment d'ubiquité: être aux extrémités et être ici à la fois... Être de l'origine, de la destinée et de la transformation.

C'est dans cette condition complexe, à l'intérieur de laquelle agit l'architecture, que renaît l'engagement humain qui présage une attitude faisant appel dans un sens large à la notion des mœurs et coutumes.

UNE HYPOTHÈSE DE TRAVAIL

Considérant cet engagement et cet état de la transformation, il devient fondamental de penser l'architecture comme écriture sur le sens de son propre temps, témoignant des coutumes et usages qui persistent ou disparaissent, qui sont modifiés ou qui apparaissent. Ainsi notre activité de projet engagerait la transformation comme état à fréquenter et connaître, et poserait la persistance, la disparition, la modification et l'apparition comme outils du projet auprès des coutumes et usages de notre lieu.

Le rapport des moyens aux besoins (le temps réel) trouverait son sens dans l'opération de poser la double condition de la ville actuelle (continuité/immédiateté) comme modèle et le lieu des coutumes et usages comme site idéal de projection et d'écriture du modèle (la cause réelle). En pratique, cette opération incarne un double appareil qui forme l'expression architecturale comme écriture sur le sens de son propre temps.

D'une part, le développement d'une visibilité matérielle ancré dans le travail d'extraction des données du territoire, celui-ci étant l'ensemble matériel où coutumes et usages interagissent. Donc un travail continu exerçant l'observation des individus et des lieux, ayant pour but de reconnaître et entretenir le connu. Et d'autre part, la formation de l'espace, apparaissant dans le travail de manipulation, exerçant l'ordre de la vision, voire l'ordre de la transfiguration du lieu, ayant pour but de sous-entendre un lieu ou créer un sous-entendu, de transformer ou former l'usage et la coutume.

Reste l'opération.

Au terme de ce travail, reste à nommer ce que l'on retiendra du temps qui précède l'usage.

Alors qu'aujourd'hui se confrontent continuité et immédiateté, profondeur et surface, la simplicité apparaît comme valeur authentique en mesure de reconnaître la complexité contemporaine et tenir à l'essentiel.

Alors que partout les signes et les messages se sont multipliés et superposés, que la parole est devenue inintelligible et le sens dilué, le silence apparaît comme le seuil d'appréciation d'une mémoire en devenir.

L'histoire ne peut naître qu'en nous, l'espoir est à la mesure de l'homme.

Le projet qui porte le temps reconnaît dans l'osmose de l'archétype, la copie et l'exception, la possibilité d'incarner l'essentiel d'une mémoire en devenir.

Puis, viennent les moyens.

PROJET TÉMOIN: LA MAISON COLONIALE

En concevant la Maison Coloniale, entre autres choses qui s'y trouvent inscrites, j'ai pensé à l'état des métiers. La solidarité qui nous lie est tellement ténue, les savoir-faire se déplacent, les langages s'autonomisent.

Témoin, la Maison Coloniale

Ici, la structure de béton n'aura pour but structural que de survivre à sa peine: le squelette est dehors, qu'est-ce que cela veut nous dire?

Partout la division, la séparation des éléments. La juxtaposition, la superposition et la répétition forment un langage dans le langage. Tout y est contact de points et de surfaces, d'appuis et d'ancrages.

Le squelette est dehors, le corps est dedans. Qu'est-ce que cela veut nous dire?

La Maison nous dit la nature de notre côtoiement. Elle nous rappelle nos liens possibles et nous apprend qu'à force de séparation, tout s'unit.

Vue de l'intérieur, la Maison Coloniale

Elle nous indique la distance qu'il nous reste à franchir pour se toucher et s'émouvoir. Elle est comme vous et moi, un peu résistante et prête à tout donner.

Elle nous dit la convivialité et nous rappelle nos moyens d'humains.

Elle nous rappelle que c'est nous qui l'avons faite!

L'ARCHITECTURE CONSIDÉRÉE SOUS L'ANGLE DU PROCESSUS – LES ENJEUX ÉTHIQUES

ROBERT PROST

The novel aspect of the recent renewal of ethical enquiry in western societies lies mainly in the many challenges posed by contemporary philosophical, scientific, and technological developments, as well as in democratic developments marked by attempts to implement the general principles underlying human rights.

Ethical debates – and, occasionally, their outcomes – are of major interest in that they reveal much about emerging values in contemporary societies. Fewer and fewer spheres of human activity, including the field of artistic endeavour, now escape this kind of enquiry.

This renewal of the ethical perspective is occurring precisely because this type of enquiry had been lost. It is important that theorists seek to reveal the complexity of contemporary philosophical debates on ethics and to show how these concerns apply to architecture. This type of contribution will make for a more realistic perspective on ethics and for a more comprehensive view of ethical issues in architecture.

Beyond these philosophical generalities, however, specific enquiries in a more pragmatic mould are also needed in order to understand the diversity, specificity, relativity, and contextuality of some ethical issues in the area of architecture. That approach, which is followed here, can be characterized as follows:

1 Our goal is to raise issues and articulate problems rather than propose solutions. In fact, the ethical enquiry in architecture, which

is still a very recent development, will require extensive debate and broadened communications among the many actors involved before it can reach the degree of maturity that will make it possible to adopt explicit, clear, and operational positions.

2 Rather than speak of architecture in a very general sense from the point of view of aesthetics or technology, for example, the focus will be on architectural production; among the actors involved in this production, the discussion will not be limited to the architect. Ethical enquiry in architecture cannot involve individual responsibility only but must consider all social actors and lead to social awareness and responsibility.

3 Finally, a third limit is introduced, in that architectural production is considered in the theoretical perspective that is suggested by the concept of process:
   - the processes that cause existing buildings to be irrevocably altered (through physical, functional, and symbolic aging), and the ethical principles inherent in the attitudes to be adopted in the face of these transformations;
   - the processes that define, generate, and result in architectural products, and the ethical challenges facing the actors responsible for their design, given the fact that their work (in progress) will inevitably undergo transformations in the future.

All of these choices restrict the scope of the enquiry that is pursued here, but the concepts of process and transformation will nonetheless make it possible to reveal specific ethical issues that are not always present in debates on architecture in contemporary western societies, which tend to focus excessively on the actual output of architecture.

RENOUVEAU DU QUESTIONNEMENT ÉTHIQUE ET ARCHITECTURE

« La nouveauté, dans la récente réintroduction du questionnement éthique dans les sociétés occidentales contemporaines, est largement inscrite, d'une part, dans les enjeux multiples attachés aux développements scientifiques contemporains et, d'autre part, dans les avancées démocratiques marquées par des tentatives de déclinaison et de concrétisation des principes généraux inhérents aux Droits de l'Homme[1]. »

L'architecture comme corps de connaissance, comme champ de pratique ou comme réalité sensible, constitue un domaine un peu en marge de ces grandes tendances: elle n'est pas, à proprement parler, installée au cœur des grands débats philosophiques, scientifiques et technologiques, comme le sont, par exemple, des secteurs tels que l'éducation ou la santé, la communication ou même les arts visuels. Elle ne constitue pas davantage un « domaine sous haute surveillance » pour les acteurs et les instances garantes du respect des droits de la personne, même si bon nombre de situations, dans des domaines aussi divers que les bidonvilles, les sans-abri ou le patrimoine, pourraient justifier une vigilance accrue et la tenue de débats non réservés aux sphères spécialisées.

Mais que l'architecture soit en marge des grands débats scientifico-technologiques et sociopolitiques – et donc à l'écart des bruits médiatiques, tout au moins dans une certaine mesure – n'enlève rien au bien-fondé d'un questionnement éthique de cette discipline. En effet, l'architecture a bien besoin, à la fin du vingtième siècle, d'échapper à toute forme d'enfermement esthétique ou technologique. Épuisée par la répétition quasi automatique d'une doctrine moderne vieillissante, par une nostalgie incurable de la désuète notion d'avant-garde, ou encore par une conception frivole du « nouveau » visant à satisfaire les goûts issus d'une idéologie du changement ou d'une quelconque rhétorique de l'innovation à la recherche d'une finalité introuvable, l'architecture doit, en effet, se reposer la question de ses fondements normatifs.

Sur un autre plan, socio-économique et sociotechnique, l'architecture est de plus en plus assujettie à l'ensemble des contraintes qui pèsent sur le secteur du bâtiment. Or, celui-ci accepte maintenant, avec retard mais détermination, les logiques technologiques et industrielles relatives à la production, à la commercialisation et à la gestion, ce qui ne manque pas de susciter des enjeux éthiques non négligeables.

Notre propos n'est pas ici d'opérer un diagnostic de l'état actuel de l'architecture ni d'établir les paramètres qui pourraient guider une telle évaluation. C'est par un argument d'un autre ordre et, si possible, loin des polémiques que nous souhaitons avancer la pertinence d'un questionnement éthique en architecture.

Si l'on considère l'architecture sous l'angle du processus, c'est-à-dire non pas comme objet (œuvre) mais suivant la manière dont les acteurs sociaux se donnent les conditions et les moyens pour penser et réaliser les œuvres architecturales, et penser et transformer ou conserver les bâtiments existants, force est de constater que la nature des problèmes d'architecture relève bien du *what ought to be* et non seulement du *what is*, pour la caractériser à partir de la distinction célèbre introduite par H. Simon à propos de la notion de design[2]. Si l'on accepte cette prémisse, on doit alors reconnaître que le questionnement éthique apporte à l'architecture une perspective que ni l'esthétique, ni la technologie ou le socio-économique ne peuvent à eux seuls lui transmettre.

MULTIPLICITÉ DU QUESTIONNEMENT ÉTHIQUE

Si nous parlons de renouveau de la perspective éthique, c'est que, précisément, ce genre d'interrogation s'était perdu. Il semble dès lors important que certains penseurs, dans la discipline qui nous concerne, cherchent, dans un premier temps, à rendre explicite la complexité des débats philosophiques contemporains sur l'éthique et, dans un second temps, à spécifier les problèmes relatifs à leurs « applications » en architecture[3]. Les apports de ce genre permettront d'actualiser le débat architectural contemporain et d'introduire des perspectives capables d'en féconder les grands pôles.

En complément à ces propos de nature générale et inscrits dans le registre philosophique, des investigations plus spécifiques nous apparaissent également nécessaires, empruntant une autre voie, plus pragmatique (*practical ethics*), afin de comprendre la diversité, la spécificité, la relativité et la contextualité de certains enjeux éthiques inhérents à l'architecture. C'est ce type de parcours que nous proposons d'emprunter ici. Plusieurs remarques permettent de caractériser le point de vue adopté.

1 Nous chercherons davantage à soulever certaines questions et à formuler certains problèmes qu'à proposer une position de nature prescriptive, ce qui ne signifie pas que nous visons à une objectivité illusoire ou une neutralité insipide. En fait, nous considérons que le questionnement éthique, encore bien « jeune » en architecture, réclame de nombreux débats et une communication élargie entre les multiples acteurs concernés avant d'atteindre une maturité qui autorise à des prises de position explicites, claires et opératoires.

2 Plutôt que de parler d'architecture au sens le plus général, à partir de propos sur l'esthétique ou la technologie, par exemple, c'est autour des œuvres architecturales que nous centrerons notre attention. Cette discussion ne saurait toutefois se limiter à l'architecte, parmi les acteurs liés à ces œuvres. En effet, le questionnement éthique en architecture ne peut se fonder uniquement sur la conscience individuelle; il réclame des con-

sidérations sur l'ensemble des acteurs sociaux, c'est-à-dire l'exercice d'une conscience, d'une responsabilité de nature collective et non seulement « professionnelle ». Cette notion de responsabilité en regard d'une œuvre architecturale est souvent très complexe. En effet, un bâtiment ne renvoie jamais qu'à ses seuls usagers ni, a fortiori, à ses « propriétaires »: en contrepoint à son statut socio-économique et juridique, il est presque toujours un élément constitutif du patrimoine collectif et le support d'expériences et de significations multiples.

3 Enfin, nous introduirons une troisième limite, qui consiste à considérer les œuvres architecturales suivant le point de vue théorique auquel nous convie la notion de processus, c'est-à-dire, d'une part, les processus qui transforment inéluctablement les œuvres existantes (vieillissement matériel, fonctionnel, symbolique) et les enjeux éthiques inhérents aux attitudes à adopter face à ces transformations; et d'autre part, les processus suivant lesquels se définissent, s'engendrent et se concrétisent les œuvres architecturales, et les enjeux éthiques que doivent relever les acteurs responsables de la conception du fait que leurs œuvres (en devenir) ne manqueront pas de subir des transformations dans le futur[4] et qu'elles sont, par ailleurs, issues de transformations (constructions intellectuelles) opérées sur le corpus existant.

L'ensemble de ces choix va, bien sûr, limiter l'envergure du questionnement que nous allons introduire, mais la notion de processus permettra néanmoins de dévoiler certains enjeux éthiques qui ne sont pas toujours présents de façon explicite dans les débats sur l'architecture au sein des sociétés occidentales contemporaines.

Enfin, bien que l'interrogation proposée se fonde davantage sur la notion de temps que sur celle d'espace, ce n'est pas une approche historique qui nous anime ici mais, plus simplement, une perspective sur les œuvres architecturales actuelles, tant celles que nous a livrées le passé que celles qui sont en élaboration, les œuvres en devenir.

ENJEUX ÉTHIQUES RELATIFS À LA TRANSFORMATION DES ŒUVRES EXISTANTES

Si l'examen des œuvres architecturales du vingtième siècle nous permet de déceler des mutations importantes (structurales, plastiques, fonctionnelles, conceptuelles, etc.), c'est dans le rapport de ces œuvres à leur contexte social qu'apparaissent les transformations les plus évidentes et les enjeux éthiques les plus représentatifs de l'époque contemporaine. Depuis plusieurs décennies, nous entrons peu à peu dans ce que certains appellent les « sociétés complexes », caractérisées entre autres par des mutations constantes, diffuses et foisonnantes. Ces transformations sociales et leurs manifestations concrètes ne sont pas homogènes et touchent à certains secteurs plus qu'à d'autres, mais leur fréquence, leur ampleur et leur interdépendance finissent par introduire une crise d'adaptation entre des formations sociales en mouvement constant et des formes architecturales irrémédiablement stables.

Si la stabilité des œuvres architecturales peut difficilement être transgressée du seul fait qu'elles sont ancrées dans la statique inhérente à toute forme matérielle, une certaine idéologie de la permanence est également à l'origine de conventions qui freinent très nettement la recherche et le développement de solutions éventuelles capables de résoudre cette crise d'adaptation. Ce n'est donc non pas seulement la matérialité inhérente de l'œuvre mais bien aussi son « statut social » qui sont au centre de notre problème. C'est en regard de ce double statut que nous cherchons à comprendre les enjeux éthiques liés à cette crise d'adaptation. Raisonnant autour de deux pôles extrêmes, il nous faut, dans un premier temps, illustrer la manière dont les acteurs sociaux peuvent prendre position face à cette situation conflictuelle.

1 Pour assurer une adaptation constante de l'espace aux besoins immédiats de la société, l'instrument le plus puissant est sans doute la possibilité de tout détruire en tout temps. À l'instar des objets usuels de la consommation, les œuvres architecturales sont, dans cette optique, considérées à partir d'une position utilitariste et instrumentaliste, très souvent fondée sur la suffisance de certains acteurs sociaux qui estiment que rien d'important n'a été fait dans le passé – on songe ici à la fameuse « tāble rase » prônée par certains courants du début du siècle – ou encore que rien ne justifie d'arrêter les exigences du « progrès ». Signalons que malgré la clarté de cette solution, la société résiste à cette notion d'éphémère, même si à certaines périodes de l'histoire on n'a pas hésité à rêver, voire à détruire des pans entiers de l'espace bâti.
2 Pouvoir toujours tout conserver constitue l'autre pôle face au problème de l'adaptation. À l'instar des œuvres d'art, les œuvres architecturales sont alors définies comme des œuvres inaltérables, qu'il faut maintenir dans l'état qu'elles avaient au moment de leur origine. Cette position, que l'on peut qualifier de conservatrice, renvoie à plusieurs attitudes, dont les unes se fondent sur la nostalgie, la valorisation a priori du passé et sur un

manque de confiance impliquant le rejet de toute production contemporaine, et les autres sur une argumentation historique et anthropologique avançant la nécessité de conserver des traces du passé: c'est la problématique du musée d'architecture étendue à l'échelle du territoire.

Dans le premier cas, les œuvres architecturales doivent s'adapter aux besoins sociaux, ce qui représente une sorte de prolongement d'une attitude quasi « fonctionnaliste », où l'adaptation se pose comme une nécessité et où n'a plus cours le rapport initial forme/fonction inhérent à une certaine conception architecturale. Dans le second cas, c'est aux formations sociales de s'adapter aux œuvres architecturales, la transformation de ces dernières n'étant pas « négociable ».

Si ces deux scénarios extrêmes illustrent certaines situations et positions bien connues, les enjeux éthiques issus de la crise d'adaptation sont généralement liés à des attitudes moins caricaturales, visant à circonscrire les conditions les moins mauvaises pour aborder le dilemme conservation/transformation. On l'aura compris, derrière ces problèmes généraux se cache toute une gamme de problèmes concrets qui vont depuis la restauration jusqu'à la démolition, en passant par la conservation, la réhabilitation, la modernisation, la reconversion, l'actualisation, la destruction, et ainsi de suite. Dans cette problématique, les constructions se situent quelque part dans la chaîne hiérarchique entre le monument et le bâtiment ordinaire et font l'objet de décisions qui dévoilent la manière dont les acteurs sociaux (et non seulement les architectes) envisagent les enjeux de la transformation. Au nom de quelles conventions, à partir de quels fondements normatifs peut-on décider si une œuvre architecturale doit rester intacte, si elle peut admettre des transformations ou si elle ne mérite que la destruction?

Mais au-delà des débats sur la valeur d'une œuvre et sur les critères capables de l'exprimer se pose également la question du système idéologique et politico-juridique au sein duquel se prennent les décisions de transformer ou de détruire. Qui doit décider de conserver, de détruire, de transformer? S'agit-il d'un problème relevant de spécialistes qui établiraient la valeur « artistique » au nom de normes culturelles établies? Ou s'agit-il, au contraire, d'un problème mettant en jeu une partie ou l'ensemble de la communauté concernée par l'œuvre? Doit-on ne considérer l'œuvre que comme un objet ou doit-on chercher à en saisir le sens en regard de son contexte? Comment, passant de l'œuvre-objet à l'œuvre dans son contexte, modifie-t-on tant les savoirs que les pouvoirs relatifs à une décision sur la transformation? Sans poursuivre la longue liste des questions inhérentes au problème de la transformation d'une œuvre, on voit bien que les rapports entre le fond et la forme sont intimement liés et que les jugements et les décisions auxquels ils donnent lieu sont très souvent complexes et ainsi ne peuvent qu'être relatifs à une situation spécifique, à un contexte donné et à une époque précise[5].

Mais plus fondamentalement, autour de ce problème, c'est toute la question du patrimoine qui se repose en cette fin du vingtième siècle. Reposer la question de l'éthique sous l'angle des enjeux inhérents à la transformation, c'est, me semble-t-il, apporter un éclairage nouveau sur le patrimoine architectural. Ce nouveau regard ne peut se contenter de poursuivre un discours anachronique centré sur la notion générale de conservation dans le domaine des beaux-arts, pas plus qu'il ne peut se suffire d'un ancrage dans une quelconque sublimation de la ruine ou dans des stratégies pseudo-progressistes d'un opportunisme à la vue courte.

Enfin, peut-on avancer dans la compréhension des enjeux éthiques sans reposer la question de la spécificité de l'architecture par rapport aux autres domaines de l'activité humaine? La notion de transformation ne serait-elle pas une des dimensions principales de cette spécificité et, par voie de conséquence, un des liens stratégiques pour la compréhension des enjeux éthiques visant le patrimoine architectural existant? En ce sens, les décisions en matière de transformation architecturale renvoient à la question plus générale relative à l'état de la pensée démocratique en matière d'architecture: elles dévoilent la manière dont les acteurs sociaux (et non seulement les architectes) pensent et conçoivent/construisent le destin des œuvres architecturales.

ENJEUX ÉTHIQUES PROPRES À LA CONCEPTION ARCHITECTURALE D'ŒUVRES « EN PROCESSUS »

La position théorique consistant à considérer les œuvres architecturales existantes du point de vue des processus qui les transforment renvoie à des enjeux éthiques qui viennent d'être esquissés. Mais cette même position amène également à s'interroger sur la conception des œuvres en devenir. En d'autres termes, ce que nous voulons introduire maintenant, c'est le problème du processus de conception des œuvres « en processus ».

Quelles attitudes les acteurs responsables de la conception de telles œuvres peuvent-ils adopter? C'est par cette question que nous aborderons les enjeux éthiques relatifs à une architecture de la transformation et non plus seulement ceux qui visent une architecture de la permanence. Cette question peut, à première vue, sembler dénuée de fondement empirique. Et pourtant, que reste-t-il, par exemple, des œuvres architecturales qui « qualifiaient/constituaient » le Montréal des années 20? En quoi ce qui est considéré aujourd'hui comme structurant la ville contemporaine ne deviendra-t-il pas secondaire dans la ville de 2040? En fait, le problème est plus réel qu'on veut bien le reconnaître et c'est pourquoi il faut admettre, me semble-t-il, que la transformation est un des problèmes majeurs de la conception architecturale et que, là encore, elle apporte une perspective permettant d'envisager certains enjeux éthiques d'un point de vue spécifique.

En évitant de tomber dans le piège d'une morale à priori quant à la nécessité de l'adaptation des formes architecturales aux formations sociales (et réciproquement), l'enjeu éthique central au plan de la conception se pose dans les termes suivants. Faut-il, au départ, prévoir les conditions, le potentiel et les dispositifs d'appropriation relatifs à la transformation d'une œuvre architecturale? Faut-il, au contraire, s'abstenir de prendre en charge la maîtrise des processus de transformation, partant du principe qu'ils comportent des incertitudes si grandes qu'il serait dérisoire de vouloir décider d'avance du destin de l'œuvre? Enfin, allant plus loin, faut-il s'opposer de manière explicite à l'idée même de transformation, soit en opérant des choix (au plan de la construction, par exemple) qui rendront toute intervention très difficile, soit en conférant aux œuvres un statut qui consacre et impose leur permanence?

L'argument relatif à l'incertitude du destin ne manque pas de poids. Mais peut-on maintenir longtemps une attitude qui consiste à nier le processus de transformation et à bâtir comme si tout allait rester toujours en place, comme si toutes les œuvres architecturale devaient irrémédiablement résister au temps? Faut-il, au contraire, accepter la possibilité que l'immeuble érigé aujourd'hui puisse disparaître arbitrairement demain? Mais ce n'est pas tant à l'égard de l'une ou l'autre de ces attitudes que se pose la question éthique, car ce qui importe davantage, c'est de poser de manière explicite la dichotomie éphémère/permanent. Par exemple, la durée de vie matérielle d'un bâtiment m'apparaît sous-tendre des enjeux éthiques qui peuvent en retour commander certains choix en matière de construction ou de matériaux. Or, ce sont là des choix que le questionnement esthétique n'aurait pu engendrer, ou plutôt qu'il aurait engendrés différemment, ou bien encore qui ne pourraient se fonder ni même être éclairés par un paramètre proprement technique (puisé dans la physique, par exemple).

Mais le choix constructif et les paramètres relevant du statut tant physique que plastique ne constituent pas le seul enjeu éthique propre à la conception d'une œuvre architecturale considérée comme processus. Le problème n'est pas nouveau, puisqu'il n'est pas rare de trouver, au cours des récentes décennies, des tentatives de mettre en œuvre – en termes instrumentalistes, le plus souvent – des dispositifs capables d'assurer à une réponse architecturale un potentiel de flexibilité, de mobilité ou d'adaptabilité[6].

Sans chercher à rendre triviale toute tentative de traitement instrumentaliste ou utilitariste du problème de la transformation, le questionnement éthique nous invite à poser sur un plan plus général la question de savoir pourquoi et comment on doit introduire dans la pensée des œuvres architecturales les conditions de leur propre transformation, les conditions d'adaptation entre les « spatialités », les « matérialités » et les « socialités ». Cette interrogation pourrait être approfondie en s'appuyant sur certaines perspectives scientifiques contemporaines qui avancent des propositions pour comprendre la complexité – en explorant, par exemple, la potentialité cognitive/active de certains concepts tels que l'auto-organisation, le temps, l'ordre/le désordre, et ainsi de suite[7].

Il faut tenter de tirer les conséquences de ce que nous venons d'esquisser.

1 Une pensée de la transformation, de l'adaptation, n'est pas chose simple en architecture, car loin des systèmes vivants, les œuvres architecturales ne peuvent échapper complètement aux contraintes inhérentes à la matérialité, quand bien même une certaine pensée de la technologie nous entraîne vers une fascination de l'immatériel. Ainsi le transfert de concepts complexes vers le design architectural reste largement à explorer, mais ce genre de travail ne manquera pas de modifier notre approche au développement d'un questionnement d'ordre éthique.
2 Une pensée de la transformation modifie le statut du créateur du fait qu'elle conjugue l'image vieillissante de l'artiste travaillant pour l'éternité/l'universel et celle de l'acteur social travaillant dans l'ombre et la complexité des rouages du système social et du système de production.
3 Une pensée de la transformation modifie également le statut de l'œuvre architecturale, car elle introduit une distinction entre le statut permanent « normalement » conféré aux œuvres d'art (conservation en l'état initial) et le statut éphémère attribué de plus en plus souvent aux objets usuels. C'est donc d'une « trace » particulière qu'il s'agirait en regard des œuvres architecturales, une trace à saisir sous la forme d'un processus, une trace dans l'espace qui se façonnerait et se métamorphoserait dans le temps, une mémoire impossible à pétrifier, un patrimoine actif, une dynamique toujours en devenir.

CONCLUSION

Enfin, les notions de transformation et de processus et, plus globalement, les problématiques issues du temps ne constitueraient-elles pas une perspective nécessaire pour éclairer, au-delà de certains enjeux théoriques et épistémologiques, une facette des enjeux éthiques – facette indispensable parce qu'elle apporte ce qui nous paraît constituer une dimension spécifique de l'architecture (une œuvre en processus) et qu'elle interdit toute forme de discours qui exclurait la place du sujet? La transformation étant le support et la résultante d'un système d'action complexe, traduisant la nécessité de prendre en compte la subjectivité des acteurs et l'impossibilité de comprendre l'architecture dans un espace positiviste, l'enjeu est bien de saisir la dynamique temporelle du rapport entre les « spatialités » et les « socialités », et la dynamique des représentations qu'un tel rapport engendre.

L'architecture peut être ainsi appréhendée comme une production sociale et historique collective qui vient complexifier constamment les œuvres qu'on a trop souvent et trop uniquement expliquées par leurs caractéristiques originales et par l'acte originel du créateur. Et c'est notamment sur une perspective de ce genre qu'il nous paraît possible et fertile d'articuler un questionnement éthique.

NOTES

1 C'est ainsi que, dans un article récent, nous introduisions notre propos sur l'architecture et l'éthique, propos plus général et plus introductif que celui qui est présenté ici; voir « L'architecture et la question de l'éthique », *Informel*, vol. 4, nº 2 (été 1991), p. 35–50. Pour une vision extrêmement riche et diversifiée des problèmes éthiques en regard du design, voir l'ensemble des quatre numéros et les actes du colloque « Prométhée éclairé » que la revue *Informel* (Université de Montréal) a consacré, sous la direction de A. Findeli, à ce genre de questionnement.

2 H. Simon, *The Sciences of the Artificial*, Cambridge, Mass., MIT Press, 1969.

3 Voir, par exemple, *The Communicative Ethics Controversy*, publié sous la direction de S. Benhabib et F. Dallmayr, Cambridge, Mass., MIT Press, 1990. Cet ouvrage réclame malheureusement beaucoup de travail pour comprendre sa portée relativement aux conditions de la mise en place d'un discours éthique sur l'architecture.

4 Nous avons longuement développé ce point de vue théorique dans deux

publications récentes; voir « L'Architecture et la question du temps », in *De l'architecture à l'épistémologie*, publié sous la direction de P. Boudon, Paris, Presses Universitaires de France, 1991, p. 323–348; et *La Conception architecturale – Une investigation méthodologique*, Paris, L'Harmattan, 1992.

5 Viollet-le-Duc est un bon exemple pour illustrer combien les conventions inhérentes à une conception de la transformation ne peuvent être intelligibles sans une compréhension du contexte et de l'époque. Si son travail « sur » Notre-Dame-de-Paris n'a pas été remis en question, il n'en est pas de même dans le cas de St-Sernin à Toulouse. À ce sujet, la position de Bruno Foucart (in *Beaux-Arts*, 1991, p. 22) révèle bien les enjeux éthiques inhérents à toute transformation.

6 Parmi les écrits et projets illustrant cette attitude, la tentative la plus significative de la deuxième moitié du vingtième siècle me paraît avoir été développée par le groupe Archigram. Au début du siècle, on trouve également dans le mouvement futuriste la volonté de créer une architecture que les générations suivantes devraient avoir la liberté de transformer.

7 Faute de fournir une bibliographie sur ce vaste champ d'enquête, mentionnons deux ouvrages récents en langue française: J.-L. LeMoigne, *La modélisation des systèmes complexes*, Paris, Dunod, 1990; et E. Morin, *La Méthode*, tome 4, Paris, Seuil, 1991 (à ce dernier tome, il faut, bien entendu, ajouter les trois précédents).

ARCHITECTURE ET TECHNO-ÉTHIQUE – CONTRIBUTION À UNE ÉTHIQUE DE L'ARCHITECTURE

ALAIN FINDELI

Much has been said about the aesthetic rupture represented by postmodernism in architecture. And yet, convincing explanations of this phenomenon are uncommon. The desire to insulate architecture in order to affirm its autonomy, coupled with the difficulty of finding room for architectural theory in a body of scientific thought that is already full (as LeMoigne puts it), in part explain the superficiality of most attempts at explanation. Just as words take on their full meaning only when used in context, similarly – and paradoxically – architecture must be taken out of its isolation in order to be understood better. Thus one cannot claim for architectural postmodernism – any more than for modernism itself – the absolute historical significance that its proponents demand.

The current concern with ethics, found in most occupations today, undoubtedly reflects a broader *Zeitgeist*, but like the aesthetic rupture just alluded to, it also reveals a more fundamental transformation in the contemporary body of knowledge – a "scientific revolution," as Kuhn calls it. It is in this context – or more precisely, in the context sometimes given the controversial designation of "techno-ethics" – that we seek to set out the links between ethics, technology, and architecture. Parallel to the "technification" of the sciences – which has led to the spreading of the designation "techno-sciences" – one can observe the growing influence of technology in areas traditionally viewed as the purview of artists – from gardening to urban design, from the decorative arts (design) to architecture, and also from

medicine to education, law, politics, and warfare. The earliest signs of this influence could be seen in the machine-age lyricism of the *avant-garde* modernist movements, but it was in the process of design itself that it became permanently established.
Here, architectural ethics is viewed as a particular form of techno-ethics. Even though, as Robert Prost stressed in a recent article, this approach raises serious theoretical, methodological, and even ethical problems, it should not be abandoned. In this paper, we consider the architect from a dual point of view – as a professional who designs technological objects and as someone who uses technological objects in the course of his professional activity. And we will see that any code of professional ethics, even when it emphasizes the ethics of responsibility (as seems to be the case these days), will continue to run into the issues raised by the private ethics of both the architect and the user. This contradiction is illustrated with examples drawn from the modern period (Wright, Bauhaus, Ulm, etc.).

« L'architecture n'est pas un problème esthétique, c'est un problème éthique avant tout... L'architecture est toujours publique car sa nature est d'être une activité sociale et collective. »

Mario Botta

LES ARCHITECTES SONT-ILS DES GENS DE PETITE VERTU?

Dès le premier chapitre de son traité d'architecture, Vitruve se préoccupe de l'éducation des architectes. La liste des « connaissances multiples et des instructions variées » nécessaires à la science de l'architecture – qui, selon lui, « naît de la pratique et de la théorie » – a de quoi déconcerter n'importe quel responsable de programme d'une école d'architecture contemporaine. Après avoir énoncé un principe fondamental dont on appréciera la remarquable modernité, à savoir qu'il convient de tenir compte de deux choses, « ce qui est signifié et ce qui signifie », Vitruve recommande que l'architecte « soit lettré, habile au dessin, instruit dans la géométrie, qu'il connaisse de nombreuses histoires, qu'il ait diligemment écouté les philosophes, qu'il sache la musique, ne soit pas ignorant de la médecine, qu'il sache les décisions des jurisconsultes, qu'il ait connaissance de l'astrologie et des lois du ciel[1] ».

Si les raisons qu'il avance pour se justifier sont parfois déconcertantes, l'ensemble laisse néanmoins rêveur. Pour notre gouverne, nous retiendrons celles qui concernent la philosophie: « Elle fait », dit-il, « l'architecte à grand caractère » et « qu'il ne soit pas arrogant, mais plutôt conciliant; équitable et fidèle sans avarice ... qu'il ne soit pas cupide et n'ait pas l'esprit occupé à recevoir des présents, mais qu'avec gravité il maintienne sa dignité, ayant bonne renommée[2] ».

Quinze siècles plus tard, Alberti réserve un chapitre de son neuvième livre aux qualités morales des architectes ainsi qu'aux sciences qui devraient leur être familières. Dans son introduction, reprise presque littéralement par l'article 3.01.01 du Code de déontologie des architectes du Québec, il précise tout d'abord que « quiconque se prétend architecte doit être un homme d'esprit raffiné, persévérant, zélé, de la meilleure éducation, possédant une expérience approfondie, et, plus particulièrement, doué de pondération et d'un jugement sûr » – en bref, conclut-il, il doit posséder « toutes sortes de qualités fondées essentiellement sur la prudence et la capacité d'une réflexion bien mûrie[3] ». Par contre, ajoute-t-il en évoquant les vertus de dévouement et de générosité, de bienveillance, de modestie et de probité, « je ne les requiers pas davantage pour l'architecte que pour tout autre homme, quelle que soit sa profession ». En effet, précise-t-il, personne ne mérite d'être appelé un homme s'il ne les possède pas. Cependant, l'architecte « doit éviter avant tout la frivolité, l'obstination, la suffisance, l'intempérance et tous ces autres vices qui lui feraient perdre la confiance de ses concitoyens et le [rendraient] odieux aux yeux du monde ».

Plus près de nous encore, dans son volumineux et « laborieux » (le mot est de lui) cours consigné en quatre tomes, Jules Guadet consacre un chapitre aux devoirs de l'architecte. Après s'être vigoureusement prononcé contre la constitution d'un ordre professionnel et pour la liberté totale de la profession, le maître propose de définir ainsi l'architecte: « [un] honnête homme habile à construire ». Autrement dit, et tout simplement, « on n'est pas architecte si l'on n'est artiste et honnête homme ». Il enjoint ainsi l'architecte, non seulement « à respecter ses engagements, à vivre comme si chaque acte de la vie avait ses témoins », mais encore à être « honnête artistiquement »; une définition qui, selon lui, découle de « tout ce qu'il [nous] a dit et redit » au long des deux mille pages de son ouvrage[4]. Le chapitre s'achève sur un ensemble de principes qui annonce nos codes de déontologie actuels, composé de trois rubriques: « Les devoirs professionnels de l'architecte envers lui-même et ses confrères, envers ses clients, envers les entrepreneurs ». Significativement, les rubriques concernant les « devoirs envers

Architecte à Rome, gravure à l'eau forte de L. Lowenstam, d'après L. AlmaTadema, 1875 (collection particulière)

lui-même et envers les entrepreneurs » ont disparu du code actuel qui, comme on le sait, prévoit les devoirs envers le public (section II), envers le client (section III) et envers la profession (section IV)[5].

On le voit sans poursuivre indéfiniment cet exercice, le souci de la conduite éthique ou morale des architectes ne date pas d'aujourd'hui, même si cette question ne fait l'objet que de mentions presque accessoires dans les divers traités. Non pas que la moralité des architectes ne valût pas la peine qu'on s'y attarde; elle semblait tout simplement aller de soi. On imaginerait mal en tout cas, à la lecture de ces extraits, qu'on pût réunir pour deux ou trois jours les membres de la profession autour du thème de l'éthique de l'architecture. Qu'y a-t-il donc de changé dans cette vénérable profession pour qu'une telle initiative soit non seulement pensable, mais encore urgente? Les architectes seraient-ils en train de perdre leur vertu en cette fin de siècle ou, au contraire, pratiqueraient-ils subitement une surenchère de bons sentiments?

Un rapide survol des rapports annuels de l'Ordre des architectes du Québec a pourtant de quoi nous rassurer: de 1974 à 1983, 200 dossiers relevant du Comité de discipline ont été étudiés, vingt-sept ont été ouverts à la suite d'informations issues du public puis ont été fermés, dans la plupart des cas parce qu'ils ne demandaient pas l'intervention du syndic (sur les 200 dossiers, soixante-neuf concernaient l'indépendance professionnelle, le désintéressement et l'intégrité). Seulement vingt-et-une plaintes comportant 116 chefs d'accusation ont été portées devant le Comité de discipline et seuls dix-huit architectes se sont vu imposer des sanctions (d'un maximum de 4 000 dollars) ou des radiations (maximum, un mois). Si, dans l'ensemble, le bilan est plus qu'honorable, les administrateurs de l'ordre se préoccupent cependant des aspects de la déontologie reliés à l'évolution de la profession. L'euphémisme « évolution » recouvre ici un ensemble de « pressions vers l'élargissement de la pratique », principalement sous l'effet de la conjoncture économique. Mais alors, s'il y a lieu, en effet, de s'inquiéter d'une éventuelle « adaptation de la réglementation » et de ses conséquences sur l'honneur de la profession, pourquoi ces questions déborderaient-elles le cadre strictement interne de la profession? Autrement dit, pourquoi convoquer non seulement des architectes, mais encore des historiens, des philosophes, des critiques, des chercheurs, des théoriciens à un symposium sur les rapports entre l'architecture, l'éthique et la technologie? Plus précisément, et ce sera notre question principale parce qu'elle touche directement au processus de conception architecturale, en quoi la question technologique modifie-t-elle les données et la problématique conventionnelles de l'éthique de l'architecture?

Nous nous efforcerons, dans cet essai, d'apporter quelques éléments qui permettront, non pas de résoudre mais de clarifier cette question en la reliant à une problématique plus large, celle de la techno-éthique, c'est-à-dire aux aspects éthiques découlant du développement technologique qui a caractérisé la société industrielle occidentale. Le texte comprend trois parties: une introduction générale résumant la genèse de la problématique, une tentative de « transfert éthique » (Prost) de la pratique biomédicale vers l'architecture, enfin un esquisse de problématisation spécifique à l'architecture.

LA GENÈSE D'UNE QUESTION – DU RAISONNABLE AU RATIONNEL

« Le terrain de l'éthique est semé d'embûches », finit par conclure l'aréopage du Symposium international sur l'éthique, la technique et la responsabilité professionnelle en design, tenu sous le titre « Prométhée éclairé », aboutissement de deux ans de réflexion sur l'éthique appliquée en design[6]. Pourquoi ces embûches? Qu'est-ce qui rend si complexe et difficile la pensée et, à plus forte raison, la pratique d'une éthique en cette époque marquée par le déterminisme technologique[7]? L'attitude la plus répandue, à cet égard, est celle d'un pessimisme généralisé, émaillé de propos alarmistes et parfois suspects, proclamant qu'il n'y a plus de guides moraux dans la société actuelle, que l'insouciance règne partout, et ainsi de suite. Nous préférons objecter que la morale traditionnelle, celle qui, malgré tout, imprègne nos comportements et nos habitudes de vie – je veux parler de la tradition judéo-chrétienne et du type d'ascèse qu'elle implique – n'offre plus de cadre adéquat pour penser et agir dans un monde métamorphosé par cinq siècles d'humanisme matérialiste. Cette morale, revue et corrigée par l'idéalisme rationaliste des Lumières, repose sur l'idée qu'il n'y a pas de liberté possible sans contrainte, ce que l'impératif catégorique kantien formulait, en gros, de la façon suivante (du moins, c'est ainsi qu'on se plaît à le présenter aujourd'hui): la seule liberté qui appartient en propre à l'individu est celle de se contraindre et de s'amender conformément à la loi. Ce qui doit nous retenir ici, ce n'est pas le caractère idéaliste de cette éthique, car il semble bien – et une étude comparative des diverses traditions éthiques qu'a connues l'humanité le montrerait vraisemblablement – qu'une éthique sans idéal n'est guère concevable. C'est plutôt son caractère individualiste et rationnel, découlant du concept de liberté, qui est nouveau, en quoi elle s'oppose à la morale collectiviste et empathique (mystique?) des grandes religions traditionnelles. Ce que nous retenons surtout, c'est qu'une telle conception de la morale est perçue comme n'étant guère compatible avec les nouvelles valeurs que véhiculent l'esprit des Lumières et l'individualisme qui le caractérise: poussée émancipatrice du libéralisme, de l'égalitarisme et de l'hédonisme, phénomène qu'il est indispensable de prendre en compte, ne serait-ce qu'à court terme, dans les problématiques contemporaines.

Lorsque nous affirmons que toute éthique est par nature idéaliste, nous voulons dire que les sentiments moraux conscients qui suscitent nos actes et nos actions sont toujours animés par un idéal. Ainsi, la tradition chrétienne a retenu de ses influences platoniciennes l'idéal du Bien, quitte à en avoir nuancé ou modifié les modalités en le « contaminant » par les idéaux du Beau et du Bon (influences mystiques) et par celui, plus scolastique, du Vrai. Le dualisme cartésien puis cartésianiste, remplaçant la polarité traditionnelle esprit/matière par celle, moins absolue et donc plus limitée, d'intellect/matière, a par la suite privilégié ce dernier idéal (le Vrai), ouvrant la voie à la science et à l'idéologie scientiste. On comprend qu'aujourd'hui, alors que l'esprit se réduit à la cognition, la dynamique potentielle de la polarité originelle se soit considérablement érodée.

Dans ces conditions, il est possible de dépasser la sempiternelle formule « il n'y a plus de morale » en esquissant une interprétation plus significative de la préoccupation éthique actuelle. L'insistance des Lumières sur le « je pense » cartésien a opéré un glissement de l'objet de réflexion: au lieu de rechercher une morale pratique, l'Occident, sous l'égide de Kant, en particulier, s'est mis à réfléchir sur les conditions d'existence rationnelles de la morale, c'est-à-dire sur la question de la morale. Il a ainsi produit une science de la morale, ce que certains auteurs définissent par l'éthique proprement dite. Il en va de même dans les champs correspondant au Beau et au Vrai: la recherche de la beauté a cédé la place à celle de ses conditions d'existence (d'où l'esthétique) et, dans une moindre mesure – car la science expérimentale se propose d'établir la vérité de façon pratique –, l'épistémologie a envahi le champ philosophique; ce qu'un jeu de mots usé traduit ainsi: la poursuite du rationnel a été privilégiée au détriment de celle du raisonnable. Notons bien que c'est précisément ce retrait du concret vers l'abstrait qui a permis le développement accéléré d'un arsenal technologique extrêmement efficace et généralisé. Dans le vide éthique (et esthétique) ainsi créé, les aspirations prométhéennes de l'humanité n'ont eu qu'à se déployer sans entraves autres que celles propres au domaine dont elles émanaient, celui de la technique. Ainsi, ce qui, traditionnellement, ne relevait que des moyens s'est imposé au niveau des fins, et désormais la technique ne répond plus qu'à elle-même. La « logomachie » (Hottois) du discours philosophique moderne, le formalisme de l'art pour l'art, l'art conceptuel (c'est-à-dire le discours sur l'art), l'« acharnement technologique » représentent quelques conséquences et manifestations notables du glissement évoqué; le « retour à l'ordre et à la figuration », le « moratoire écologique » et le « reflux vers le local » constituent quelques aspects de la réaction post-moderne à ce phénomène.

Cette réaction n'est que la suite d'un processus historique entamé depuis plus d'un siècle. C'est en effet sous la poussée de la philosophie romantique, d'une part, et à la suite du relativisme épistémologique qui s'est emparé des sciences physiques, d'autre part, que le vingtième siècle a peu à peu remis en question les dogmes de l'idéalisme rationaliste. Nous savons désormais, et il y a derrière cette affirmation tout le poids et le prestige dont jouissent ces magiciens contemporains que sont les physiciens, que *la* vérité n'existe pas, qu'il n'y a de vérités que relatives, que *la* réalité – si elle existe – est aléatoire, chaotique, stochastique, etc. Dès lors, la fameuse relation d'incertitude de Heisenberg (1927) prend une signification qui s'étend au fondement même de la pensée occidentale. Le relativisme a fini par s'emparer non seulement de l'épistémologie, mais encore de l'esthétique et, bien entendu, de l'éthique.

Un nouveau paradigme est donc en voie de s'installer sous des noms divers: phénoménologisme, constructivisme, systémisme, holisme, intuitionnisme, etc. Puisqu'il n'y a plus de Vérité, à chacun et chacune la sienne; on « personnalise » à tout va, comme on dit dans l'industrie des biens de consommation. De même, à chacun sa conception de la beauté et de la morale, avec toutes les conséquences et les incertitudes qui s'ensuivent. L'angoisse contemporaine (« *No future* ») rappelle singulièrement celle qui s'empara de l'humanité lorsque les astronomes lui révélèrent que la Terre n'était plus au centre du monde; or si le Vrai, le Beau et le Bien ne siègent plus dans la raison de l'humanité mais sont éparpillés dans la multitude des consciences individuelles, à quel point fixe doit-on se référer pour en apprécier la valeur absolue?

Même si son tour s'est fait attendre, l'architecture n'a pas été épargnée. Le modernisme architectural s'est distingué, en effet, par l'apparition d'un abondant discours sur l'architecture – son origine remonte aussi loin qu'à la fondation de l'Académie en France en 1671, mais il s'est surtout déchaîné au début du vingtième siècle –, par la recherche privilégiée de la structure (au plan conceptuel mais aussi matériel) et de l'essence au détriment de la substance, par la poursuite de l'universel (d'où le style dit « international ») sous la houlette du fonctionnalisme. On connaît bien les divers « retours à » que le relativisme post-moderniste nous a proposés en réaction aux excès de cette abstraction rationalisante. Tout se passe comme si, apeurée par la complexité théorique mais aussi sociale et technique de la situation, l'architecture s'était repliée frileusement et orgueilleusement sur elle-même pour se réfugier à l'autre extrémité de la polarité fondamentale (forme/fonction) qui l'anime. Sans anticiper sur la dernière partie de cet essai, disons tout de suite qu'on ferait fausse route si l'on cédait aux exhortations souvent intolérantes de ces divers « retours à » (et à leurs dangers), qui ne font que dénoter l'impuissance et le manque d'imagination d'une époque incapable d'affronter les problèmes ou même de les poser de façon adéquate. L'approche proposée par l'épistémologie post-moderniste n'est, hélas, qu'une illusion, car elle renferme, à peu de choses près, les mêmes structures mentales fondamentalement dualistes qui sont à l'œuvre dans le modernisme; à cet égard, modernisme et post-modernisme ne sont que les deux faces d'une même médaille et c'est à un effort conceptuel bien plus radical qu'il conviendrait de faire appel pour dépasser les apories qui empoisonnent l'esprit occidental et parvenir, comme dit Piaget, à « une totalité nouvelle dont les propriétés dépassent les deux systèmes jusque-là distincts et séparés ». Il nous apparait plus constructif de rechercher un cadre général permettant de guider la réflexion et la pratique; à cet égard, la question éthique, qui n'est qu'un symptôme parmi d'autres de ce malaise, nous semble constituer une plate-forme intéressante et féconde. Il conviendra cependant de bien veiller à ne pas imposer des préceptes ou un nouveau dogme car, nous l'avons dit, une telle attitude irait à l'encontre de l'individualisme qui, pour le moment du moins, caractérise notre époque. C'est plutôt un outillage conceptuel, une façon de présenter la problématique que nous souhaiterions suggérer, l'initiative devant toujours demeurer, en fin de compte, entre les mains des individus.

ÉTHIQUE ET TECHNIQUE

C'est dans le domaine des sciences de la vie et de la médecine que la préoccupation éthique actuelle s'est manifestée tout d'abord, avant de s'étendre à toutes les disciplines et aux professions. Elle a donné lieu, au début des années 70, à la bioéthique ou éthique biomédicale[8]. On s'est aperçu, en effet, que les professions médicales, encadrées comme toute profession par des codes de déontologie (ou d'éthique), faisaient face à un ensemble de problèmes nouveaux mais bien concrets, engendrés par un accroissement considérable des possibilités d'intervention rendues possibles par le développement technologique. Au demeurant, ces possibilités nouvelles touchaient autant la recherche en sciences biologiques que les actes professionnels des divers praticiens – médecins, chirurgiens, pharmaciens, infirmiers, etc. Les conséquences de ces actes se sont accrues en « intension » (efficacité thérapeutique locale plus grande) et en extension (conséquences sociales globales plus étendues). Or, les codes de déontologie existants, fruits d'une très longue tradition, ne permettaient plus de prendre en compte ces aspects nouveaux; on les considéra dépassés, débordés, obsolètes. D'où la nécessité d'en réviser le contenu, de les actualiser, de les « moderniser ».

À ces conséquences pratiques de l'effet amplificateur de la technique s'en ajoutent d'autres, aux niveaux épistémologique et méthodologique, c'est-à-dire au niveau de nos façons générales de penser les problèmes et de les ré-

Un entrepreneur-architecte au dix-septième siècle, gravure sur bois d'après A. Brun pour le « Magasin pittoresque », 1882 (collection particulière)

soudre. L'attitude techniciste de type *problem solving* s'est répandue au point de devenir ce que les sociologues appellent un *habitus*: notre rapport au monde, dans toutes sortes de situations (famille, milieu professionnel, loisirs, éducation, vieillesse et mort, etc.) se pense désormais sous forme de problème à résoudre. Il en va de même de l'éthique, ce que trahit d'ailleurs singulièrement le terme récent d'éthique « appliquée ». Or, nous avons tout lieu de craindre l'extension de cet habitus au champ de l'éthique, car les problèmes qui s'y posent sont par nature irréductibles au paradigme technicien, c'est-à-dire à une logique de type instrumental. Nous n'évoquerons que deux raisons pour cela, qui mériteraient d'être élaborées davantage.

Tout d'abord, l'attitude technicienne qui, comme l'a très bien signalé Donald Schön[9], s'est calquée sur la démarche de la science expérimentale, exige que l'acteur se détache de son objet, qu'il l'examine de la façon la plus objective possible, sans engager sa propre subjectivité. Ce sont là des exigences constitutives du postulat épistémologique fondamental de l'ascèse scientifique occidentale. La décision éthique, au contraire, exige un engagement total de la personne; celle-ci ne peut, par conséquent, se dissocier de son objet, de son problème. L'objection est majeure car, dans les milieux des professions où, sous diverses formes, s'est imposée la méthodologie générale du *problem solving*, on s'orienterait volontiers, et en toute bonne foi, vers la recherche de méthodes analogues de résolution des problèmes éthiques (de préférence assistées par des systèmes experts informatisés); autrement dit, on court le risque d'assister à une réduction de l'éthique à la technique[10].

La seconde raison s'appuie sur une observation qu'ont effectuée plusieurs philosophes des sciences et de la technique, et qui fait aujourd'hui l'unanimité; on ne parle plus, en effet, de science, mais plutôt de technoscience[11]. À cet égard, si l'on a surtout insisté sur le fait que la science est devenue techniciste (ou technicienne), on a bien moins souvent relevé le fait que la technique est devenue de plus en plus scientifique (par opposition à « casuistique »), dans la mesure où elle repose aujourd'hui, dans de nombreux domaines, sur des méthodes et des principes plus sûrs, plus fiables, plus efficaces, plus universels. Elle s'en trouve ainsi considérablement valorisée en raison du prestige dont jouit la science dans l'imaginaire contemporain: si elle est scientifique, la technique est nécessairement plus vraie, donc incontestable. De cette constatation, un pas est vite franchi, celui qui envisage l'élaboration d'une éthique scientifique de type déductif. Mais là encore, il existe un obstacle majeur: alors que le domaine propre de la science est celui de l'universel, du général, de l'abstrait, l'éthique ne s'occupe, en dernière instance, que de cas particuliers, individuels, concrets. La décision éthique est toujours à recommencer dans chaque cas particulier; elle relève toujours de l'initiative et de la volonté individuelles, même si celles-ci peuvent être guidées par des principes généraux, dont l'impératif catégorique de Kant est l'exemple le plus connu. Donc, si une éthique formelle semble concevable et même nécessaire, elle ne saurait en aucun cas être suffisante.

Les deux réserves énoncées ci-dessus se résument ainsi: la question éthique ne peut être abordée de façon satisfaisante avec l'outillage intellectuel qui nous est le plus familier. Ou, en termes encore plus précis, il est vain d'espérer et d'entreprendre la mise sur pied d'un cadre théorique et méthodologique général qui nous permettrait de nous affranchir à bon compte de l'encombrante question éthique, car celle-ci exige nécessairement un investissement personnel dans chaque décision, c'est-à-dire un dialogue toujours recommencé de chacun et chacune avec soi-même.

LE « PARADIGME BIOÉTHIQUE »

Dans un article récent, Robert Prost distingue trois itinéraires possibles pour « explorer les conditions de mise en place d'un questionnement éthique en architecture[12] ». Il écarte d'emblée les deux premiers pour ne s'attarder qu'au troisième avec une pertinence qui force la conviction. Le second itinéraire, qui se proposerait d'analyser les discours, les pratiques et les œuvres architecturales pour en extraire les présupposés éthiques, est renvoyé à une étude plus approfondie qui aurait débordé du cadre de son article. Quant au premier, il consisterait à « prendre appui sur les débats [...] existant dans d'autres domaines (bioéthique par exemple) » afin de les appliquer à l'architecture. Il écarte cependant cette avenue pour des raisons diverses: importance moindre du déterminisme technoscientifique en architecture, et, surtout, problèmes méthodologiques. Sans mésestimer ces derniers, qui sont considérables, nous croyons cependant que le « transfert » de la bioéthique vers l'architecture peut s'avérer fécond. Les bioéthiciens, dont l'expérience accumulée est appréciable, ont en effet développé un ensemble de concepts, de méthodes et de pratiques qui peuvent s'avérer précieux à condition que l'analogie s'effectue sur les principes et non sur la substance. Illustrons ceci par quelques exemples.

1 La première question qui se pose aux praticiens de la santé est celle de la définition même du terme, autrement dit de l'objectif principal de leur pratique. Voici la définition adoptée: « la santé est un état complet de bien-être physique, mental et social et ne consiste pas seulement en une absence de maladie ou d'infirmité ». La transposition à l'architecture, qui peut être immédiate si l'on remplace « santé » par « confort » dans la première partie de la phrase, ne manque pas d'être évocatrice. Son adoption aurait des conséquences considérables sur les impératifs de la profession. En effet, si l'on conçoit sans peine la notion de confort physique et la contribution qu'y apporte l'architecture, l'exercice s'avère bien plus difficile – mais non moins instructif – lorsqu'il s'agit de définir, en termes archi-

tecturaux, l'idée de confort mental et social. On peut, soit penser que l'architecture ne peut contribuer que faiblement à ce dernier, soit imaginer dans quelle mesure elle le pourrait. Quoi qu'il en soit, nous sommes bien éloignés de « l'art de construire les édifices » (définition de l'architecture dans le *Petit Robert*) ou de la « personne diplômée, capable de tracer le plan d'un édifice et d'en diriger l'exécution » (définition de l'architecte, in *ibid.*), et nous nous rapprochons davantage d'une définition du design architectural proposée par Amos Rapoport: « Le but de la conception [architecturale] est de fournir un cadre approprié aux caractéristiques et besoins bio-sociaux, psychologiques, culturels et autres des personnes diverses à qui elle s'adresse[13] ». La question centrale est donc celle des besoins et il nous faut bien convenir qu'une théorie des besoins appropriée nous fait encore défaut, bien que les récents développements des méthodes phénoménologiques et systémiques appliquées au domaine bâti constituent sur ce point des apports appréciables[14]. Ainsi, dans un chapitre intitulé « Qu'est-ce qu'habiter? Les besoins », Peter Schmid distingue dix-sept points à l'attention des architectes, après avoir posé l'acte d'habiter dans un espace complexe de nature psychosomatique et psychospirituel exigeant des architectes une approche « biologique » intégrale[15]. Les répercussions d'une attitude aussi radicalement nouvelle seraient considérables sur, par exemple, la conception de bâtiments publics tels que hôpitaux, écoles, lieux de travail, etc. La notion même de projet en serait métamorphosée.

2 Il est un autre aspect qui retient l'attention en bioéthique, celui d'acharnement thérapeutique. C'est aussi un concept central de l'éthique de la technique, que l'on pourrait traduire par « acharnement technologique » et qui dénote le fait que le développement technique ne trouve sa raison d'être qu'en lui-même. Il en résulte que « tout ce qui est techniquement possible doit être réalisé ». C'est une déviation bien connue des esthéticiens, celle de « l'art pour l'art », avec toutefois une différence notable au plan des conséquences. L'architecture n'est pas à l'abri de « la technique pour la technique » (ni de « l'art pour l'art », du reste), dont les exemples abondent actuellement (*high-tech*, industrialisation, domotique, etc.). Ce dérapage de type prométhéen n'est pas justifiable éthiquement, car il revient à prendre les moyens pour des fins, avec toutes les conséquences humaines, sociales et économiques dans lesquelles nous entraîne ce vertige technophile. Il est bien entendu qu'une époque matérialiste comme la nôtre a tendance à juger les actes à leurs conséquences et non à l'intention qui les motive. Or, si une morale de l'intention n'est concevable que dans

des civilisations où les conséquences des actes posés n'atteignent pas la démesure et les dangers des décisions techniques actuelles, ce n'est pas une raison suffisante pour ne plus se poser la question des intentions, donc de la finalité. Pourtant, c'est bien ce qui se produit actuellement: il n'y a pas de débat sur le fond, seules les finalités techniques importent, on évacue la phase du *problem setting* au profit du seul *problem solving*, prétextant que la première n'est pas du ressort de l'architecture et des architectes. En outre, cette attitude tient pour acquise la neutralité axiologique de la technique. Aussi, on ne se demande pas quelles sont les valeurs impliquées dans l'outil informatique, par exemple. Robert Prost a peut-être raison de sous-estimer cet aspect en architecture – après tout, s'est-on demandé quelles étaient les valeurs impliquées dans l'usage du té ou du rapidographe? –, mais la question mérite d'être posée si l'on considère l'informatisation non seulement de l'objet, mais du processus même de la conception architecturale.

3 L'éthique pharmaceutique nous a paru particulièrement évocatrice en raison des liens qu'entretient cette profession avec l'industrie et des enjeux économiques considérables qu'ils recouvrent. L'industrialisation croissante de la construction et l'activité bancaire que nourrit la consommation de logements (les transactions immobilières et hypothécaires représentent l'activité principale des banques) incitent à examiner cette analogie de plus près. Les notions centrales de l'éthique pharmaceutique sont celles d'efficacité et d'innocuité des produits. Les deux propriétés doivent être établies et prouvées par les concepteurs, ce qui est souvent bien plus aisé pour la première que pour la seconde. Or, l'importance des enjeux économiques pousse à raccourcir le plus possible la durée de mise à l'essai des nouveaux médicaments; il s'ensuit une tendance, qui se manifeste quotidiennement pour tous les autres produits industriels, à utiliser le marché, donc les consommateurs, comme laboratoire d'essai de ces produits. Le nombre croissant des accidents domestiques, pour ne citer que les conséquences les plus visibles d'un mauvais design, incite à prendre ces notions au sérieux dans le domaine de l'architecture. En raccourci, cela revient à la recommandation suivante: au lieu de ne penser qu'aux problèmes que résoudra une décision au niveau du design, tâcher de penser également aux nouveaux problèmes qu'elle engendrera, et ceci sur des registres dépassant aussi largement que possible la pure technicité du projet. C'est une autre façon de poser la question centrale suscitée par l'acharnement technologique: telle solution technique est-elle socialement et éthiquement acceptable?

4 Mentionnons également un autre postulat de la bioéthique, qui est aussi le premier: c'est le respect de l'autonomie, de la liberté et de la dignité des personnes concernées par l'acte professionnel, dont on doit s'assurer du « consentement libre et éclairé ». On comprend qu'en pratique biomédicale, certaines décisions professionnelles décident de la vie ou de la mort d'une personne. C'est exceptionnellement le cas en architecture, bien que les accidents, déjà évoqués, soient nombreux. Ce n'est pas une raison, cependant, pour évacuer ces notions de l'acte architectural. Considérer une personne, un client, comme un consommateur ou même un usager n'est pas la même chose que de l'envisager comme un être intelligent, sensible, volontaire, animé par des aspirations morales, esthétiques, cognitives et, ajouterait Simmel, religieuses (au sens non confessionnel). Respecter en priorité ces aspirations et ces qualités peut avoir des conséquences considérables en architecture; il suffit, encore une fois, de songer à la commande publique: hôpitaux, écoles, bâtiments administratifs et ainsi de suite. Ne tenir compte que des besoins biologiques ou matériels des individus, c'est nier la dignité humaine, c'est réduire un complexe d'aspirations au simple désir, c'est transformer le désir de l'architecture en architecture du désir. Le postulat du respect de la dignité nous enjoint donc, avant même de songer à sa matérialisation en termes d'architecture, de nous forger une anthropologie, c'est-à-dire une conception de la personne humaine et, partant, de nous-même, ainsi que de la contribution que l'architecture pourrait offrir au développement et à l'émancipation d'une telle personne.

5 Pour terminer, l'un des points que le Comité consultatif national d'éthique chargé de rédiger le rapport *Sciences de la vie et société* a soulignés nous semble particulièrement pertinent à l'architecture: il s'agit de la nécessité et de l'urgence d'encourager la recherche en chronobiologie[16]. Cela revient à inclure systématiquement la variable temporelle, à diverses échelles (une journée, une saison, une vie, plusieurs générations), dans la conception architecturale. Certains efforts dans ce sens ont déjà été effectués, mais nous sommes loin d'une application systématique. Tout comme ses habitants, l'environnement construit devra donc être considéré comme un organisme vivant soumis à des biorythmes intrinsèques et extrinsèques et non plus comme une réponse idéale à des besoins absolus énoncés par des usagers abstraits.

Ce sont là quelques observations que suggère cet exercice de transfert de la bioéthique vers l'éthique de l'architecture. Nous nous sommes bien éloignés, apparemment, du quotidien des architectes: plans et devis, contrat, dessins d'exécution, surveillance de chantier, etc. Tâchons donc, pour terminer, de résumer les considérations précédentes de façon un peu plus concrète.

VERS UNE ARCHITECTURE ÉCLAIRÉE

Nous avons déjà signalé l'écueil que constituait le glissement d'une recherche pratique de l'action morale vers l'intellectualisation de cette question. Un tel glissement laisserait entendre que la connaissance d'une science abstraite de la morale serait préalable à l'attitude morale proprement dite ou, pis encore, s'y substituerait. Kant lui-même nous met en garde contre une telle déviation: « Il n'est besoin ni de science ni de philosophie pour savoir ce qu'on a à faire pour être honnête et bon, même sage et vertueux. L'on pouvait même bien supposer déjà d'avance que la connaissance de ce qu'il appartient à tout homme de faire, et par conséquent encore de savoir, doit être aussi le fait de tout homme, même le plus commun[17] ».

Cette précision rappelle l'observation d'Alberti, qui n'estimait pas nécessaire de distinguer les vertus des architectes de celles de n'importe quel honnête homme. Dans ces conditions, il n'y aurait pas lieu de poursuivre la recherche d'une éthique particulière à l'architecture, car l'observation du code de déontologie existant serait suffisante. Bien plus, l'inclusion dans ce code de dispositions particulières empiétant sur la morale « privée » serait perçue comme une atteinte à la liberté des architectes. De fait, c'est de façon plus qu'ambiguë que le premier article du code de déontologie spécifie les devoirs et obligations de l'architecte envers le public: « Dans l'exercice de sa profession, l'architecte doit tenir compte de ses obligations envers l'homme et son environnement (section II, article 2.01) ».

Signalons au passage que, selon les renseignements dont nous disposons, aucune plainte n'a jamais été déposée auprès du Conseil de la discipline de l'Ordre des architectes au titre de cet article. Le moins qu'on puisse dire, c'est que son interprétation est équivoque, tant sur l'acception du terme « obligations » que sur la définition de l'« environnement ». Dans la mesure, en effet, où l'acte architectural agit sur le milieu physique et social, l'environnement de l'objet architectural est à la fois matériel et symbolique. Les obligations envers le premier relèvent de l'écologie physique et industrielle; pour ce qui est du second, elles relèvent de l'écologie humaine et sociale, constituée majoritairement, comme l'ont bien montré les psychologues et les sociologues, de représentations et de significations.

Ces remarques montrent toute la complexité dans laquelle s'inscrit le design architectural et soulèvent l'importante question de la responsabilité des architectes envers l'étude et la connaissance de cette complexité. Or, et Robert Prost l'a bien noté, les connaissances et la recherche dans ce domaine sont plus qu'indigentes; nous ne savons pas grand-chose, en effet, des conséquences à moyen et à long termes des décisions architecturales qui sont prises à l'occasion d'un projet: « Il est bien délicat », dit-il dans l'article déjà cité, « de prouver par exemple qu'une façade vous donne des boutons quand vous passez devant ». S'il n'est pas envisageable, contrairement à ce qui s'est effectivement produit en recherche biomédicale, d'imposer un moratoire sur les projets, il n'en demeure pas moins que les méthodes traditionnelles de projétation (design) nécessitent une remise en cause sérieuse et un effort de recherche particulier.

À cet égard, et pour plus de clarté, nous effectuons une distinction entre une pratique compétente et responsable, la seconde englobant la première. Il est évident que tout architecte doit bien posséder son métier, ses outils, ses méthodes – et en connaître les limites. Tout ceci relève des techniques propres à toute profession, du savoir-faire, de la compétence. En revanche, nous avons défini ailleurs la responsabilité comme « la capacité d'un professionnel de saisir la configuration (*Gestalt*) dynamique globale, c'est-à-dire les aspects non seulement techniques, mais encore économiques, sociaux, politiques, écologiques, juridiques, éthiques ou symboliques de ses actes professionnels et leurs conséquences ». À l'appui de cette formulation, nous émettions l'hypothèse que rendre ainsi conscient un acte et sa rationalité sous-jacente, c'était déjà les dépasser, donc se créer la possibilité de les modifier. En bref, poursuivre une pratique éclairée constitue un premier pas important vers l'éthique. Depuis quelques années, c'est dans la discipline nouvelle des sciences de la conception ou « sciences de l'artificiel » (Herbert Simon) que s'effectue l'analyse épistémologique du design: ingénierie, architecture, urbanisme, etc. À leur façon, Robert Prost et Jean-Louis LeMoigne contribuent par des analyses pénétrantes à cette discipline fondée par H. Simon qui, en 1969, déclarait: « En tant que concepteurs ou que concepteurs de produits de conception, nous avons à être explicites comme jamais nous n'avions eu à l'être auparavant sur tout ce qui est en jeu dans l'acte de concevoir et dans la mise en œuvre des processus de conception-création[18] ». Poussées plus avant, de telles études montreraient, entre autres, que l'acte de design évolue généralement au sein d'un paradigme profondément dualiste, dont l'une des manifestations principales se révèle sous la configuration forme/fonction, qui n'est qu'un avatar du dualisme matière/esprit plus fondamental. Dans la pratique professionnelle quotidienne, tout se passe comme s'il s'agissait, au pire, de choisir pour l'un ou l'autre des deux termes et, au mieux, de trouver un moyen terme arithmétique entre les deux à l'aide d'une analyse classique coût/bénéfice.

C'est là, au demeurant, l'une des conséquences du déterminisme techniciste et de l'un de ses corrélats principaux, le déterminisme économique ou économisme. Comme ces deux déterminismes s'inscrivent, par nature, dans la sphère matérielle, ils évoluent toujours dans un espace clos, limité, fermé: si l'on ajoute d'un côté, il faut enlever de l'autre. Par contre, en transposant le problème dans le domaine éthique, on se retrouve dans un espace illimité (car il n'y a pas de limite a priori dans la recherche du Bien), dans lequel la décision ne relèvera plus d'une rationalité de type arithmétique mais procédera de l'invention, comme l'ont si bien remarqué les éthiciens[19]. Le saut conceptuel et épistémologique qu'implique le passage du premier espace dans le second est considérable et modifie radicalement l'attitude envers le design et la prise de décision. L'acte qui s'ensuit n'en demeurera pas moins essentiellement conflictuel ou, comme le dit Guy Durand, « à double effet ». Mais le conflit technique y cède la place à un conflit de valeurs, irréductible à un processus formel de type déductif car exigeant l'implication totale du sujet, comme nous l'avons souligné dans notre seconde partie.

Le questionnement épistémologique, on le voit, conduit ainsi au questionnement sur les valeurs. Cet aspect est fondamental, dans la mesure où il exige des architectes, non seulement qu'ils s'interrogent sur leur propres valeurs, mais encore qu'ils explicitent et hiérarchisent, pour chaque projet, les valeurs mises en jeu. C'est ici que le code de déontologie s'avère insuffisant, car il ne statue pas sur les intentions, qui relèvent d'une éthique de type téléologique. La liste des valeurs peut être longue, depuis les valeurs fondamentales (liberté, égalité, altruisme, justice, solidarité, autonomie, bien-être, etc.) jusqu'à des valeurs plus spécifiques (nouveauté, compétitivité, durabilité, progrès, possessivité, confort, individualisme, austérité, distinction, etc.). Dans cette perspective, la notion de projet, omniprésente en architecture, prend une dimension nouvelle qui exige qu'on s'extraie du domaine restreint de la profession (où se cantonne le code de déontologie, à l'exception de l'article cité) pour s'inscrire dans un contexte et une signification plus larges. La « Déclaration de Montréal », rédigée à l'occasion du dernier congrès de l'Union internationale des architectes, qui s'est déroulé en 1990 sous le titre « Pour des politiques nationales de l'architecture », va dans le sens indiqué et exprime le souci des architectes de se diriger vers une pratique plus éclairée et plus responsable[20]. Dans le même esprit, mais avec des accents corporatistes et technicistes bien plus marqués, le document de l'Institut américain des architectes (AIA), *Vision 2000: The Challenge of Change*, vise à ressourcer la pratique architecturale à la veille du troisième millénaire[21]. On signalera enfin, au chapitre des valeurs, une proposition originale émanant du designer et théoricien italien Ezio Manzini, qui s'applique à promouvoir une nouvelle éthique du « faire » à l'intérieur de « scénarios socio-environnementaux »: se fixer comme valeur de référence la lenteur pour s'opposer à l'accélération croissante du mode de vie sous l'effet de la poussée techno-économique[22].

CONCLUSION

Nous souhaiterions conclure par quelques considérations sur l'enseignement et la recherche en architecture, sans lesquelles les principes exposés ci-dessus resteraient lettre morte. Pour la recherche, le constat sera vite fait: l'architecture semble loin d'être à la hauteur de la tâche. Il est vrai que les obstacles méthodologiques sont énormes. Ce n'est pas une raison pour démissionner là où les sociologues, psychologues et psycho-sociologues continuent de piétiner. Nous avons désormais à notre disposition un ensemble de méthodes dites « qualitatives » et de modèles théoriques qui, convenablement adaptés, seraient d'un grand secours à une recherche architecturale qui continue de se chercher elle-même.

La situation de l'enseignement semble aussi alarmante, à la suite du vide créé par le rejet du paradigme moderniste. On conçoit aisément l'effet dévastateur d'un relativisme généralisé et mal interprété sur les méthodes et les contenus de l'enseignement et on comprend – sans l'approuver, bien entendu – le retrait et le retour vers le métier et la table à dessin (ou l'ordinateur) auxquels on assiste actuellement. À cet égard, l'adoption d'une éthique de type « ouvriériste » ne nous apparaît pas en mesure d'affronter les responsabilités qui s'imposent. S'il est sage d'« agir localement », on semble avoir oublié de « penser globalement ». Il faut dire que toute allusion à une préoccupation théorique qui menacerait d'éloigner les étudiants de l'atelier – là où se fait l'architecture, c'est bien connu – rappellerait avec trop d'insistance le mouvement moderne et ses excès, évoqués plus haut. Mais le rejet – légitime – du style moderne ne cache-t-il pas, comme l'ont souligné certains auteurs, un rejet plus profond, celui d'une conception sociale de l'architecture[23]? Les débats sur la forme sont aussi vains que les débats sur la technique. Nous retrouvons, dans le dualisme théorie/pratique, le même syndrome que nous avons identifié auparavant et qu'à sa manière bien particulière, Goethe nous décrit ainsi: « Tout ordre donne finalement naissance à la pédanterie; afin de se débarrasser de cette dernière on détruit l'ordre, et il se passe un certain temps avant qu'on ne se rende compte qu'il faut le rétablir de nouveau[24] ». Or, l'éthique semble venir à point pour hausser – et non supprimer – le débat, comme en témoignent plusieurs contributions récentes à la question de l'enseignement de l'architecture[25].

Nous avons tenté ici de justifier la présence de l'éthique dans la réflexion et la pratique architecturales. Une ultime question demeure cependant: comment enseigne-t-on l'éthique? Si l'on nous a bien suivi, et pour paraphraser Moholy-Nagy, l'un des maîtres du Bauhaus, on aura compris que l'éthique, tout comme l'art, ne peut s'enseigner: on ne peut qu'indiquer la voie qui y mène car, aimait-il répéter, « le design n'est pas une profession, mais une attitude[26] ».

NOTES

1 *Vitruve,* présenté et traduit par Auguste Choisy, Paris, T. de Nobele, 1971, tome I–II, livre I, chap. I, p. 5–6. Les précautions d'usage concernant la féminisation des textes n'étaient pas requises à l'époque de Vitruve. Dans la suite de notre texte, il va de soi que la forme masculine utilisée doit être considérée comme une simple commodité de langage et non pas comme une volonté d'exclure les nombreuses femmes architectes de notre propos.

2 Ibid., p. 9.

3 L. B. Alberti, *L'architettura. (De re aedificatoria)* (1485), présenté par P. Portoghesi et traduit par G. Orlandi, Milan, Il Polifilo, 1966, tome II, livre IX, chap. 10, p. 854 (trad. de l'auteur).

4 J. Guadet, *Eléments et théorie de l'architecture*, Paris, Librairie de la Construction Moderne, n.d. [1909], tome IV, « Additions », p. 565.

5 *Code de déontologie des architectes*, Loi sur les architectes (L.R.Q., c. A-21) et Codes des professions (L.R.Q., c. C-26, a. 87), Gouvernement du Québec.

6 « Prométhée éclairé », Symposium international tenu du 8 au 11 mai 1991 à l'Université de Montréal, compte-rendu en préparation.

7 Plusieurs auteurs se sont attachés à définir les termes « éthique » et « morale », qu'il conviendrait, en toute rigueur, de distinguer. Ainsi, pour Raymond Abellio, la morale se préoccupe de l'action et demande, « que faire? » alors que l'éthique va au-delà et s'interroge, « comment vivre? », distinction où l'on retrouve celle, classique, qu'Aristote établit entre faire (*poiesis*) et agir (*praxis*). Pour notre part, et par souci de simplicité, nous considérons ici les deux termes comme interchangeables.

8 Le champ pluridisciplinaire de la bioéthique a produit des centaines d'ouvrages depuis lors. Un résumé concis mais bien utile est proposé par G. Durand dans « De la déontologie médicale à la bioéthique », *Informel*, vol. 3, n° 2 (été 1990), p. 30–35.

9 D. Schön, *Educating the Reflective Practitioner*, San Francisco, Jossey Bass, 1987.

10 L'épistémologue montréalais M. Bunge a proposé, sous le terme de « techno-éthique », une telle approche, contre laquelle G. Hottois s'élève vigoureusement dans « Vérité objective, puissance et système – D'une éthique pour l'âge technoscientifique », *Informel*, vol. 3, n° 2, p. 16–29, en particulier p. 20.

11 Nous renvoyons au texte cité dans la la note précédente pour un exposé bref et remarquablement bien argumenté de cette question. Rappelons au passage que le mot « technique », que l'anglais traduit par *technology*, doit être distingué de « technologie », également traduit par *technology*, un peu comme « morale » se distingue de « éthique ». Cette question terminologique divise encore parfois les philosophes de la technique, ainsi que le déplore Jacques Ellul dans *Le système technicien*, Paris, Calmann-Lévy, 1977. Pour plus de précision, voir Gunther Ropohl, « La signification des concepts de "technique" et "technologie" dans la langue allemande », *De la technique à la technologie*, Cahier STS n° 2, Paris, CNRS, 1984, p. 30–41.

12 R. Prost, « L'architecture et la question de l'éthique », *Informel*, vol. 4, n° 2 (été 1991), p. 35–50.

13 A. Rapoport, « Architectural Education: There Is an Urgent Need to Reduce or Eliminate the Dominance of the Studio », *Architectural Records* (octobre 1984), p. 100 et 103.

14 Les travaux promouvant la méthode phénoménologique pour l'analyse des besoins sont nombreux. Citons, entre autres, J. E. Pirson, *La structure et l'objet*, Bruxelles, Mardaga, et Liège, Métaphores, 1984; C. Norberg Schulz, *The Concept of Dwelling*, New York, Rizzoli, 1985; M. Mangematin, « Mode, modernité et architecture », *Le Carré bleu*, n° 1 (1988), p. 55–59. Pour une approche plus originale, voir S. Rossbach, *Fenq Shui – L'art de mieux vivre dans sa maison*, Paris, Souffles, 1988.

15 P. Schmid, *Bio-logische Architektur*, Cologne, Muller, 1988, chap. 10, « Was ist wohnen? Gebrauch ».

16 T. Gros, T. Jacob et P. Royer, *Sciences de la vie et société*, Paris, La Documentation Française et Le Seuil, 1984.

17 E. Kant, *Fondements de la métaphysique des mœurs* (1783), Paris, Delagrave, p. 106–107.

18 R. Prost, « L'architecture et la question de l'éthique ». J.-L. LeMoigne s'est exprimé de nombreuses fois sur la question de l'épistémologie du design

(ou sciences de la conception). Citons entre autres: « Les paradoxes de l'ingénieur », *Culture Technique* n° 12 (mars 1984), p. 327–335; « Questions sur l'épistémologie des sciences de la conception », *Informel*, vol. 5, n° 1 (hiver 1992); « Quelle épistémologie pour une science des systèmes naturels "qui sont avec cela artificiels"? », *Revue internationale de systémique*, vol. 3, n° 3 (1989), p. 251–272. Enfin, rappelons l'œuvre pionnière de H. Simon dans ce domaine, *The Sciences of the Artificial*, Cambridge, MIT Press, 1969; la citation est tirée de la traduction de J.-L. LeMoigne sous le titre *La science des systèmes, science de l'artificiel*, Paris, Épi, 1974.

19 Sur ce point, voir le chapitre intitulé « L'invention éthique » dans l'article de J. Ladrière, « Biologie et morale », *La bioéthique*, Cahiers de bioéthique, Québec, Presses de l'Université Laval, 1979, p. 59–80.

20 « Déclaration de Montréal », *Esquisses*, vol. 1, n° 3 (juin–juillet 1990), p. 8–9.

21 American Institute of Architects, *Vision 2000: The Challenqe of Change*, Washington, D.C., AIA, 1988. Par ailleurs, l'association des écoles universitaires d'architecture a organisé un colloque sur le thème des valeurs en architecture, dont les actes ont été publiés sous la direction de W. G. Gillard et D. Woodcock; voir *Architectural Values and World Issues*, Washington, D.C., Association of Collegiate Schools of Architecture, 1984.

22 E. Manzini, « Écologie de l'artificiel et responsabilité du designer », *Informel*, vol. 4, n° 2, p. 21–29; « Les nouveaux artefacts et le rôle du designer – Fluidification de la matière, accélération du temps et habitabilité du monde », *Actes du Symposium « Prométhée éclairé »*, Université de Montréal, en préparation.

23 A. Kopp, « Une culture du désespoir – Réflexion sur l'architecture "post-moderne" », *Le Carré bleu*, n° 3 (1988), p. 57–58; H. Harris et A. Lipman, « Eine Kultur der Verzweiflung », *Archithese*, vol. 18, n° 3 (mai–juin 1988), p. 12–16, 32.

24 J. W. Goethe, *Écrits sur l'art*, Paris, Klincksieck, 1983, p. 274.

25 L'excellent *Journal of Architectural Education* publie régulièrement des articles d'auteurs préoccupés par ces questions. Citons A. Rapoport, « On the Cultural Responsiveness of Architecture », vol. 41, n° 1 (automne 1987), p. 10–15; J. Mehta, « Good, Useful and Beautiful », ibid., p. 30–33; B. Burnham, « Specialized Knowledge, Professionalism and the Discipline of Architecture », vol. 41, n° 2 (hiver 1988), p. 53–55; D. Schön, « Toward a Marriage of Artistry and Applied Science in the Architectural Design Studio », vol. 41, n° 4 (été 1988), p. 4–10; J. Mayo, « Critical Reasoning for an Enlightened Architectural Practice », ibid., p. 46–57; J. Albrecht, « Architecture and the Disproportionate Development of Human Faculties », vol. 43, n° 3 (printemps 1990), p. 20–25.

26 L. Moholy-Nagy, *Vision in Motion*, Chicago, Paul Theobald, 1947, p. 42. Une lecture attentive des textes des maîtres du Bauhaus – de Gropius (1919–28) et de son successeur à Chicago (1937–46) – révèle qu'il est plus qu'abusif de réduire le projet du Bauhaus aux excès du rationalisme moderniste (Hannes Meyer) ou à ses dérivés maniéristes (Mies van der Rohe). On y trouve, en particulier, des propositions au plan de l'éthique qui méritent une sérieuse reconsidération et des intuitions épistémologiques et méthodologiques dont la pertinence ne fait que se révéler actuellement.

PART THREE / TROISIÈME PARTIE

THE ARCHITECTURAL USES OF HISTORY AND NARRATIVE IN A TECHNOCRATIC WORLD

LA FONCTION DE L'HISTOIRE ET DU RÉCIT EN ARCHITECTURE DANS UN MONDE TECHNOCRATIQUE

THE ETHICS OF NARRATIVE AT TRENT UNIVERSITY

RICHARD HENRIQUEZ AND GREGORY HENRIQUEZ

Dans le passé, le concept occidental de l'espace rituel était ancré dans la foi religieuse et s'incarnait dans les gestes partagés qui définissaient et unifiaient ses adhérents. Ces rapports intersubjectifs se concrétisaient dans des institutions qui fournissaient le cadre nécessaire pour lier la Terre mortelle au cosmos divin. L'orientation survenait lors de la « mise en scène », au cours de laquelle était révélée la présence d'une continuité divine au sein de la vie. L'architecture servait de véhicule à une interaction rituelle et dynamique – une danse symbolique qui circonscrivait une « aire sacrée » dans le domaine de l'homme. Cet « espace d'action » permettait à l'humanité de se sentir chez elle dans un monde hostile et sans limites. Le rôle joué par le rituel dans un monde qui a rejeté l'action collective et la conception du divin est l'une des principales questions qui se pose à l'architecture contemporaine, à supposer que l'âme de l'architecte soit demeurée la même dans notre contexte existentiel actuel.

S'il est même possible d'envisager une réconciliation avec le reste du monde vivant, il nous faut éduquer nos enfants en remettant en question ce qui est présentement considéré comme une action éthique et collaborer avec eux à la création d'une nouvelle mythologie. L'interrogation sur la nature de l'action éthique dans une société démocratique et polythéiste est donc inhérent à cette position. Nous pouvons néanmoins être certains que ces nouveaux mythes uniront un jour l'humanité ou alors que la cupidité et le vertige suscités par la technologie et la croissance économique détruiront une grande partie des œuvres de la création. En tant que citoyens, nous pouvons tenter d'amorcer le changement par l'intermédiaire de nos commettants; mais en tant qu'architectes, nous pouvons et nous devons jouer un rôle actif pour stimuler davantage la mémoire collective de nos communautés.

Dans notre pratique, nous nous servons de la narration parce qu'elle permet de stimuler l'imagination et de susciter la participation. Le but ultime de notre travail est de situer l'individu dans un continuum historique sensible. Notre tâche est de concevoir des narrations qui établiront une correspondance avec l'histoire d'un lieu spécifique – une histoire qui intégrera à la fois l'univers naturel et l'univers construit, le passé réel et le passé fictif, et qui fera en sorte que les citoyens pourront projeter leur existence dans l'avenir.

The incomprehensible rate of change of the modern world brings with it the more-than-uneasy feeling that the outcome is unknown, that doom is at hand. The threat once represented by the atomic bomb has been replaced by the ecological crisis and the population explosion. We need somehow to restore faith in the future of the human race, both physically and spiritually. Canada's multicultural society could be called an excellent example of the homogenization of values and, at the same time, a mosaic of values. As we await the new unified "world order" of trade and tariffs, our country's cultural values swim in a sea of existential possibilities. There is no ritual, myth, or collective action in Babel, but inherent in human nature is the potential for reconciliation with the rest of the living world.

In the past, western man's concepts of ritual space were anchored by a religious faith, bodied forth by the shared actions that defined and unified its members. These intersubjective relationships found concrete form through institutions that provided the framework necessary to link the mortal earth and the divine cosmos. Orientation was experienced during the "enactment" that unveiled the presence of the divine continuity within life. Architecture embodied a dynamic ritual interaction that was a symbolic dance that circumscribed a "holy place" within the human realm. This "space of action" allowed humanity to feel at home in a hostile and endless world. One major question for contemporary architecture, presupposing that the core of the architect's being remains intact in our present existential context, is the role of ritual in a world that has abandoned collective action and the concept of the divine.

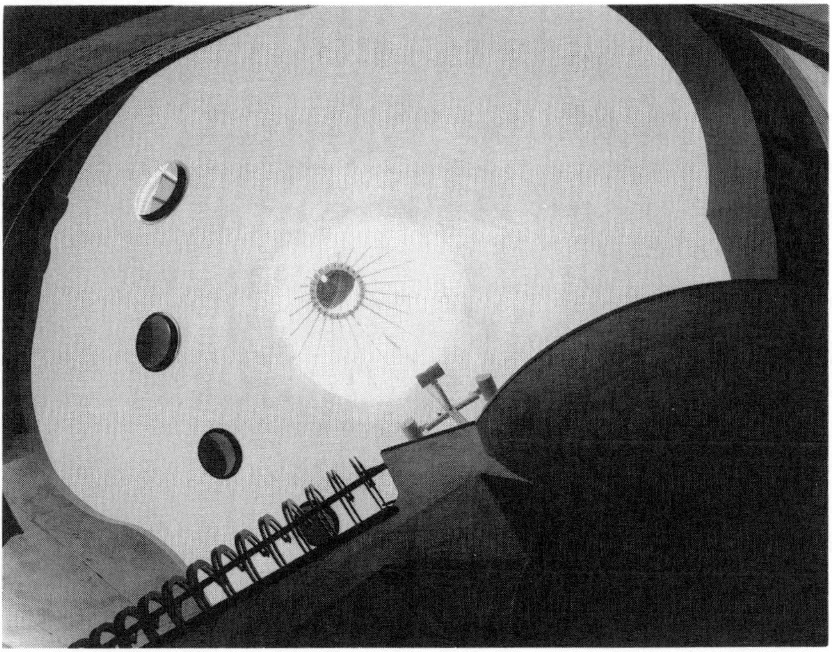

Ceremonial sphere

If reconciliation with the rest of the living world is to be even contemplated, we must educate our children through the questioning of what is presently deemed ethical action, and work with them to create a new mythology. Inherent in this position is the questioning of what is ethical action in a democratic, polytheistic society. Yet one can be certain that these new stories must one day unite humanity or that the "greed and speed" of technology and economics will destroy much of creation's work. As citizens, we can attempt to effect change through those whom we elect; but as architects, we can and must play an active role in helping to stimulate the collective memory of our communities.

In our practice, the tool of narrative is used for its power to stimulate the imagination and engage participation. The overall goal of our work is to situate the individual in a perceived historical continuum. The task is to create narratives that resonate with the history of a specific place – a history that includes both the built and the natural world, the real and the fictional pasts, and that enables citizens to project their lives into the future.

Trent University's Environmental Sciences Building in Peterborough, Ontario, is a significant attempt to develop specific narratives for an archi-

View of Environmental Sciences Building from across the Otonabee River

tectural project.[1] The new building fulfils a number of roles within the university. On a pragmatic level, it provides expanded new facilities for the earth science departments, including faculty and graduate student offices, as well as research and teaching laboratories. On the scale of the campus as a whole, the building fills a missing link in the pedestrian circulation system of the university, connecting a large footbridge crossing the Otonabee River to the plaza level adjacent to the existing Chemistry Building. This link is accomplished by means of an outside walkway that is integrated with the east side of the new building.

The Environmental Sciences Building combines specific elements from other buildings on the campus, designed by Ron Thom with previously unrealized fragments from his original master plan, in a manner that suggests an overlay of time. The Animal Care facility is one such fragment. Here, a barn-like building linked to the site's agricultural past is overlaid on the footprint of Thom's intentions.

A sloped wall, the "groundhog ramp" leading to the roof garden, is another fictional reconstruction from Thom's master plan that rises to meet the roof of the new building. The wall carries the earth upward and on to the roof itself. The roof is reserved not for people but for the small creatures that were displaced by the construction. The ramp symbolically gives access to the roof garden, where grass, clover, and wild flowers will grow. This gesture has to do with humanity's need to abdicate its domination of nature and instead consider itself a member of a community of living things.

Adjacent to the groundhog ramp is the front entry canopy of the new Environmental Sciences Building. The roof of the building is hinged upward by an inclined concrete beam from which emerges the skeleton of a dragon. This canopy is intended to be the gateway to the future east side of the campus.

On the west side of the building, the abandoned railway line appears to split the new building from the old, recalling the power of this now-dormant "line of force." A part of the railway line is raised to platform level and may be used in a proposed graduation ceremony, a ritual developed by the architects. To the south of the building is a proposed spiral forest of oak trees that relates to this ceremony and could become the focal point of the future East Campus.

The architects have proposed that every year an extended graduation celebration take place for students in Environmental Sciences. The purpose of this proposed new ritual is to reinforce the importance of the sacred relationship that humanity shares with the earth. Through this ritual, a spiral forest will emerge, with a tree being added each year and with space for new trees to be added 350 years into the future. The intention is that this unbroken chain will enable the participants to connect themselves to their descendants.

The graduation ceremony starts with a procession moving from the main plaza in front of the Bata Library, across the river on the footbridge and into the rotunda of the Environmental Sciences Building. Two representatives, one male and one female, are selected from the graduating class. The female student receives an oak sapling from the nursery located west of the building. Adjacent to the nursery are two "parent" trees from which subsequent trees will be grown. The male student carries the water vessel housed in the centre of the rotunda of the Environmental Sciences Building.

Together in the ceremonial sphere, they ascend the stairs beneath an ocular skylight. Walking east through the corridor, they emerge at the gap between the old and the new buildings. To the right is the rail car, the carriage that descends the trestle into the spiral forest. The railway car comes to a stop in the forest at the granite walkway. Four students approach the car-

Skeleton of a dragon gateway

riage and carry the water vessel to the "water pavilion," where it is lowered into the Otonabee River. The filled vessel is returned to the point where the car has stopped and is carried along the spiral walkway behind the female student and her oak sapling. At the appointed spot for that year, the young

Ceremonial water vessel

tree is planted and watered. A commemorative stone and time capsule are laid in the monument at the centre of the garden. At the first ceremony, twenty-five trees are to be planted by benefactors of the university – one for each year that the university has been in existence. A "ritual committee" of senior students and faculty could elaborate, organize, and pass on knowledge of this ritual.

There are five major architectural components to the graduation ritual. At the heart of the project are the Central Sphere and Ceremonial Stair. The sphere is the public forum where students meet and ceremonies take place. It is metaphorically both an eye and a cosmic sphere. This will be the focal point where the graduation ritual begins.

The Railway Car and Trestle Ramp make special reference to the human history of the site and act as a symbolic carriage that would transport the male and female representatives along with the oak sapling and the water vessel. The railway car, restrained by ropes held by six students, will descend

Stairs and ceremonial sphere

Where the railway line meets the building

by gravity from the platform level into the ceremonial courtyard. At the end of the ceremony, the car will be raised to the top of the trestle ramp and bolted into place under a protective canopy to await the following year's ceremony.

The Granite Walk to which the railway car descends forms a spiral shell in plan and initially defines the edge of the spiral forest. It is the sacred wall, a memory walk that terminates at the tower monument at the centre of the forest. Along this wall important contributors and founders of the university would be remembered. At the edge of the Otonabee River, there is to be a Water Pavilion from which the ceremonial water will be drawn for the seedling in the planting ceremony.

The forest of the Ceremonial Grove will be on a small scale at first; the future will be marked by tree grates awaiting occupation. Seventy (three score years and ten) grates mark one human life on earth; they will follow the growth of the forest, advancing one year at a time. At the centre of the forest is the Tower of Continuity, a monument to remind us of man's reconciliatory role on earth. The tower would grow over time, as it is formed by stone blocks housing time capsules constructed by each graduating class. These capsules would be opened seventy years later.[2]

The architects have proposed an environmental college for the next phase of the university, where a mature, multi-disciplinary approach to ecological studies could be applied. Economists, behavioural scientists, artists, religious scholars, poets, philosophers, and historians could live and work together on an approach to life in keeping with nature. This is the challenge facing architects interested in the issue of continuity. In a world that is being swept away by seemingly incomprehensible forces of change every day, architects must create places where individuals are able to perceive the historical relevance of their daily actions. The intention is that this work may help to stimulate others, in the hope that one day enough momentum may develop to allow new mythologies to be written and nurtured for future generations. The dream is for humanity to once again lock arms in the chora of life.

NOTE

1 The Environmental Sciences Building was designed by Richard Henriquez Architect, OAA, in a joint venture with Laszlo Nemeth Associates.

2 The architects' proposals for the graduation ceremony and ceremonial garden are as yet unrealized.

"THE PROBLEM WITH THE ARCHITECT AS WRITER ...": TIME AND NARRATIVE IN THE WORK OF ALDO ROSSI AND JOHN HEJDUK

LILY CHI

Nous savons pertinemment qu'on ne peut exorciser le spectre de la technologie au moyen de polémiques simplistes opposant l'homme à la machine, par exemple. La technologie, en effet, laisse des traces aussi profondes que celle d'une existence; Heidegger, du reste, nous prévient que l'interrogation sur l'essence de la technique revient à s'interroger sur *soi*. La narration architecturale met en lumière un aspect essentiel de la connaissance de soi, c'est-à-dire son caractère temporel. Chose curieuse, la narration est un sujet controversé en architecture, mais ce sont précisément les polémiques, les possibilités et les problèmes que soulève son emploi qui permettent de pénétrer dans le monde silencieux de l'intentionnalité dans l'architecture moderne.

Les méditations d'Aldo Rossi et de John Hejduk ouvrent la voie à une double réflexion. La première explore la complémentarité entre le récit et la création de lieux que les deux architectes invoquent dans leurs projets. En mettant en parallèle « programme » et « mascarade », Hejduk fait intervenir toute une constellation de formes narratives empruntées à la mascarade. Dès lors, comment et dans quel contexte le programme architectural constitue-t-il une approche au théâtre, au rite, au mythe – bref, à la construction d'un « récit »? L'attention que Rossi porte à « l'heure et l'événement », aux « gestes habituels » et aux « sentiers tout tracés » illustre une conception théâtrale de l'architecture. Elle se distingue en cela des conceptions programmatiques et fonctionnelles plus traditionnelles.

L'hypothèse adoptée ici est que le récit et l'architecture visent tous deux à moduler et à situer l'action humaine, c'est-à-dire, dans une perspective temporelle, ce qui apparaît comme la plus fragile et la plus éphémère des capacités de l'homme. Pour les deux architectes, le temps et son énigmatique entrelacement avec les gestes et les souffrances des hommes réapparaissent comme une préoccupation à la fois mouvante et inexorable. Le temps sert aussi de point de départ à ce qui serait, chez les philosophes Paul Ricœur et Hannah Arendt, une interrogation sur le rôle de la narration. Tous deux distingueraient la temporalité du récit de celle des processus matériels, démarche qui offre une approche critique aux analogies temporelles du fonctionnalisme moderne et du vocabulaire et des pratiques qui y sont associés. La critique que fait Arendt de la fonctionnalisation de l'histoire et des actions humaines dans la pensée et la politique modernes trouve sa contrepartie architecturale la plus inventive dans le jeu qui s'établit entre l'inscription orthographique et le texte narratif dans le projet *Victims* de Hejduk.

La deuxième partie de cette expérience aborde la question de la narration architecturale du point de vue opposé, c'est-à-dire qu'elle examine les facteurs qui font de la narration un élément problématique sinon controversé de l'œuvre architecturale. On peut trouver un excellent exemple de cette opposition dans la différence évidente qui sépare les rites sociaux de la Venise du quinzième siècle et les architectures théâtrales de Rossi et de Hejduk. Dans le premier cas, la narration et la représentation brossaient *simultanément* un tableau de l'ordre moral et spatial, de telle sorte que les frontières de la géométrie et du mythe étaient en somme complémentaires. Aujourd'hui, il nous est impossible de postuler une telle correspondance au départ. Le fossé est plus profond que jamais dans les scènes muettes mais lourdes d'attente de Rossi, dans-le combat entre le dessin et le texte qui encadre les mascarades de Hejduk. Pour nous autres modernes, issus de ce moi qui doute de soi et se déclare l'*arkhe* de l'univers, la relation entre les mots et les objets, entre le physique et le moral, entre la géométrie et les actions humaines est une question qu'il faut toujours reposer. Cette tâche est tout à la fois la source du problème et notre responsabilité irrévocable. En d'autres termes, les *problèmes* de la narration dans l'architecture contemporaine se résument dans cette image déconcertante de l'architecte comme écrivain – à la fois fabricant de textes et auteur de narrations humaines. Si l'on peut définir l'esprit de la modernité par la foi en cette qualité d'auteur, c'est-à-dire par la croyance que l'homme peut inventer une fin plus heureuse au récit de l'humanité, le fait de reconnaître cette position solipsistique de l'auteur, c'est-à-dire que le récit peut être une fiction, pourrait bien caractériser le dilemme de la post-modernité.

Les œuvres de Hejduk et de Rossi, qui manifestent au départ des préoccupations réciproquement sympathiques, manifestent par la suite des attitudes divergentes vis-à-vis la responsabilité de l'auteur. L'extra-historique chez Rossi et le fictif chez Hejduk proposent deux pôles, pas forcément antipodiques, qui incarnent de façon dramatique les responsabilités antinomiques de l'architecture contemporaine.

PROLOGUE 1

What can an architect mean when he writes of his own discipline that "in order to be significant, architecture must be forgotten?" The author of *A Scientific Autobiography* makes this assertion repeatedly. Indeed, he admits, "Forgetting Architecture comes to mind as a more appropriate title for this book, since while I may talk about a school, a cemetery, a theater, it is more correct to say that I talk about life, death, imagination."

It would be difficult to miss Aldo Rossi's polemic: what is at stake in architectural work ultimately eludes the most familiar terms and techniques that circumscribe it as a discipline. "How can one measure the quality of Lord Jim's fall," he asks, "when it is a fall from which he will never rise again? How can one measure buildings, if an amphitheater can become a city, and a theater a house?" What does seem to be at stake for Rossi is given in the enigmatic image of a seashell, the "daughter of stone and the whitening sea" who drew him to architecture. The shell, wrote Bachelard, is "a witch's cauldron in which bestiality is brewing." A figure of dazzling poetic clarity, the shell is also a long-time riddle – an impossible intimacy between stone and fluid animality, an inscrutable commensuration of geometric "firmity" and liquid movement. Shell-like is Rossi's language as he muses on the matter of architectural work. The *Autobiography* betrays a gaze driven to distraction by the enigma of "habitual gestures and paths," by unfolding actions and stages lying in wait. For the architect who seeks "the continuation of the *insula*, the space of the people," theatricality offers a provocative

The Melancholy and Mystery of a Street, by G. de Chirico

analogy for architectural work and experience. So too, apparently, do literary and cinematic metaphors. Stories spill out where architectural discourse falters, like the sounding-out of some vague, inscrutable distance.

28 *Librarian*
She has made the following observations:

There are three kinds of detectives
1 *The Fiction Detective*
2 *The Non-fiction Detective*
3 *The Semi-fiction Detective*
  1 The Fictional Detective departs from his office in search of the fictional murderer.
  2 The Non-fictional (Real?) Detective departs from his office in search of the non-fictional (real?) murderer.
  3 The Semi-fictional Detective (Newspaper Reporter) departs from his office in search of the semi-fictional (newspaper report) murderer.

ALL are Detectives and Murderers.

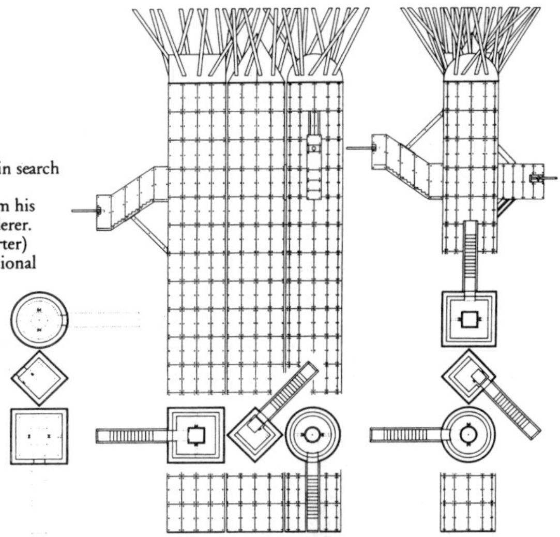

"Librarian Structures – Book Towers/Fiction/ Semi-Fiction/Non-Fiction," Structure 28, *Victims*, by John Hejduk

PROLOGUE 2

John Hejduk writes in *The Mask of Medusa* that he objects to sculptors who pretend to be dealing with architecture because their spaces are, in fact, empty. He confesses, furthermore, that for him, "architectural space has to have a function." While this sounds at first like an all-too-familiar modernist chauvinism, Hejduk's most recent projects point to a very strange functionalism indeed. The catalogue of sixty-seven structures in the *Victims* project, for example, is ostensibly a multi-use program, a master plan in format. Closer scrutiny, however, reveals disconcerting elements in the list of provisions: among the offices and shops, "record hall," and studios, prescriptions are also given for "a Time Keeper, a Child, a Judge ... The Dead." The functionalist dictum of integrity between form and function takes on an absurd gravity in *Victims*, where it is no longer possible to tell what is being "designed" – the habitat or the inhabitant, the container for the institution or the moral/political characters themselves. "Planning," here, slides insidiously towards "plotting."

Like Rossi, Hejduk seems to emphasize the eventfulness of architectural settings. The polemical tenor in Hejduk's work is more explicit, however:

the architectural project is radicalized as an overtly temporal construction with the insertion of narrative text into the space of orthographic drawing. Hejduk's intentions are made clear in the concluding remark to the *Berlin Masque*, a forebear of *Victims*: "As it was necessary for the highly rational-pragmatic city of 15th c. Venice to create masques, masks, masses for its time in order to function, it would appear that we of our time must create masques (programs???) for our times."[1]

The juxtaposition of "program" and "masque" suggests that, far from forgetting architecture (as Rossi urges), Hejduk wants to inculpate the very terms and techniques of modern practice. Program and function are two preconditions for the architectural project as it is understood today; ordinarily, both are but silent players relegated to the pre-design margins of professional procedure. The masque, on the other hand, is a Renaissance court spectacle, a tradition of civic ritual with tenebrous ancestry in pagan mysteries and sacrificial rites. The question begged is this: How could a pre-design technique developed to clarify and organize desire be seen in the light of dark and terrible stories whose eternal re-enactment dismembered kingly bodies and intoned their resurrection?

NARRATION AND THE ARCHITECTURAL PROJECT

This essay explores the inferred complicity – or complementarity – between narration and architectural making as it is invoked by Aldo Rossi and John Hejduk. In their respective meditations on architectural work, the former points to unfolding dramatic plots to explicate a non-visual, ineffable dimension of lived spaces; the latter names "function" and "program" as locales wherein this complementarity is played out in contemporary practice. In both is found the same scepticism about the capacity of existing practices to deal adequately with their concerns; in both are also found intimations of a critical agenda for bringing narrative into the architectural domain. Rossi sidesteps architecture to speak of "life," while Hejduk implicates but then polemicizes function and program as ironic formulations with disquieting implications.

I would like to explore this suggested complementarity with the following questions: In what way can the subject of storytelling be an architectural concern? What could be occluded by modern practice to the point that narration could be seen as compensation for disciplinary discourse? On the other hand, are there limitations to narration as a viable strategy for contemporary architectural production? In what way could the notion of the architect as writer be a problematic one in our age?

The secret but pivotal word in this discussion is "time." First, because for both Rossi and Hejduk storytelling seems to voice an elusive but persistent preoccupation – that with the temporal dimension of architectural quality. Second, because temporality itself names a context of discussion, an accountability that forces the questions posed above beyond a benign argument of disciplinary boundaries.

This essay is written at a time when the heroic model of the architectural project – and of the architect as gnostic visionary – has lost much force as a guiding idea. In a world of increasingly indifferent "difference," it may not be just fatigue that renders us insensitive to each new promise of redemption, each proclaimed reserve of original authenticity upon which to recover architectural ideals in a post-industrial age. And monotonous is not the only description for the plethora of discourses, forms, strategies, and technologies that have debated this ambition. Boredom is not an option, for the apparent relativism of ideas – already acutely discerned by the nineteenth century – menaces with a much more sinister cast in the afterglow of two world wars. In the age of psycho-babble, this paralysing aimlessness, which not only architects experience, might be called a monumental collapse of self-confidence. And "post-modernity," if this term be used, might describe an ambivalent state of mourning for the waning vigour of this self.

At stake is not just the loss of a self-identity but the concordant destitution of a whole mode of operation, a whole tradition of thinking, acting, and making. Modernity was (and is), in practice, a way of living through time. Octavio Paz has put this most succinctly:

> What distinguishes our modernity from that of other ages is not our cult of the new and surprising ... but the fact that it is a rejection, a criticism of the immediate past, an interruption of continuity. Modern art is not only the offspring of the age of criticism, it is also its own critic. The new is not exactly the modern, unless it carries a double explosive charge: the negation of the past and the affirmation of something different ... The modern age is the first to exalt change and convert it into a foundation. Difference, separation, otherness, plurality, novelty, evolution, revolution, history – all these words can be condensed in one: future ... time which is not yet and which always will be to come ... Our perfection is not that which is, but that which will be.[2]

Modernity is here characterized by a heroic task taken on by thought and action in the last two centuries – that of wresting human destiny from the province of transcendent authority. Avant-garde art and poetry share this timely ambition with modern politics and technology – to rupture an (inauthentic) present by positing a (more) authentic past in order to champion a better future.

Although it seems more difficult today to maintain the same faith in the goodness or reason of this will to Creation, the task of achieving one's own destiny remains – the single fruit of a paradoxical tradition of revolution. In architecture, this task appears in the form of the program – something that only became a concern for architects at the end of the eighteenth century. (Ledoux's *L'Architecture considérée sous le rapport de l'art, des mœurs et de la législation*, published in 1804, was one of the first treatises to problematize architecture's socio-political agenda.) The question of narration in architecture must be situated in this cultural context, for "writing" has, in this way, been a responsibility of the architect for nearly two centuries. Thus the question at stake in this essay is not whether architecture should emulate literature but rather how an architect can best confront the task shared by all makers of modern life – that of constructing, without recourse to a higher authority, whole futures and destinies. How does narration differ from programming as a way of construing that destiny? Can storytelling be compared to the retelling of binding myths in earlier times, when no such transcendent myth can today be taken for granted but must itself be posited? Given the ominous overtones of this fearful freedom, what positions do John Hejduk and Aldo Rossi take in this matter?

These thoughts are developed in two parts. The first compares programs and plots as ways of formulating human affairs. To this end, I will consider the role of narrative in the political philosophies of Paul Ricœur and Hannah Arendt. Both authors deal with narration within a broader context of questions about language, action, and thought in modern times – a context that architecture shares with both philosophy and politics. Most interestingly, both see in narration a mode of self-understanding that has significant ethico-political implications for contemporary culture. The second part of the essay draws these themes together to focus on the potential problems and limits in the idea of the architect as writer.

PROGRAM, FUNCTION, PLOT

Although its framework was well in place by the nineteenth century, architectural programming became a formal discipline only very recently. The

Wall painting, Pompeii

basic premise of programming practice, as it was outlined in the 1970s, is the application of method to a critical juncture in the path from life to building, to the point at which life situations must be translated into workable terms and techniques of the architectural profession. However this may have been done traditionally, the modern program reifies this transference as a discrete technique. The essential aim of programming procedure is to articulate clear goals (or "ends") out of the tangle of human affairs in order to facilitate the question of form (the "means"), hitherto defined as the specific and proper province of design activity itself.

This partition of architectural work into program and design was called for only in part by the increasing material complexity of a technologically equipped world. The science of programming emerged as a discipline in its own right when it appeared that the architect's personal knowledge and experience no longer sufficed for understanding, and operating upon, a rapidly changing social landscape. Strategies were gleaned from the managerial, social, and behavioural sciences to formulate workable ends out of the tangle of interpersonal relations that comprise the social realm. Programming subsumes and clarifies less reliable, traditional modes of client/architect interaction. Goal-oriented, accountable technique takes the place of "playing it by ear," the intent being to minimize the area of uncertainty embodied by design work itself.

This program-design structure of practice was itself made possible by the formulation of functionalism that emerged with the social sciences in the nineteenth century. Sociology, as it was developed by Auguste Comte, was to political practice as programming is to architectural design – an application of positive method to the apparent ambiguity and relativity of human affairs. The "science of society" was born in the wake of the French Revolution amidst the turmoil of an emergent industrial and historically conscious world. In the chaotic aftermath of the revolution, it became clear that true reform could only come about with a global reorganization of the social realm based upon more fundamental, and therefore more lasting, principles of social harmony and order. For Comte, the framework of the natural sciences promised new legitimacy for the domain of moral and practical inquiries. Sociology had the task of grounding these matters anew in "positive knowledge," distinct from the merely speculative framework of metaphysics or the "fictitious" base of theology: "I shall treat the Social Organism as definitely composed of the Families which are the true elements or cells, next of the Classes or Castes which are its proper tissues, and lastly of the Cities and Communes which are its real organs."[3] Comte sets out here the outlines of the organicist analogy: the socio-political body is a living system consisting of parts. Functionalism looks at the operative relationships between these parts.

In conceiving of society as an organism to be modified, functionalism was but one of two approaches to the search for foundations of social order in nineteenth-century France. It was, however, only superficially different from the rival conception of society as a history to be interpreted. Both were developed to grapple with the new problem of time and change; both dealt with this problem through the dual concept of "pathology" and "norm."[4] Indeed, for Comte, "no conception can be understood otherwise than through its history" – but a history that exhibits intelligible laws. A normative structure for assessing and modifying social phenomena is inscribed into the very premise that human relations can be described in terms of laws derived from the explanation of means and ends. The ultimate, normative end of a social system in this view is social integration and equilibrium.

A number of problems intrinsic to this analogy became apparent almost immediately. Émile Durkheim's attempt to distinguish between the cause of a social structure and its function (need, end) – hence between causal and functional analysis – pointed to one ambiguity that nonetheless continued to haunt functionalist analysis. If the existence of a social phenomenon can have pathological or normal causes, the ultimate ambiguity between cause and end (the latter being the ultimate concern of social science and technique) would obviously prove to be a problem of grave proportions. Furthermore, the apparent tension between the concepts of visible organ and invisible function proved to be a tenacious dilemma for nineteenth-century city planners. The functionalist analogy revealed itself to be at heart a non-representational problematic that offered no ready explanation for dealing with the world of visible form.[5] That this would become a serious ambiguity for architects is shown by the complete lack of clarity and consensus about how function could operate as a normative basis of architectural form in twentieth-century criticism.

How does this formulation of human affairs compare with that of narration? *Mythos*, Aristotle wrote, is the imitation of action; storytelling is also a way of making sense of human relations. For Paul Ricœur and Hannah Arendt, this is also the locus of a great difference between plotting and programming. The difference lies in the respective ways in which each articulates and thus resolves the temporality of human acting and suffering.

Ricœur begins with the question of what narrative articulates as a mode of discourse distinct from others. He sees narration as a poetic construction similar to the workings of metaphor – that is, as a conjoining of unlikely terms that *thereby* offers indirect articulation of the inscrutable (in this case, the paradoxical experience of time that philosophy has never succeeded in defining). For Ricœur, our pre-reflective experience of time is embedded in the common world of acting and suffering. An analysis of everyday speech reveals that the most simple action statements bring into relation modes of futures, pasts, and presents, as well as ethical stances through the concrete and culturally mediated articulation of goals, agents, causes, circumstances, and so on. Stories emerge from this linguistic and communal, or practical, texture of temporal experience, but they remain inextricable from it. As a poetic construction, storytelling in effect makes dynamic temporal wholes (durations) out of disparate events in human life. At the same time, stories locate their characters within specific temporal orientations. These paradigms for understanding and living time constitute the shared, ethical horizon of a community. Conversely, the stories by which an individual or community articulates its self-understanding commit their subject as an agent of such actions, as well as a sufferer of others' actions. Ricœur develops this into the notion of "narrative identity" – an understanding of subjectivity that sees selfhood as arising out of practical relations or polity first. In

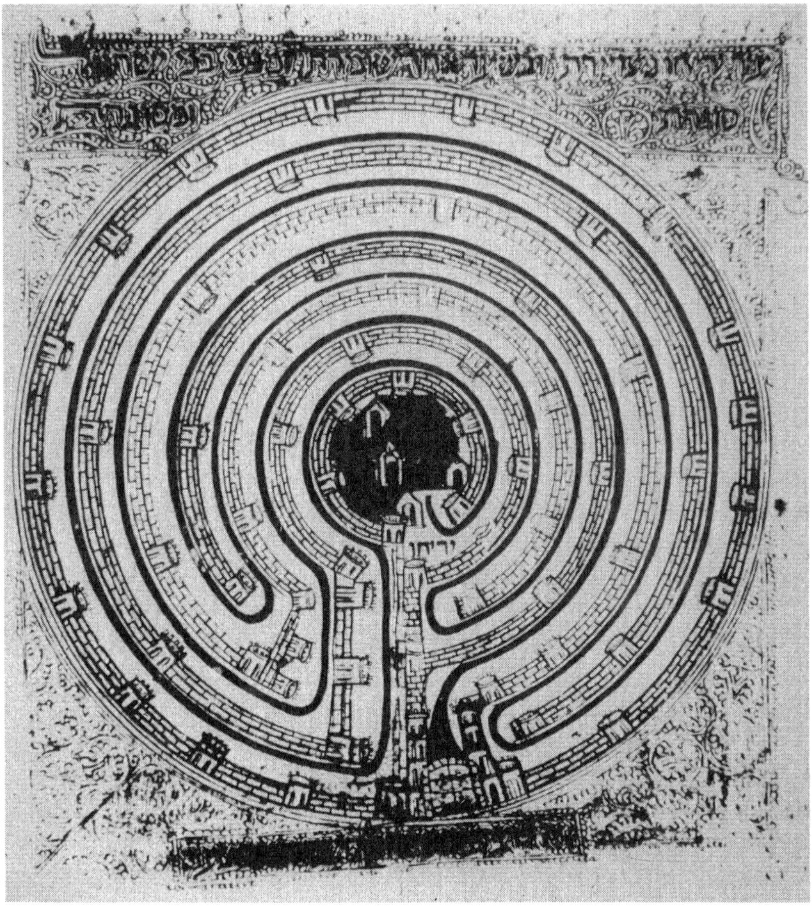

The city of Jericho as a labyrinth, Hebrew Farhi-Bible, c. 14th century

other words, before reflecting upon our substantive nature, we are already constituted by everyday language as politically engaged and hence ethically responsible beings by the sheer fact that we are embedded in the world as actors and sufferers of human action.

Ricœur's account of narrative offers an intriguing prospect on an old architectural question – that of the apparent tie between myth and geometry, between constructional procedure and cultural ritual in many pre-modern cultures. If narration is a linguistic mode of self-orientation in time and with respect to others, stories would constitute, in effect, temporal and ethical maps to the world of human affairs. Perhaps this is a clue to the complementary dynamic between *placing* and *plotting* that Joseph Rykwert describes in *The Idea of a Town.*[6] His thesis there is that ancient founding rituals and civic dramas set architectural making within the context of tales retold about origins and destinies. There is no possible separation of geometry and myth, space and time. Building-as-retelling appeared to work a double-mapping – an enduring plot in the double sense of site and *mythos* – so that henceforth, mortal dealings could take place in the light and topography of ultimate deeds and their consequences. As a symbolic portent, shelter would thus have a temporal component as some form of duration (the ordered time opened up between the birth and death of the cosmos, for example); and architecture, as the shelter of human affairs, would be as much an embodied story against their loss in chronology as it was physical endurance.

History abounds with such tales of invincible walls whose strength lies not in stone and mortar but in the spells of collective mythology: the wall that girded Romulus's city was raised upon the plaitings of a labyrinthine dance; Jericho's seven skins were woven through with the same incantatory binding. Something of this resonates still in the nineteenth century when Gottfried Semper locates the essence of walls not in structural support but in the act of enclosure that first takes form in the ephemeral depth of woven, coloured tapestry. A more wistful expression is given in Étienne-Louis Boullée's proposal for a city gate:

> Lorsqu'ils ont construit des Entrées de Ville, les ingénieurs se sont contentés de faire les murs assez épais pour mettre les habitans à l'abri de l'artillerie. Ils ont, en cela, rempli leurs ministère; mais ils n'ont pas joint ce qui appartient à l'architecture civile, en présentant, comme je crois qu'ils le devoient, l'image de la force ...
>
> L'entrée de Ville dont j'offre le dessein, présente des murs qui semblent indestructibles. Sur le stylobate qui sert à la décoration de ces murs, est placée une file de guerriers que l'on a lieu de croire invincibles. J'ai cherché ... à rappeler l'héroïsme du courage de ces Lacédémoniens qui, en voyant une ville déffendue par des murs, se demandoient quels étoient, les hommes assez pusillanimes pour employer de tels moyens de deffense. En plaçant, sur les murs de la ville, tous les guerriers armés pour la déffense, j'ai cru que cet emblème diroit aux spectateurs. Ces murs ne sont rien; Redoutez le courage des habitans.[7]

Wistful, because for Boullée at the end of the eighteenth century, the case for locating the idea of a city wall in the space of the social imagination must be made in terms of a polemic – engineering versus "image" and decoration; physical wall versus dramatic wall; the wall that wards off artillery versus the wall that stages brave citizens.

This same social imagination figures prominently in Hannah Arendt's political philosophy. Like Ricœur, she privileges narration as a mode of articulating and engaging in this "public realm." As a poetic construction, narration constitutes a way of understanding human relations that is not without political significance in late modern society.

For Arendt, the antecedent of the *polis* was early Greek poetry. Stories of singular words and deeds offered mortal beings the possibility of a unique form of immortality not found anywhere else in the universe: the potential dignity of individual life lies not in perpetual return but in the capacity to accomplish the extraordinary, to initiate the unprecedented through action and speech. Thus human greatness resides, for Arendt as for the early Greeks, "in the least lasting of activities." In word and deed lies the human capacity to initiate beginnings and, therein, the essence of human freedom. Stories could be compared with *polis* because the one reveals in remembrance what the other is set up to maintain – the "space of appearance," the stage opened up between human beings whereupon freedom (the capacity to begin through action and speech) is enacted. In making human multiplicity a condition of meaning, Arendt stresses the primacy of what is shared because it is created by many and owned by no single person. This is Arendt's understanding of "worldedness," of the primacy of the public realm as a medium of human reality. Plot and *polis* both exhibit action in its own medium – the "web of human relations," that flesh of the world wherein human life can be said to amount to more than vain passage or mere biological existence.

But the web of human relations is also what confines speech and action to an intrinsic uncertainty. Falling always into an existing world of other actors, the consequences of action are always both unpredictable and irreversible. This inherent frailty in the realm of human affairs accounts for the historical fact that political actors and philosophers alike, in Arendt's opinion, have always attempted to find a substitute for action. The theory and practice of "making history" in the nineteenth and twentieth centuries are just such a substitute and constitute the culmination of western metaphysics as a fundamentally anti-political tradition.

Both Marx's practical philosophy of history and Comte's positive science of society find their antecedent in Plato's philosopher-king, who governs through cognition. Knowledge as command, action as obedience and execution, *eidos* and *techne* – this was the ancient model of fabrication. Transposed into the realm of word and deed, whose medium is human multiplicity, this is essentially a model of instrumental action. In Arendt's estimation, modern historicism, technology, and totalitarianism in politics all share this same temporal model of human affairs – that of action conceived as a process with intelligible ends, and hence with identifiable means. For Arendt, this is the meaning of "justified violence" in the modern age. (Operation "Desert Storm" comes to mind as an impressive example in recent history.) While the full manifestation of this model of action could only come about with the inversion of the thinker's traditional disdain of worldly affairs – that is, with the ideal of practical knowledge attendant to modern technology – its framework was already present in Descartes's epistemology with the identification of thinking, acting, and making. David Lachterman offers the following observation of this modern *ethos*:

> The "idea" giving significant shape to modernity, *in both its revolutionary and projective modes*, is the "idea" of construction or, more broadly, the "idea" of the *mind* as essentially the power of making, fashioning, crafting ... To be modern now means at least this: to have a share in or to be caught up by a *project*, an enterprise *projecting* itself by anticipation into an unbounded future. It means, therewith, to gauge any present by its bearing on the future, however dark the details or design of the latter might now be.[8]

The reader may recognize here a temporal attitude – resentment of the present – that Octavio Paz had called "the spirit of critique." Arendt disparages this same temporal disposition for its corrosive effect on our belief in, and commitment to, an enduring public world.

"Dark times" is her term for an age in which the most horrific crimes are not the work of hatred by the best, not the effect of monstrous ideologies, but a capacity inscribed into the very heart of the routine. This is the "banality of evil" in the modern age. Totalitarianism is generally understood as the subjection of thought and action to the tyranny of one-man rule, but Arendt believes that this also describes democracies wherein "behaviour" has replaced "action." For Arendt, the "worldless" society of labourers is tyranny in the form of no-man rule. Dark times is the state of affairs where words and deeds have no meaning but as means to ends. The impotence of modern polity stems from a twofold "loss of the world." When everything, natural or human, can be conceived of as a process, everything can be seen as a means or end capable of being acted upon. Ultimately, all things are reduced to transitory value in an endless dynamic of consumptive instrumentality. Transience thus infects not only the natural universe but effaces, too, the worldliness of human artifice when works and matters of public life become but "exchange value."

Arendt sees in poetic narration the effect of a lament: "Narration, which shapes history, solves no problems and assuages no suffering"; and yet it seems to offer the only hopeful form of responsive engagement between barbaric intervention and corrosive indifference. Storytelling offers a mode of understanding through reliving. Remembering through retelling weaves the web of re-engagement: for Arendt, the essence of tragedy lies in the moment when the actor suffers the consequences of his/her own action. For the author who continued to maintain faith in beginnings, works of poetic narration are *worldly* fragments.

THE ARCHITECT AS WRITER

Arendt's analysis gives rise to serious questions about the process analogy inherent in the framework of functions and programs. It is not just the effects of mass destruction that are most disturbing about the *fact* of Auschwitz or Hiroshima. It is the "banality" of the evil therein – an obscuration of polity against which the theory and ethics of the day appeared impotent. This should serve as an ironic reminder of what may be further occulted by the contemporary disdain of the area of work named by these two terms.

On the other hand, if the estimation of narration by Arendt and Ricœur tempts us to replace program with plot in architecture, there remains the legacy of our modern heritage, which should serve as a sobering caution. Narration cannot play the same role that it did for fifteenth-century Venice, for Homer, or for Herodotus, now that Man has assumed the role of God as maker of nature and history. I believe that the best demonstration of this is offered by Hejduk's *Victims* project.

Presenting itself equivocally as a proposal for both container and contained, the sixty-seven structures cross and recross a line made proverbially fine between building and institution, between institution and inhabitant, and between inhabitant and habitude/ethos/moral character. By thus implying that architectural work may be as much temporal construction (plot)

Two Children Are Threatened by a Nightingale, by Max Ernst

as object prescription, Hejduk seems to inculpate two innocent bystanders into the web of stories that is the world's "Berlin." Not just the citizen whose banal routines are inscribed into the very brick and mortar of the constructional details, but also the one character whose ubiquitous presence remains unnamed – that is, the architect, pen in hand. To elaborate, I will focus on

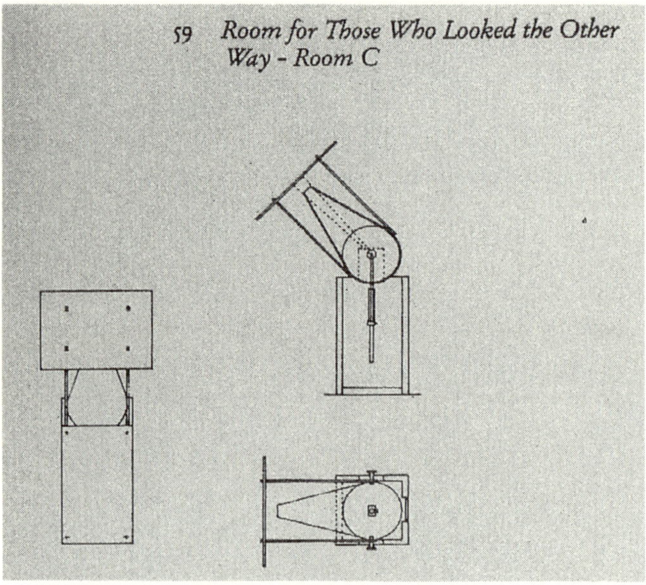

Structure 59, *Victims*, by John Hejduk

the apparent tension between narrative text and orthographic drawing in Hejduk's *Masques* projects.

Hejduk's drawings are rather conventional in themselves; that is, they seem to work on the same three planes of projection that construction drawings do. The distance between these plans, sections, and elevations and a complete set of working drawings is only a matter of time: the *Masques* drawings are imminently buildable. Rather, it is language – in the form of names, fables, or riddles – that warps the plane of projection from drawing to building. Textual inscription renders the projects speculative, fictive in status, a paper architecture in contrast to a buildable one. Moreover, the structures that comprise the *Masques* constitute a too-literal narration of architectural "character" and thus brazenly overstep the limits of architectural design. Hejduk does not seem content to set the stage, as Rossi would do, but presumes to prescribe movement, direct action, write life stories.

This apparent opposition between narrative text and projective drawing is, however, deceptive. One need only recall the status of the construction document and the legal contract after the nineteenth century. In modern architecture, drawing *is* building. For any architect, the most sinister implication of the *Victims* project comes rather from the cruel conjoining of two expressions of futurity – the imminent "will" of constructive technique and the prophetic "will have done" of narrated acting and suffering. Hejduk's narrative text makes explicit an ignominious task inscribed in the very heart of the modern architectural project: the program of life is to be rewritten with every new design proposition. This is the major difference between Hejduk's *Masques* and the archaic rituals that inspired him: for modern culture, neither the order of, nor the relationship between, the moral and the physical can be presumed. For us moderns, born of the doubting ego that declared itself *arkhe* of the world, the relationship between words and things, between the physical and the moral, between geometry and action, is a question to be constructed ever anew. Every project must begin by declaring its plot.

In Hejduk's work, narration has a critical role; but as Arendt would say, it offers no solution. The work of architecture remains bound to a provisional public realm opened up in private reckoning. The dark humour of projects such as *Victims* feeds upon deeper ironies that provoke the following thoughts. It seems that whether she or he draws or writes, the modern architect lives in a God-less world and is thereby condemned to playing both author and scribe, caught between "writing" (i.e., making) history or transcribing (mere) "fiction." I refer here not to the tired distinction between history and fiction as truth claims but to two modes of conceiving and engaging with the world of human affairs. To put it another way, the practical conundrum that plagues the question of ethics at "the end of modernity" is a disturbingly equivocal choice between *active change*, which perpetuates the same world-effacing resentment, and *passive resistance*, which risks indifference and withdrawal. The dilemma is best summarized by a question posed by the novelist Milan Kundera: What is the role of the writer to be "in the absence of the Supreme Judge" when *mimesis* is no longer possible and reason is suspect? The difficulty with the idea of the architect as writer is thus a problem that extends well beyond the apparent opposition of drawing and building, of body and text. It is a problem that is part and parcel of architectural making as a political act, as a form of engagement in the realm of interpersonal affairs and in language.

EPILOGUE

These speculations on architectural narration were provoked by the meditations of Aldo Rossi and John Hejduk. I would like to now consider how these two architects might have responded to the burden of authorship.

Courtyard, Elementary School, by Aldo Rossi

Rossi's *Autobiography* is filled with musings upon the interrupted scene. For Rossi, the suspended moment seems to bear the secret of stillness – not eternity, perhaps, but some sort of release from the will to succession. Time, as understood by the word *tempo,* signifies both atmosphere and chronology. Atmosphere is caught up in things as an in-between condition that haunts objects and for which the notion of function is but a poor substitute. In this context, the making of architecture becomes a preparation of settings, of modest "instruments." The instrument has the character of a trap – a *wrinkle in time* whereupon memory and possibility may coincide and crystallize. Limitation "amplifies memory, objects, events": the more detailed is the stage, the richer is the catch. Tuning his ear to the same sonority that Hejduk would set to words, Rossi could be thus very specific about its tone and tenor, about the instruments that await its resonance, and yet still never claim authorship of its manifestations. Working upon conventions, architectural work becomes a discipline in the most concrete sense – a rigorous "repetition" and patient waiting: "Today if I were to talk about architecture, I would say that it is a ritual rather than a creative process. I say this fully understanding the bitterness and the comfort of the ritual. Rituals give us the comfort of continuity, of repetition, compelling us to an oblique forgetfulness, allowing us to live with every change which, because of its inability to evolve, constitutes a destruction."[9]

Comments such as this indicate that Rossi's conception of architectural work is a historically motivated position. His words articulate a melancholy at the end of time – that is, at the waning of modernity as a way of living time. His tone is not one of despair, however. Rossi seems to find solace in the conviction that architectural intention can no longer be futural in the old way. The architectural project can no longer claim unselfconsciously to be the bearer of utopian tidings, the purveyor of a better ending for the story of Man. The temporality of an architectural work is given instead in the analogy of holiday time, oblivious to historical causality, to succession. When "great deeds are no longer possible," passage does not bring change; it is not the new that expectation awaits, but the will have been. Summer encampments, the plots of the *teatrini* "whose dramas made no progress" – these are alternative expressions for this sense of fateless happening:

The widows' walks on the houses in New England recall the Greek ritual of scanning the sea for what does not return – a substitution of ritual for pain, just as obsession is a substitution for desire. Similarly the repetition of the form of the tympanum on a building does not cause the event itself to recur ... I am more interested in the preparations, in what MIGHT happen on a midsummer night. In this way architecture can be beautiful before it is used; there is beauty in the wait, in the room prepared for the wedding, in the flowers and silver before High Mass.[10]

Architectural work continues under the sign of repetition. And for Rossi, there seems to be a kind of freedom in repetition.

Rossi's disillusionment with, and abandonment of, the "u-topic" model of revolutionary creation for a position that privileges the commonplace of social convention invites us to reconsider the resentful stance of the modern gnostic, for whom worldly convention and custom are inauthenticities to be overcome. This would be radical advice indeed for North Americans for whom revolution *is* tradition, and a frenetic cycle of production and consumption is the everyday reality. Would Rossi's reconciliatory approach to the familiar and well-worn extend to video imagination, fashion, and the marketplace in North America? This is perhaps the most provocative implication of his position, and it merits careful consideration.

Hejduk, who *is* American, seems to follow another trajectory altogether – not a melancholic memory but an ominous exacerbation of the productive, projective force of the modern imagination. The *Masques* can be read and discussed as architecture even in their present form as drawing-texts, because the ancient gap between thinking and making, between intention and actualization – which could only be spanned by a divine intelligence, whereas mortals required two different approaches in dialectic and rhetoric – has narrowed to near simultaneity in projective space. A poignant, if unwitting, corroboration of this mode of imagination is given by Ricœur himself: "Reading," he writes, "is itself already is a way of living in the fictitious universe of the work."

As I write this paper, more and more progenies of the "Berlin Masque" continue to appear in print – and now, in constructed form. Hejduk is very proud of the construction of these machinations, which he likens to the building of a cathedral by a community. This is a highly provocative comparison, for it begs the question of who or what, in these *Masques*, replaces the divine Word, the Christian *mythos*, the Biblical heavenly city that ultimately delimited all medieval building enterprises. The answer can only be: a lone American architect; fragments of children's stories, modern history, novels; and Berlin/Lancaster/Hannover/Vladivostok ... Irony is a very fine line indeed – but perhaps one capable of projecting Man the author to the point of his eclipse on the far side of history.

The "extra-historical" in the work of Rossi, the fictional in that of

Hejduk: neither would claim to be conclusive responses to Hannah Arendt's question of the public realm, for just as Hejduk's *Masques* are bracketed by their own textual space, Rossi's theaters remain silent like the stages of the *teatrini* between festivals. This is perhaps the most poignant limit of all, a final modesty imposed upon the architect as writer: public space, as Hannah Arendt would caution, cannot ultimately be created by one, but by many.

NOTES

1 From *Mask of Medusa* (New York: Rizzoli 1985), 152.

2 Octavio Paz, *Children of the Mire* (Cambridge, Mass.: Harvard University Press 1974), 3*ff.* For similar accounts of modernity, see also G. Vattimo, *The End of Modernity* (Baltimore: Johns Hopkins University Press 1991); and G. Grant, *Time as History* (Toronto: University of Toronto Press 1969).

3 Comte, *System of Positive Polity*, v. 2.

4 See Paul Rabinow, *French Modern: Norms and Forms of the Social Environment* (Cambridge Mass.: MIT Press 1989); see also Michel Foucault, *Discipline and Punish* (New York: Random House 1979).

5 An early problem confronted by the new discipline of urban design in nineteenth-century France was the question of how social "functions" could be translated into spatial locality and visible form. Paul Rabinow maps the issues opened up by this problem in *French Modern.* For Hannah Arendt, this intrinsic denial of the world of appearance constitutes the fatal flaw of the functionalist model of social science; see *The Life of the Mind* (San Diego, New York, London: Harcourt Brace Jovanovich 1971). Aldo Rossi makes a similar point in rejecting functionalism: "For if urban artifacts present nothing but a problem of organization and classification, then they have neither continuity nor individuality; they do not 'say' anything to us"; see *The Architecture of the City* (Cambridge, Mass.: MIT Press 1984).

6 J. Rykwert, *The Idea of a Town* (Princeton, N.J.: Princeton University Press 1988).

7 É.-L. Boullée, *Architecture – Essai sur l'art*, ed. H. Rosenau (London: Alec Tiranti 1953).

8 David Lachterman, *The Ethics of Geometry* (New York: Routledge & Kegan Paul 1989), 2.

9 A. Rossi, *Scientific Autobiography* (Cambridge, Mass.: MIT Press 1981), 30.

10 Ibid., 55.

LA VILLE TELLE QUELLE – UN THÈME DE REPRÉSENTATION ARCHITECTURALE

IRENA LATEK

The critique of scientific positivism has enabled contemporary architectural theory to distance itself from the modern discourse in a way that goes beyond mere criticism of modernism and its impact: it opens up the possibility of undertaking a critical analysis of the modern paradigm from an architectural perspective. The idea of architectural representation that is at the core of this approach shapes the "exterior" to which architecture refers and, at the same time, reveals the depth of the processes thus undertaken, pointing to new possibilities. On the other hand, the dominant paradigm of contemporary architectural practice, overflowing with scientific data, is that of production. In a system where only immanent values count, the architect, reduced to being a tool of technopolitical entrepreneurs, becomes alienated in a discipline where the widening gulf between theory and practice signals a crisis or integrity.

To escape from this situation, alternative approaches to the architectural project obviously have a particular relevance. The project, the transformation of the idea into its execution, is revealed as the essential tool for critical approaches that mark a true departure from dogmatic instrumental thinking and bring with them the promise of a new outlook and a possible reorientation of interests.

This paper, illustrated with examples from school projects, proposes an approach based on the notion of architecture as a perpetual celebration of its original and universal state – the city. It explores the possibility of using the urban narrative to shape the architectural object and investigates the relationship between reality and its representations in drawing. In this view, the drawing is the *locus* of architectural figurations and constructions, a potent medium aimed at leaving traces on the built object.

La critique du positivisme scientifique émergeant du discours philosophique du vingtième siècle a permis à la pensée architecturale contemporaine de se distancier du modernisme tout en dépassant la simple critique de ses retombées. Cette pensée ouvre la possibilité d'entreprendre, à partir de la perspective architecturale, une analyse critique du paradigme moderne. La notion de la représentation au centre d'un tel discours renoue avec le symbolique en tant que sens permanent de l'architecture; elle révèle également la profondeur des processus engagés dans le questionnement des modes de représentation de l'espace à travers l'histoire. Le discours qui se penche, entre autres, sur la dissociation du vrai et du bon à travers l'invention de l'esthétique, confronte le questionnement de la science et de la technologie à celui de l'éthique. L'analyse des noyaux conflictuels des concepts de la pensée moderne (symbolique/instrumental, objectif/subjectif...) démontre la difficulté de cerner la notion de vérité. Dans cette perspective, le rôle de légitimation de la vérité se déplace de la science vers l'énoncé poétique[1].

La pratique architecturale contemporaine, en revanche, dominée par la techno-politique, s'édifie sur la doctrine positiviste; son paradigme constructif est une aspiration à instaurer sur des données scientifiquement vérifiables une société apte à assurer le bonheur des masses. Dans ce système où seules les valeurs immanentes sont prises en compte, tout ce qui le dépasse est considéré comme douteux et non valable. Sous l'emprise de cette réalité, l'architecture au quotidien fait face à des problèmes multiples: la conception quasi exclusive de la culture comme culture technologique, la conception restrictive de la liberté comme liberté de l'entreprise mènent à la crise de l'écologie naturelle et urbaine et à une architecture animée par la spéculation foncière. Par suite, le fossé entre la théorie et la pratique, entre le penser et le faire, se creuse de plus en plus. Ce clivage, que l'architecture a de la difficulté à assumer, met en doute son intégrité. Dans des situations extrêmes (caractéristiques, dans le contexte nord-américain), tout questionnement sur l'objet de l'architecture conduit aux vicissitudes de l'interrogation sur le sens du métier. Qui est l'architecte? Celui qui met en forme les projets des investisseurs de la construction ou celui qui, lors d'un travail théorique, ne cherche que le sens de l'architecture? Ces questions ne sont évidemment pas de celles auxquelles on peut répondre. Elles témoignent néanmoins, confortées en cela par la qualité de la production courante, de la profondeur de la crise actuelle.

Dans cette perspective, les approches alternatives au projet architectural présentent un intérêt évident. Comme un moment de l'existence humaine où l'idée, dans un rapport *sine qua non*, conjugue le faire, le projet se révèle alors un objet clé. Son rôle privilégié s'amplifie tout naturellement au sein d'approches critiques qui s'écartent fermement de la pensée instrumentale dogmatique. Il aspire ici au statut d'objet architectural en soi, se libère des limites qu'impose son rapport ancillaire avec l'objet bâti dans l'espace réel, et trace ainsi le parcours vers le symbolique. Ce parcours n'est cependant pas direct. Il est clair que la condition moderne qui est la nôtre rend impossible le simple retour à la tradition symbolique pré-moderne et nous engage plutôt dans une perspective de l'architecture critique qui fait de l'esprit moderne et de l'esprit traditionnel à la fois son objet d'étude. Une telle démarche n'aspire pas à la conception de l'architecture nouvelle mais cherche à rééduquer les regards et réorienter les intérêts.

L'approche architecturale et pédagogique présentée ici se situe dans ce contexte. Elle postule que le projet architectural peut émerger de la narration – de la narration urbaine – et elle choisit le dessin comme agent actif et lieu de sa genèse. Son exposé est illustré au moyen de projets effectués par les étudiants de notre atelier à l'École d'architecture de l'Université de Montréal[2].

La conception première de l'espace architectural est celle d'un système de lieux qui véhiculent un sens, qui témoignent de l'existence humaine: maison, usine, école, rue, place ou parc, dispersés dans une étendue sans fin. C'est la réaffirmation du sens de l'espace antique, médiéval et moderne, c'est-à-dire l'amalgame des notions aristotéliciennes et newtoniennes; cela correspond à ce qui domine toujours notre lecture de l'espace, et les théories contemporaines de l'espace ont peu d'impact sur cet état de choses[3]. À ces emplacements, il faut donc trouver des figures reconnaissables, idéales. Sur le plan théorique, ces idées épousent le concept rossien de la mémoire collective[4]. Elles annoncent une structure dans laquelle se situe chaque fait architectural et urbain. Un lieu particulier est cependant un phénomène à la fois concret, unique et irréductible à un cas typique. Le dévoilement de son sens repose donc aussi bien sur son appartenance à un système plus large que sur la découverte de la richesse de sa propre expérience. C'est pourquoi la forme d'un nouveau lieu est anticipée par les couches successives des inscrip-

La maison d'un journal montréalais. Relevé de la ruelle (fragment), Katherine Paré; mine de plomb et crayon de couleur sur vélin

tions qui s'y trouvent déjà. L'archéologie des différents moments de l'existence du lieu, la lecture des impressions de ses tracés formels et l'interrogation de leurs contenus permet, par addition et synthèse, de constituer des archétypes. La découverte des superpositions et transparences oblige cependant aussitôt à laisser les idéaux en suspens et à procéder à leur mise en abîme. Par l'injection du temps dans le projet et à travers la construction du récit urbain dans l'espace fictif du projet se développe une situation particulière qui devient alors concrète.

La recherche historique est donc pour le projet l'interprétation du passé à partir d'une position que l'on pourrait qualifier d'architecturale et non scientifique; elle exclut, par ailleurs, toute nostalgie du passé. Elle postule

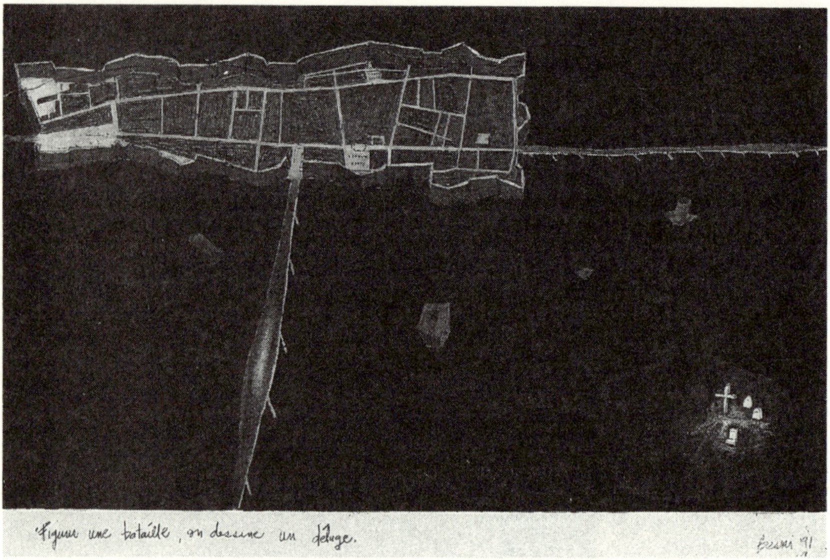

La maison d'un journal montréalais. Étude: « La ville », Dominique Besner; crayon de couleur sur photocopie sur vélin

plutôt que l'essence d'un lieu est détenue par le lieu présent et réel et par toutes ses représentations produites dans le passé; dans cette recherche, la réalité trouve ainsi son fondement dans la fiction, et réciproquement. Cette démarche, ne cherchant pas à établir la vérité historique factuelle mais à dépister des permanences du sens et de la forme du lieu, capte à travers les documents une narration évocatrice de sa spécificité. Autrement dit, elle cherche à édifier un nouveau lieu par le dévoilement de la vérité du lieu déjà existant. Cette découverte, inséparable du projet postulé comme œuvre, rend légitime ce dernier comme seule et unique forme de la vérité en question.

Dans cette démarche, on accorde aussi de l'importance à la spécificité du lieu au sens large, à la spécificité collective. Le milieu dans lequel nous vivons est porteur de charges culturelles que l'individu intègre nécessairement parce qu'ainsi s'opère son apprentissage affectif de l'espace. L'« espace réel » transformé en lieu signifiant sert, selon Piaget, d'ancrage psychologique[5]. Si l'érosion de la différence semble être aujourd'hui un phénomène qui s'élargit et prend de l'ampleur, il ne s'agit pas là d'une condition stable ou permanente. Le contre-courant, aussi constant que global, se manifeste dans l'essor des cultures régionales et, d'une façon plus générale, dans le besoin de

La maison d'un journal montréalais. La mémoire du lieu, Katherine Paré; mine de plomb et crayon de couleur sur vélin

La maison d'un journal montréalais. Mur des machines, façade sur la rue St-Antoine (fragment), Katherine Paré; photocopie, mine de plomb et crayon de couleur sur vélin

formes de plus en plus nouvelles d'expression des spécificités. La leçon de l'œuvre de Melvin Charney sur la « montréalité » de Montréal a fourni la clé à la compréhension de la forme de Montréal en re-présentant sa culture architecturale; cette œuvre a également construit un imaginaire qui s'y imprime de façon permanente[6]. Dans les projets des étudiants, les maisons en rangée, les escaliers extérieurs sont des conditions premières à partir desquelles tout commence, le moment où le développement de la représentation devient possible.

Logements collectifs à Pointe Saint-Charles. L'Idée du projet (fragment), Katherine Paré; mine de plomb sur vélin

Logements collectifs à Pointe Saint-Charles. Entre le banal et le monumental, Katherine Paré; mine de plomb et crayon couleur sur vélin

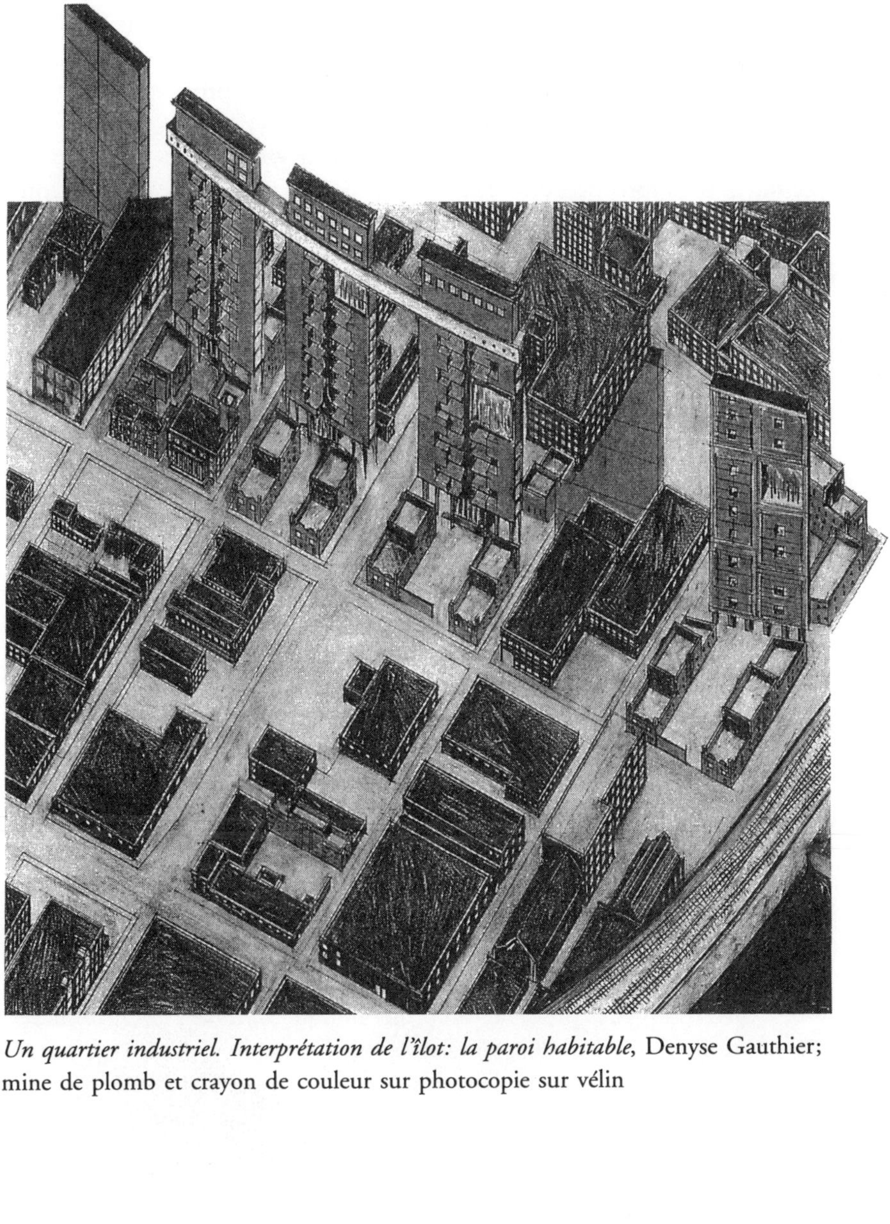

Un quartier industriel. Interprétation de l'îlot: la paroi habitable, Denyse Gauthier; mine de plomb et crayon de couleur sur photocopie sur vélin

Aussi faudrait-il admettre que ce sont là les figures du quotidien, les choses mineures, la « réalité pauvre de nos vies », très précisément les choses qui renferment les mythes essentiels[7]. Pour le bâtiment, cette notion est importante car elle permet, malgré la prédominance de la production de masse et du statut commercial de l'objet bâti, d'engager les procédés subversifs qui peuvent convertir sa valeur marchande en valeur symbolique. Sur un point décisif, notre approche s'écarte cependant de certaines procédures récentes (qualifiées de post-modernes) qui consistent essentiellement en une « esthétisation » du banal, c'est-à-dire en une conversion en œuvres originales d'objets produits en masse au moyen de procédés industriels. Ici, l'objectif est autre; les frontières entre l'original et le banal disparaissent parce que c'est dans le banal que l'on découvre le pathétique et le grandiose. La défamiliarisation que fait subir la machine au projet déplace son rôle habituel vers un récit architectural nouveau; il est désormais situé dans l'ensemble des relations dues au lieu, au thème et à la condition unique du projet.

À ce point, il est important de signaler la double nature ambiguë du projet. C'est un espace virtuel et temporaire en attente d'un objet bâti; il s'affirme cependant comme objet en soi, un espace d'articulation de l'architecture comme idée, comme chose mentale. L'exploration de ce dernier aspect, qui renvoie l'architecture à l'univers conceptuel en la libérant momentanément de la contrainte matérielle, implique désormais l'expérimentation des entraves dues à la complexité des jeux des idées et de leurs représentations. Le dessin, la photo, la maquette ne sont donc plus des instruments innocents; ils sont clairement distincts et distants de la réalité qu'ils représentent, ils contiennent leur propre espace codé, la convention qui indique la façon de construire le lieu.

Le dessin s'est ici proclamé comme instrument privilégié. C'est un médium tactile – l'expérience corporelle du monde y est transférée –, mais aussi un outil souple, conceptuellement manipulable. Il marque une affinité avec le projet comme œuvre en chantier permanent. Il y a une continuité dans le dévoilement du monde et le dessin est proche de cette condition provisoire. Ainsi compris, le processus du projet indique au travail de création une organisation selon un double versant: les représentations des divers aspects de l'objet, issues des lectures de son univers conceptuel, constituent de véritables constructions par couches qui interposent l'objet dans l'espace réel, mais en même temps chaque représentation de l'objet à construire est aussi une couche non pas qu'on fabrique mais qu'on soulève pour découvrir un peu plus de la profondeur de la chose.

Étude du lieu pour « La maison d'un journal montréalais ». Parcours, Manon Levesque; mine de plomb et crayon de couleur sur vélin

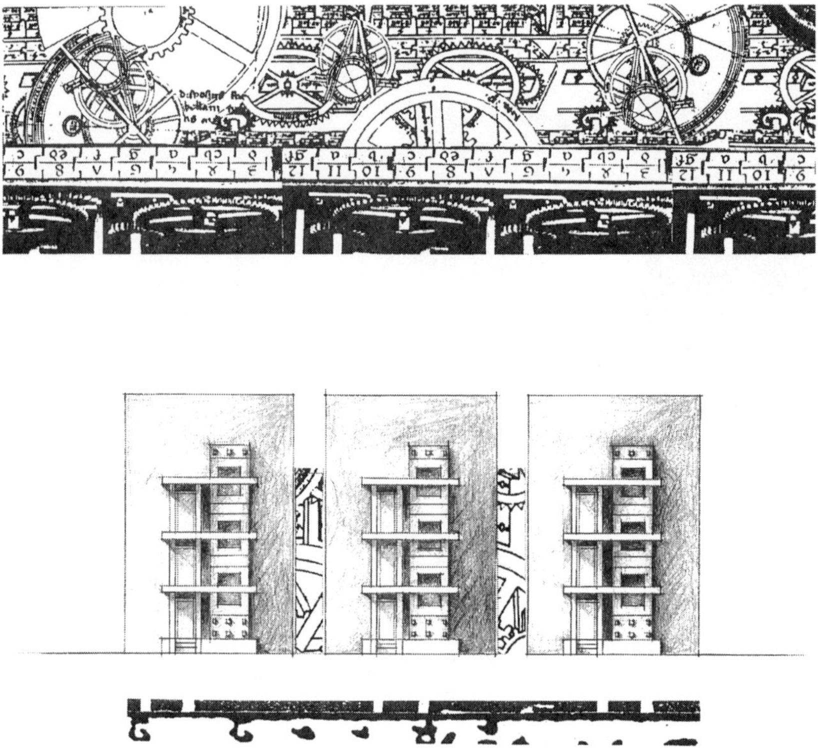

La maison d'un journal montréalais. Côté rue et Côté ruelle (deux fragments), Catherine Labrecque; montage de photocopies sur vélin et mine de plomb sur vélin

Paradoxalement, cette démarche, bien qu'éloignée des principales idées de l'architecture moderne, notamment par son aspiration à désintégrer le pragmatisme de la construction et sa prétention à dominer et à contrôler l'espace, est un produit de la pensée moderne. La découverte de la profondeur historique du concept de l'espace et son rapport avec la représentation n'est qu'une articulation parmi d'autres du dilemme contemporain que Koyré appelle la « tragédie de l'esprit moderne qui "résolut l'énigme de l'Univers" mais seulement pour la remplacer par une autre: l'énigme de lui-même[8] ». Cette critique ouvre cependant une nouvelle perspective: l'architecture n'a désormais plus raison de croire que l'unique fondement de sa modernité repose sur la vision utopique de la science positive et elle peut ainsi se dégager du paradigme productif.

La seconde modernité de cette approche, c'est la ville. Celle-ci est au centre de la démarche architecturale parce que l'humanité civilisée est insé-

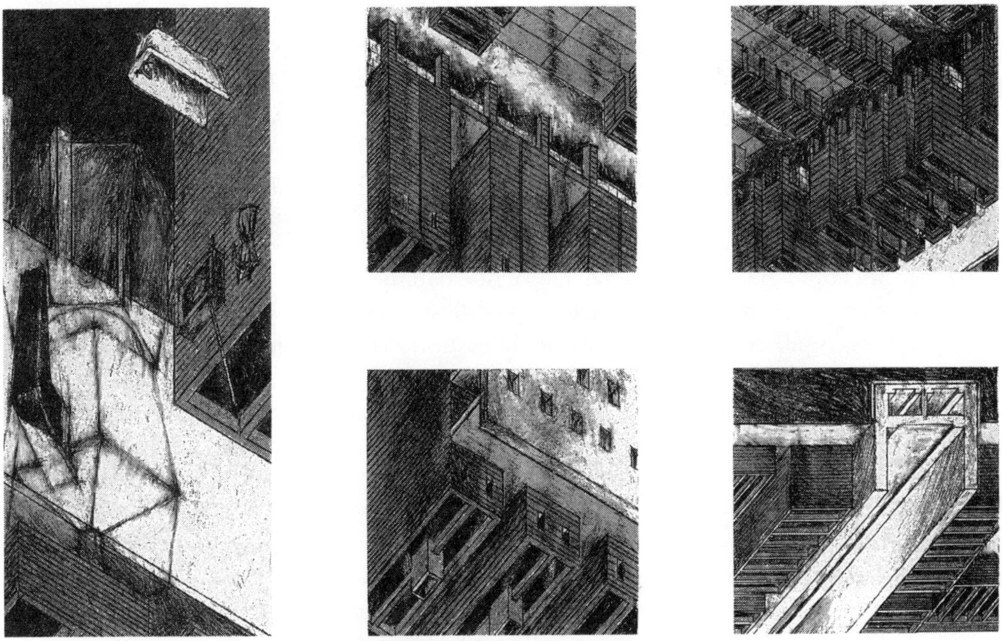

Housing on Toronto's Main Street. Idées urbaines fondatrices du projet, Alain Carle, Denyse Gauthier et Nicolas Roquet; mine de plomb et crayon de couleur sur vélin

parable de l'urbanité. À mon avis, on ne peut parler de la mort de l'architecture que dans les sociétés où l'idée de ville est devenue obsolète. La ville est pourtant un lieu d'émancipation moderne toujours possible et disponible, lieu des frottements et de la communion, de l'échange et de l'isolement, du support social et de l'autonomie. Si l'architecture en question comprend la ville comme son univers privilégié, elle ne veut pas forcément exalter la ville dite « traditionnelle », mais elle postule en revanche l'urbanité comme valeur.

Trop complexe pour être définissable, l'urbanité ne peut se décrire en termes de modèle: elle est notre mode de vie. Aussi faudra-t-il admettre que la seule ville de son objet est *la ville telle quelle.* Le regard qui la guette est un commentaire critique, la réalité qui l'occupe est contradictoire. Ce regard cherche les idéaux et expose les discontinuités, mais il évite cependant l'apologie du chaos. Il doit saisir l'essence d'une totalité non fragmentée par les modèles modernes mais aussi la vérité de son érosion par les nouvelles machines et les nouvelles vitesses.

Si l'architecture est considérée ici comme la perpétuelle célébration de la ville en tant qu'état originel et universel, la notion de lieu vient en première position. Or, le pluralisme, aujourd'hui souvent porté au rang de nouveau dogmatisme parce qu'il interprète la réalité comme une multitude des fragments aux origines diverses, a pour effet de dissoudre la notion de lieu[9]. La défense de cette notion doit donc non seulement intégrer la compréhension de la fluidité et de la mobilité de la mémoire du lieu mais également imposer le degré de stabilité permettant de s'y reconnaître.

Si l'approche présentée envisage la ville comme système de signes reconnaissables, elle se distancie quelque peu de ses mauvais usages. La création architecturale n'est pas un système de gestion des signes mais leur analyse critique et la remise en jeu, dans une perspective unique, de la vérité de chaque énoncé nouveau qu'est le projet. Cette impossibilité de soumission à la règle superficielle révèle les limites de la pensée normative.

Dans cette recherche des approches que l'on appelle volontiers alternatives, il ne s'agit pas de nier que la ville est toujours et perpétuellement à construire et à reconstruire. Le renouvellement du projet urbain correspond à notre existence constamment nouvelle, au mouvement de la condition temporaire de l'humain; il constitue aussi un acte de re-découverte continuelle de l'essence de cette condition par l'intermédiaire des formes idéales des archétypes. La ville est à construire mais, paradoxalement, cela ne peut constituer pour le projet une fin ultime. Ce qui importe vraiment, c'est de découvrir au sein du projet que la ville est le lieu de notre existence; elle est son contexte, sa situation et sans doute sa condition moderne. L'architecture a donc quelque chose à révéler sur tout cela. C'est ici que se résout la question de la place qu'occupe l'espace architectural – réel ou virtuel, dessin ou bâtiment. C'est aussi pour cette raison que la lutte contre le paradigme productif et la technologie se dessine comme une tactique très partielle. La technologie affiche une compatibilité avec la société de consommation et c'est là l'argument principal en sa faveur. Le vrai drame est que cela constitue le seul moyen de vérification en usage (surtout en Amérique du Nord). C'est pourquoi une idée d'Antoine Artaud, énoncée au seuil de l'instauration de ce système à l'échelle mondiale, me semble de plus en plus actuelle. « Le plus urgent », dit Artaud, « ne me paraît pas tant de défendre une culture dont l'existence n'a jamais sauvé un homme du souci de mieux vivre et d'avoir faim, que d'extraire de ce que l'on appelle la culture, des idées dont la force vivante est identique à celle de la faim[10]. »

La recherche de ces idées est aujourd'hui la question essentielle. La quête post-moderne de la forme signifiante a rejoint cette condition de la société démocratique que Beaudrillard a caractérisée comme fonctionnant à la fois à la base du produit de masse mais aussi à la recherche du renouvellement constant des signes visibles de l'émancipation et du bonheur. La récente reprise automatique et aveugle des éléments historiques dans une architecture technologique correspond bien à cet état de choses. La véritable réitération de la fonction symbolique de l'architecture ne peut donc pas ignorer les questions concernant les contenus de la représentation architecturale, car de leur force vivante dépendent entièrement les modalités de la mise en place de cette fonction, c'est-à-dire les moyens de la mise ou de la remise du symbolique dans la culture architecturale.

Par ailleurs, situer l'architecture dans l'espace virtuel peut être, provisoirement, la seule possibilité concrète de renouer avec son rôle historique d'agent culturel. Par cette politique éducative, on met l'accent sur le fait que l'architecture peut maintenant survivre parce qu'il y aura des individus créateurs qui pourront la transposer dans une société qui, un jour, aura besoin d'elle et érigera des structures qui permettront sa production dans l'espace.

NOTES

1 Je fais ici surtout allusion à la théorie élaborée par Dalibor Vesely dans « Architecture and the Conflict of Representation », *AA Files*, n° 8 (janvier 1985) ainsi qu'à l'ensemble des idées de son article « Architecture and the Poetics of Representation », *Daidalos*, n° 25 (15 septembre 1987).

2 Entre la présentation de cet exposé lors du symposium sur les rapports entre l'architecture, l'éthique et la technologie, et sa réalisation sous forme d'essai pour le présent recueil, une monographie a été publiée sur le travail de l'Unité d'architecture urbaine, atelier dont il est ici essentiellement question; voir *Ville, métaphore, projet – Architecture urbaine à Montréal, 1980–1990*, Montréal, Éditions du Méridien, 1992. Le présent texte vise surtout à mettre en évidence les aspects de ce discours qui apportent un témoignage pertinent sur la question de l'architecture et de la pensée instrumentale, de l'architecture face au progrès technologique et à la croissance illimitée.

3 Je fais référence à Michel Foucault, « Of Other Spaces », *Diacritics*, vol. 16, n° 1. Le problème de la conception de l'espace, dans la perspective de l'histoire des sciences, est présenté de manière exhaustive dans l'ouvrage d'Alexandre Koyré, *Du monde clos à l'univers infini*, Paris, Presses Universitaires de France, 1962.

4 Aldo Rossi, *Architecture de la ville*, traduit de l'italien, Paris, L'Équerre, 1981.

5 Jean Piaget et B. Inhelder, *La représentation de l'espace chez l'enfant*, Paris, Presses Universitaires de France, 1977.

6 Voir Melvin Charney, « The Montrealness of Montreal: Formation and Formalities in Urban Architecture », *Architectural Review*, vol. 167, n° 999 (mai 1980).

7 Je fais référence à la notion de « réalité pauvre » travaillée par Tadeusz Kantor et son théâtre Cricot 2. Kantor cherche l'essence de l'être dans le geste le plus quotidien et dans les comportements les moins empreints de conventions. De façon analogique, l'essence du lieu s'affirme avec force dans les gestes constructifs souvent les plus humbles. Pour le travail de Kantor, voir *Scènes*, n° 2 (avril 1986).

8 Par le biais de l'examen épistémologique, Alexandre Koyré célèbre l'apparition d'une nouvelle modernité, celle qui se tourne vers elle-même; voir *Études newtoniennes*, Paris, Gallimard, 1968, p. 42–43.

9 Voir notamment la discussion de l'idée de la sublimation du pluralisme et de son impact sur l'art dans Hall Foster, « Against Pluralism », *Recordings: Art Spectacle, Cultural Politics*, Port Townsend, Wash., Bay Press, 1985.

10 Antoine Artaud, *Le théâtre et son double*, Paris, Gallimard, 1964, p. 11.

THE AUTHORS / LES AUTEURS

LILY CHI received a Bachelor of Architecture degree from Carleton University and a Master of Philosophy degree in Architectural History and Theory from Cambridge University. She has worked in architectural firms in Ottawa and Buffalo, and has taught at schools of architecture at the State University of New York at Buffalo, at Carleton University, and at the Technical University of Nova Scotia. Ms Chi is presently engaged in doctoral studies at McGill University in the History and Theory of Architecture program.

ALAIN FINDELI est actuellement professeur titulaire à la Faculté de l'aménagement de l'Université de Montréal, où il enseigne l'histoire, la théorie et l'esthétique du design. Ingénieur diplômé de l'Institut national des sciences appliquées de Lyon, il a obtenu une maîtrise en aménagement de l'Université de Montréal en 1975 et un doctorat en esthétique et sciences de l'art de l'Université de Paris VIII en 1989. Ses travaux de recherche ont porté sur le Bauhaus et la création du New Bauhaus à Chicago par Laszlo Moholy-Nagy. De 1989 à 1991, M. Findeli a été rédacteur-en-chef invité de la revue *Informel* pour quatre numéros spéciaux consacrés au symposium « Prométhée éclairé » (1991) sur le thème de l'éthique, la technique et la responsabilité professionnelle en design. Sa recherche actuelle porte sur les conséquences éthiques du développement technologique et sur la responsabilité des concepteurs.

JEAN-PIERRE HARDENNE a obtenu un diplôme d'architecte en 1970, avec grande distinction, de l'École nationale supérieure d'architecture et des arts visuels de la Cambre en Belgique, et un doctorat en sociologie en 1978 de l'Université de Montréal. À l'Université du Québec à Montréal, il a fondé le département de Design, dont il occupe la direction depuis plus de sept ans. Il a occupé pendant trois ans la direction du module Design de l'environnement et a assumé pendant quatre ans la direction du Secteur des arts comme vice-doyen. II a aussi fondé le laboratoire ARCHINFO où, à travers une pratique du projet, il a travaillé sur les questions que l'infographie pose à l'architecture. M. Hardenne collabore régulièrement à de nombreux séminaires et colloques au Canada et à l'étranger qui ont pour thème les moyens de représentation en architecture et les techniques nouvelles, dont l'infographie. À travers une pratique de recherche-création, il a obtenu plusieurs distinctions et prix dans des expositions et concours internationaux. Enfin, il participe régulièrement à des jurys de diplôme en architecture, design, graphisme et arts visuels, tant au Canada qu'en Europe.

GREGORY HENRIQUEZ graduated from Carleton University, where he studied architecture. He also studied at McGill University in History and Theory of Architecture. He is currently working with his father, with whom he co-authored the chapter in this volume.

RICHARD HENRIQUEZ, MAIBC, RCA, FRAIC, has been practising architecture in Canada since 1966. Born in Jamaica, he studied architecture at the University of Manitoba, graduating in 1964. He continued his studies at the Massachusetts Institute of Technology, receiving a Masters degree in Architecture and Urban Design in 1967. Since then, he has been based in Vancouver, British Columbia, working on projects throughout Canada and overseas. His most notable projects include the Environmental Sciences Building at Trent University in Peterborough, Ontario; the winning entry in the United Nations Peacekeeping Monument competition in Ottawa; the Sinclair Centre Federal Office Complex, an award-winning heritage restoration of an entire city block in Vancouver; three luxury condominium towers in Vancouver; and various municipal buildings. He has received numerous awards and prizes, including the Governor General's Award for Architecture (1990), the Lieutenant Governor of British Columbia Medal in Architecture (1990, 1992), the Canadian Architect Award for Excellence (1990, 1985, 1976, 1972), and the City of Vancouver Heritage Award (1988, 1986). He is regularly invited to participate as a competition juror, guest critic and lecturer. His projects and design approach have been featured in many magazines in North America and Europe.

DAN HOFFMAN studied architecture at the Cooper Union School of Architecture, where he received his Bachelor of Architecture degree in 1976. After teaching three years at Carleton University, he practised in the office of Edward Larrabee Barnes Associates, in New York. During the past six years, he has been artist-in-residence at the Cranbrook Academy of Art, Michigan, where he chairs the Department of Architecture. Mr Hoffman has lectured extensively across the United States and Canada and is often invited as a visiting critic to architecture schools. His architectural projects have been published and are widely recognized as provocative ground for speculation on the nature of architectural practice, poetics, and representation.

IRENA LATEK est née en Pologne, où elle a reçu sa formation d'architecte. Elle détient un diplôme de maîtrise en architecture et en design urbain de l'École polytechnique de Varsovie. Elle a pratiqué l'architecture en Pologne et à Paris, et depuis 1978, elle a collaboré avec quelques cabinets montréalais sur de nombreux projets et études à dimension urbaine. Madame Latek est actuellement professeure adjointe à l'École d'architecture de l'Université de Montréal, et elle a reçu en 1991 le New Faculty Teaching Award décerné par The Association of Collegiate Schools of Architecture en reconnaissance de l'excellence de l'enseignement pendant les premières années dans la carrière de professeur d'architecture. Elle est co-auteure du livre *Ville, métaphore, projet* (Montréal, Éditions du Méridien, 1992) et est également l'auteure de plusieurs articles, dont « La ville comme récit sans fin » (ARQ, n° 57), « Architecture comme idée » (ARQ, n° 50, 1989), et « Lieux et Figures » (*Section A*, vol. 3, 1985). Ses recherches portent sur la notion de représentation dans l'architecture contemporaine.

PHILIPPE MADEC, né en 1954 à Brest, est architecte de métier. Sa première réalisation architecturale a obtenu une mention au Prix de la première œuvre du Moniteur en 1991. Ses travaux sont publiés dans de nombreuses revues françaises d'art et d'architecture. De 1983 à 1984, il a été professeur invité en stage de recherche à l'école d'architecture de Columbia University à New York. De 1985 à 1991, il a enseigné à l'École nationale supérieure du paysage de Versailles. Il a été tour à tour professeur invité au module Design de l'environnement de l'Université du Québec à Montréal (1989), à l'École d'architecture de l'Université de Montréal (1991) et à l'école d'architecture de paysage de Harvard University (1991–1992). M. Madec est également l'auteur d'ouvrages sur l'architecture, dont *Boullée* (publié en 1986 aux éditions Fernand Hazan, en cours de traduction en Espagne) ; « Le jeu subtil et poétique des espaces sous la lumière » in *La Cité des Sciences* (Calman-Lévy/ E.C.S.I., 1988) ; *Le Matin de l'architecte* (M.U.L.T.M., 1991). Deux ouvrages sont en préparation: *Pour que la vie ait lieu* (version intégrale) et *L'État des lieux*, en collaboration avec l'écrivain Julien Cendres.

ALBERTO PÉREZ-GÓMEZ was born in Mexico City in 1949. He has taught at universities in Mexico City, Houston, Syracuse, and Toronto, and at the Architectural Association in London. He was director of the Carleton University School of Architecture from 1983 to 1986. Mr Pérez-Gómez obtained his undergraduate degree in architecture and engineering in Mexico City, did postgraduate work at Cornell University, and was awarded M.A. and Ph.D. degrees by the University of Essex in England. His articles have been published in *Journal of Architectural Education*, *AA Files*, *Arquitecturas Bis*, *Section A*, *Architectural Design*, and other periodicals. His book, *Architecture and the Crisis of Modern Science* (MIT Press, 1983), won the Alice Davis Hitchcock Award in 1984. In 1987, Mr Pérez-Gómez was appointed Saidye Rosner Bronfman Professor of the History of Architecture at McGill University, where he is currently director of the Master's program in the History and Theory of Architecture. From 1990 to 1993, he was also been director of the Institut de recherche en histoire de l'architecture. Two new books, *Polyphilo or the Dark Forest Revisited* (MIT Press), inspired by Francesco Colonna's *Hypnerotomachia Poliphili*, and a translation and introduction to Claude Perrault's *Ordonnance des cinq espèces de colonnes* (Getty Center for the History of Art and the Humanities) were published in 1993.

ROBERT PROST est ingénieur et architecte. Diplômé en architecture de l'École spéciale d'architecture de Paris en 1967, il obtient un doctorat en urbanisme à l'Université de Montréal en 1972. Après avoir été professeur à l'Université de Montréal, maître de conférence à l'Institut Auguste Comte, chargé de mission au CESTA et professeur invité à la School of Architecture and Urban Design du Massachusetts Institute of Technology, il est depuis 1985 professeur et chercheur à l'École d'architecture Paris-Tolbiac. Auteur de nombreux travaux de recherche sur la planification et les « sciences de l'action », il se consacre depuis plusieurs années aux « sciences de la conception ».

PETER ROSE founded the firm Peter Rose Architect in 1974. Before opening his own practice, he was in partnership with Rose, Burr, Righter Associates (1970–74) and a designer for ARCOP Associates in 1972. Mr Rose has received several design awards, and his work has been featured in many major publications. Educated at Yale, Peter Rose has taught at McGill, Princeton, and Harvard, where he is currently Adjunct Professor in Architecture. The drawings and models prepared by his office have been exhibited in a number of Canadian and American museums and galleries. He recently completed the Canadian Centre for Architecture. Currently, his office is working on several projects: the restoration and renovation of Windsor Station in Montreal; an addition to the Brookside School at Cranbrook Educational Community, north of Detroit; renovations and additions to the Westmount Library in Montreal; a master plan for the expansion of, and the addition of a recital hall to, the Agnes Etherington Art Gallery at Queen's University in Kingston, Ontario. He has worked on a master plan for the mile-long, historic waterfront of the Old Port of Montreal and on an urban design for a thirty-acre, nine-million-square-foot, mixed-use project in the old Canadian National Railway yards in downtown Montreal.

JACQUES ROUSSEAU, diplômé de l'École d'architecture de l'Université de Montréal, développe depuis plus de dix ans un travail de recherche, d'enseignement et de réalisation dans les domaines urbain et architectural. Associé principal de la société d'architecture Fortin et Rousseau, il est également professeur invité au module Design de l'environnement de l'Université du Québec à Montréal depuis 1982, où il occupe la direction des ateliers d'architecture urbaine. Il a reçu le Prix de Rome du Conseil des Arts du Canada en 1988. Sa compétence et son expérience résident dans son aptitude à articuler des concepts intégrant plusieurs niveaux de complexité et à produire des opérations spatiales audacieuses, judicieuses et simples. Ses recherches théoriques sur la ville et l'architecture, sa participation à divers projets et concours soutiennent l'idée d'une architecture civique et conviviale alliant urbanité et humanité.

DR MARGARET A. SOMERVILLE, AM, was both a practising pharmacist and lawyer in Australia before beginning her academic career in Canada. She was awarded an AUA (Pharmacy) by the University of Adelaide in 1963, an LL.B. (Hons I) and the University Medal by the University of Sydney in 1973, and a Doctor of Civil Law degree by McGill University in 1978. Her current appointments at McGill University are as Gale Professor of Law (she holds the first Chair in Law awarded to a woman in Canada); professor, Faculty of Medicine; and director, McGill Centre for Medicine, Ethics and Law. Dr Somerville is very active in presenting medicine, ethics, and law in all forms of media and undertakes a heavy schedule as a speaker and writer both nationally and internationally in wide-ranging subject areas. She acts as a consultant to governments and non-governmental bodies – in particular, the Global Programme on AIDS of the World Health Organization. Her work also includes issues related to aged persons and aging, reproductive technology, genetics, medical malpractice, human medical research, etc. Dr Somerville was recently appointed to the Order of Australia and elected a fellow of the Royal Society of Canada in recognition of her work in law and bioethics.

LUKAS SOSOE est professeur agrégé au département de Philosophie de l'Université de Montréal. Docteur de l'Université de Fribourg, il a d'abord poursuivi sa carrière universitaire dans cette université avant d'assumer en 1989 les fonctions de professeur d'éthique et de philosophie du droit à l'Université de Montréal. Parmi ses plus importantes publications, signalons son livre sur *La critique du naturalisme et l'autonomie de l'éthique*, publié en allemand aux Éditions Karl Alber de Fribourg-en-Brisgau en 1988, et l'ouvrage *Philosophie du droit*, rédigé en collaboration avec le professeur Alain Renaut de l'Université de Caen et paru aux Presses universitaires de France. Les travaux du professeur Sosoe ont pour caractéristique essentielle d'intégrer l'apport de méthodologies et d'analyses appartenant aux traditions philosophiques française, allemande et anglo-américaine, et d'ouvrir des perspectives originales sur les thèmes centraux de la philosophie morale.

DALIBOR VESELY was born in Prague, Czechoslovakia. He studied engineering, architecture, art history, and aesthetics in Prague and Munich, and obtained his Ph.D. from Charles University in Prague. While in Czechoslovakia, he was a member of the "Continualist" group, closely associated with the surrealists. He has lived in England since 1968, where he taught first at the Architectural Association and presently at Cambridge University. Dr Vesely has developed an original position introducing a serious consideration of phenomenology in architectural discourse. This position has led to increasing attention being paid to the poetics of architecture and to the fallacies of reductive thinking. Dr Vesely has been a visiting professor in a number of American and European universities and has produced many influential articles in journals such as *AA Files* and *Daidalos.* Dr Vesely's meticulous study on the genealogy of the poetry of architecture is now close to completion.